Beyond Failure

Other Titles of Interest

Failure Mechanisms in Building Construction, edited by David H. Nicastro (ASCE Press, 1997). Addresses the question of why buildings fail by presenting a wide range of failure mechanisms and case studies. (ISBN 978-0-7844-0283-2)

Guidelines for Forensic Engineering Practice, edited by Gary L. Lewis (ASCE Committee Report, 2003). Recommends guidelines for five areas important to engineers with a forensic practice: qualifications, investigations, ethics, legal issues, and business. (ISBN 978-0-7844-0688-5)

In the Wake of Tacoma: Suspension Bridges and the Quest for Aerodynamic Stability, by Richard Scott (ASCE Press, 2001). Surveys changes in the design of suspension bridges evolving from the 1940 collapse of the first Tacoma Narrows Bridge. (ISBN 978-0-7844-0542-0)

Lessons from the Oklahoma City Bombing, by Eve E. Hinman and David J. Hammond (ASCE Press, 1997). Documents the incident from a structural standpoint, including the building's construction as well as hazard mitigation steps taken during the rescue/recovery process. (ISBN 978-0-7844-0217-7)

The New Orleans Hurricane Protection System: What Went Wrong and Why, by the ASCE Hurricane Katrina External Review Panel (ASCE, 2007). Focuses on the direct physical causes and contributing factors that led to the failure of the New Orleans hurricane protection systems in 2005. (ISBN 978-0-7844-0893-3)

The Pentagon Building Performance Report, by Paul E. Mlakar, Donald O. Dusenberry, James R. Harris, Gerald Haynes, Long T. Phan, and Mete A. Sozen (ASCE Committee Report, 2003). Presents the results of an exhaustive examination and analysis of the Pentagon's performance during and immediately after the 2001 terrorist attack. (ISBN 978-0-7844-0638-0)

Beyond Failure

Forensic Case Studies
for Civil Engineers

NORBERT J. DELATTE JR., PH.D., P.E.

Hartness Library
Vermont Technical College
One Main St.
Randolph Center, VT 05061

Library of Congress Cataloging-in-Publication Data

Delatte, Norbert J.
 Beyond failure : forensic case studies for civil engineers / Norbert J. Delatte, Jr.
   p. cm.
 Includes bibliographical references and index.
 ISBN 978-0-7844-0973-2
 1. Forensic engineering—Case studies. 2. System failures (Engineering)—Case studies. 3. Civil engineering—Case studies. 4. Structural failures—Investigation—Case studies. I. Title.

TA219.D45 2009
624.1'71—dc22

2008045372

Published by American Society of Civil Engineers
1801 Alexander Bell Drive
Reston, Virginia 20191
www.pubs.asce.org

Any statements expressed in these materials are those of the individual authors and do not necessarily represent the views of ASCE, which takes no responsibility for any statement made herein. No reference made in this publication to any specific method, product, process, or service constitutes or implies an endorsement, recommendation, or warranty thereof by ASCE. The materials are for general information only and do not represent a standard of ASCE, nor are they intended as a reference in purchase specifications, contracts, regulations, statutes, or any other legal document.

ASCE makes no representation or warranty of any kind, whether express or implied, concerning the accuracy, completeness, suitability, or utility of any information, apparatus, product, or process discussed in this publication, and assumes no liability therefor. This information should not be used without first securing competent advice with respect to its suitability for any general or specific application. Anyone utilizing this information assumes all liability arising from such use, including but not limited to infringement of any patent or patents.

ASCE and American Society of Civil Engineers—Registered in U.S. Patent and Trademark Office.

Photocopies and reprints. You can obtain instant permission to photocopy ASCE publications by using ASCE's online permission service (http://pubs.asce.org/permissions/requests/). Requests for 100 copies or more should be submitted to the Reprints Department, Publications Division, ASCE (address above); e-mail: permissions@asce.org. A reprint order form can be found at http://pubs.asce.org/support/reprints/.

Copyright © 2009 by the American Society of Civil Engineers.
All Rights Reserved.
ISBN 978-0-7844-0973-2
Manufactured in the United States of America.

17 16 15 14 13 12 11 10 09   1 2 3 4 5

*For my mother, Virginia McClary Delatte,
who inspires my teaching.*

Contents

FOREWORD   xi
by Kenneth L. Carper

PREFACE   xiii

1 Why Case Studies?   1
   Organization of This Book   3
   Notes to the Student   4
   Sources for Case Materials   5

2 Statics and Dynamics   7
   Hyatt Regency Walkway   8
   Tacoma Narrows Bridge   26
   Aircraft Impacts   38
   Other Cases   42

3 Mechanics of Materials   51
   Quebec Bridge   51
   Point Pleasant Bridge Collapse   70
   Comet Jet Aircraft Crashes   82
   Other Cases   86

4 Structural Analysis   89
   Agricultural Product Warehouse Failures   90
   Ronan Point   97

	L'Ambiance Plaza Collapse	107
	Cleveland Lift-Slab Parking Garage	121
	Kemper Arena	124
	Other Cases	126
5	**Reinforced Concrete Structures**	129
	Air Force Warehouse Shear Failures	130
	2000 Commonwealth Avenue	133
	Skyline Plaza in Bailey's Crossroads	144
	Harbour Cay Condominium	149
	Bombing of the Oklahoma City Murrah Federal Building	155
	The Pentagon Attack	162
	Other Cases	167
6	**Steel Structures**	173
	Hartford Civic Center Stadium Collapse	174
	Mianus River Bridge Collapse	184
	Cold-Formed Steel Beam Construction Failure	188
	The World Trade Center Attacks	195
	Pittsburgh Convention Center Expansion Joint Failure	206
	Minneapolis I-35W Bridge Collapse	211
	Other Cases	215
7	**Soil Mechanics, Geotechnical Engineering, and Foundations**	221
	Teton Dam	223
	Vaiont Dam Reservoir Slope Stability Failure	234
	The Transcona and Fargo Grain Elevators	249
	Other Cases	255
8	**Fluid Mechanics and Hydraulics**	257
	Johnstown Flood	257
	Malpasset Dam	267
	Schoharie Creek Bridge	277
	New Orleans Hurricane Katrina Levee Failures	287
	Other Cases	299

9 Construction Materials — 301
- Austin Concrete Dam Failure — 303
- Liberty Ship Hull Failures — 310
- Willow Island Cooling Tower Collapse — 316
- Boston's Big Dig Tunnel Collapse — 325
- High-Alumina Cement — 330
- Other Cases — 330

10 Management, Ethics, and Professional Issues — 333
- Citicorp Tower — 333
- Space Shuttle *Challenger* — 345
- Sampoong Superstore, Korea — 352
- Misuse of the Professional Engineer License — 357
- Property Loss Investigations — 359

APPENDIX A: Notes to the Professor — 361

APPENDIX B: The ASCE Code of Ethics — 377

APPENDIX C: Some Cases on Video and DVD — 385

REFERENCES — 391

INDEX — 403

ABOUT THE AUTHOR — 407

Foreword

FOR THE PAST 35 YEARS, MY ACADEMIC CAREER HAS BEEN devoted to teaching structural engineering concepts to students in architecture and in construction management—a challenging assignment indeed. Throughout that time, I have integrated lessons from forensic engineering case histories into my classroom instruction with a considerable degree of success. Exposure to landmark engineering failures and the subsequent investigation reports gives students a better theoretical understanding of the nature of materials. They become aware of the manner in which engineering design strategies evolve over time. They also gain a greater appreciation for the inherent professional responsibilities of their chosen professions. In addition to technical factors, the study of engineering failure case histories provides opportunities for classroom discussion of important nontechnical topics: ethics, professional liability, human factors, and the critical interpersonal skills and interdisciplinary relationships required for the delivery of a successful project.

I first met Norbert Delatte in 1999, when he became a member of the ASCE Technical Council on Forensic Engineering (TCFE). Since then, we have worked together on several TCFE projects, including a number of initiatives by the Committee on Education. A principal goal of that committee has been to encourage the use of forensic engineering case histories in undergraduate and graduate professional engineering education. On the basis of committee surveys, the two most prominent obstacles to furthering this goal were identified as (1) the crowded engineering curriculum that makes it nearly impossible to introduce a new required course devoted to forensic engineering topics and (2) the perceived lack of case histories suitable for introduction into the classroom without extensive effort by the educator.

The TCFE Committee on Education has produced materials to support stand-alone courses and capstone design courses related to forensic engineering. However, the focus of most committee activities is on the development of materials and strategies for integrating case histories efficiently into existing courses throughout the undergraduate curriculum. Norbert Delatte has been at the center of these initiatives. His own undergraduate students have responded enthusiastically to his experiments with a variety of teaching strategies, using materials jointly developed by Delatte and his students. Several comprehensive case history papers written by his undergraduate students have appeared in ASCE journals. Some of these have been the recipients of prestigious paper awards.

It is most gratifying to review this major contribution by Norbert Delatte. It addresses all of the issues identified by TCFE surveys and provides a truly workable set of resource materials directly suited for use in the classroom. This volume includes not only the classic, well-known landmark engineering failures, but also recent events. The discussion is concise, yet sufficient detail is given, along with specific references for further in-depth study. Suggestions are given as to the most appropriate course in which to introduce each case history. Some of the more instructive cases are recommended for re-visits at several stages of the curriculum, as students develop in maturity. The purpose of this book is clear. It does not introduce new information about the case histories, but rather synthesizes the available information in a form that is readily accessible. I expect that engineering educators will find this to be a most valuable publication.

Finally, while the book is directed to engineering educators and their students, it contains much information that is relevant to practitioners. It should be required reading for all members of the design and construction community.

KENNETH L. CARPER
College of Engineering & Architecture
Washington State University

Preface

THIS BOOK REPRESENTS THE CULMINATION OF MORE THAN a decade of effort, but I couldn't have done it alone. Many of my friends and colleagues on the American Society of Civil Engineers Technical Council on Forensic Engineering (ASCE TCFE) have helped me put case study materials together.

I would like to thank the following people and institutions, who contributed photos and illustrations and granted permission for their use:

- Nicholas J. Carino, retired from the National Bureau of Standards/National Institute of Standards and Technology (NBS/NIST)—Hyatt Regency, Figs. 2-1, 2-2, and 2-4; L'Ambiance Plaza, Figs. 4-9 and 4-12; Skyline Plaza in Bailey's Crossroads, Figs. 5-5 and 5-6; Harbour Cay Condominium, Figs. 5-7 and 5-8; and Willow Island Cooling Tower, Figs. 9-7 and 9-9;
- Anthony M. Dolhon of Wiss, Janney, Elstner Associates, Inc.—Hyatt Regency, Fig. 2-3;
- Lee L. Lowery, Jr. of Texas A&M University—Hyatt Regency, Figs. 2-5 through 2-7;
- University of Washington Special Collections—Tacoma Narrows Bridge, Figs. 2-10 and 2-11;
- Library and Archives Canada—Quebec Bridge, Figs. 3-1, 3-2, and 3-5;

- Cleveland State University Special Collections and the *Cleveland Press*—Cleveland Pigeonhole Parking Garage, Figs. 4-16 through 4-18;
- Michael J. Drerup, Exponent—Oklahoma City Murrah Building, Figs. 5-9 through 5-12;
- *Pittsburgh Tribune-Review*—Pittsburgh Convention Center, Fig. 6-10;
- Howard F. Greenspan, Howard F. Greenspan Associates—Schoharie Creek Bridge, Figs. 8-9 and 8-10;
- Library of Congress—Austin Dam, Fig. 9-3; and
- Robert Pitt, University of Alabama—Citicorp Tower, Figs. 10-1 and 10-2.

I would like to acknowledge grant funding from the National Science Foundation (NSF) that has supported this work. Much of this material is based on work supported by the National Science Foundation under Grants No. EEC-9820484, DUE-0127419, and DUE-0536666. Any opinions, findings, and conclusions or recommendations expressed in this material are those of the author and do not necessarily reflect the views of the National Science Foundation.

My former students Rachel Martin, Suzanne King, Stacey Solava, Chris Storey, Cynthia (Rouse) Pearson, Dan Miller, and Constantine Kontos originally helped me research many of the case studies in this book. Their work in locating references and preparing illustrations was particularly valuable.

I am indebted to the ASCE TCFE staff contacts Verna Jameson and John Segna. The ASCE TCFE and the NSF have supported six faculty failure case study workshops (2003–2008). Workshop instructors Paul Bosela, Ken Carper, Kevin Rens, Kevin Sutterer, Oswald Rendon-Herrero, Jack Gillum, and Michael Drerup helped develop and refine the workshop content. The faculty participants in the workshops have provided valuable feedback to help improve the case studies.

My employers—the U.S. Military Academy (West Point, New York), the University of Alabama at Birmingham, and Cleveland State University—have also been supportive of this work with resources and small grants. The librarians at UAB and CSU have been particularly helpful. I would like to thank in particular Theresa M. Nawalaniec and Lynn M. Duchez Bycko of the CSU library who helped me locate references and illustrations.

I owe a considerable debt to my professors at The Citadel, the Massachusetts Institute of Technology, and the University of Texas at Austin. Many of these teachers discussed failure case studies and the importance of engineering responsibility and professionalism.

I apologize to anyone I have forgotten to thank—there have been so many. Of course, it goes without saying that the inevitable errors and omissions in this book are mine and mine alone.

Finally, I would like to thank my wife Lynn and our children Isabella and Joe for their patience and understanding while I spent way too much time working on this book.

Norbert Delatte
Cleveland, Ohio
July 21, 2008

1

Why Case Studies?

ENGINEERS DESIGN. ENGINEERING DESIGN MAY BE SEEN AS an attempt to use science, mathematics, and other principles to prevent failures. Most of the time the attempt is successful, but the times it is not successful can provide useful lessons for students and practitioners. The lessons learned from failures have often led directly to changes to engineering codes and procedures. Students are more likely to appreciate advances in design and analytical procedures if they are placed in a historical context.

Leonards (1982) defined failure as "an unacceptable difference between expected and observed performance." Feld (1964) noted that although structural collapses are rare, if failure is defined as "nonconformity with design expectations" then there are many failures. The latter is, admittedly, a broad definition.

I would propose a simplified two-part definition of design:

1. Figure out everything that can possibly go wrong.
2. Make sure it doesn't happen.

To figure out what can go wrong, it is necessary to know how structures, facilities, and systems fail. This knowledge of failure I define as "failure literacy." It is a knowledge of potential failure modes and limit states, informed by a historical perspective on reasons for past failures and patterns of failure.

Luth notes,

> Buildings are not like automobiles and airplanes. It is not possible to build and test prototypes to work out the bugs. Nor is it possible to design and draw every detail without overly constraining the construction activities. We are faced, then, with a system that requires building from less than complete drawings, working in less than ideal conditions, with laborers of uncertain skills. Such a system cannot help but produce more failures unless there are conscientious professionals working on all sides all the way through the project. To suggest otherwise is simplistic and unrealistic. The fact that there are so few failures is a credit to the professionalism of both the construction and the design professionals that design and build modern projects. (2000, p. 61)

The introduction to *Construction Disasters: Design Failures, Causes, and Prevention* states, in a preface entitled "Why *Engineering News-Record* Reports Failures,"

> "ENR policy is to report both failures and successes . . . for the same reason: to give readers the information they need in their own businesses, so that they can avoid the failures and emulate the successes. . . . Some weeks we also have to report the failures of people and companies and the things they plan, design, make or build; financial failures such as bankruptcy, personal failures such as crime and corruption, planning failures such as downtown renovation that doesn't work, disasters that are caused by natural phenomena like earthquakes and floods. There are also structural failures, such as dam collapses, and system failures, such as hotel fire-safety assignments that fall apart in a fire." (Ross 1984, p. xi)

ENR reports on accidents that kill and injure people, as well as functional failures—potholes, cracks and leaks in facades and roofs, and other problems.

According to James Amrhein, as cited by Carper (1989), structural engineering may be defined as "the art and science of molding Materials we do not fully understand, into Shapes we cannot precisely analyze, to resist Forces we cannot accurately predict—all in such a way that the society at large is given no reason to suspect the extent of our ignorance." That is true of most other subdisciplines of civil engineering as well. I would recommend failure literacy and a healthy dose of humility for both engineering students and practicing engineers.

To make failure cases more useful and relevant, it is necessary to link them to specific courses and course topics. It has been demonstrated how specific cases may be linked to engineering mechanics course topics (Delatte 1997). Based on this earlier work, a more comprehensive master plan was published in 2000 (Delatte). The plan was developed further, adding more topics and cases, and the revised version was published in 2002 (Delatte and Rens).

For example, the Teton Dam case in Chapter 7 deals with engineering geology and geotechnical engineering topics, including the suitability of foundation and borrow materials, the importance of compaction, and movement of water within rock and soil masses. The Schoharie Creek Bridge collapse in Chapter 8 illustrates points in hydraulic engineering, such as stream velocity and scour, as well as structural engineering topics, including the advantages of continuity and redundancy of structures.

Organization of This Book

This book consists of chapters of case studies that parallel typical courses in the civil engineering and engineering mechanics curriculum. The cases are reviewed in detail, and the technical elements that relate to the course's chapter are outlined. Often, failure case studies in other chapters will be relevant. These are listed under the heading of "Other Cases" at the end of Chapters 2–9.

Typical elements of the failure case studies included in this book are:

- an introduction;
- a description of the design and construction of the project;
- a narrative describing the failure;
- a discussion of any investigations undertaken and the results, which may include a review of who the investigators were, who hired them and why, and any limits on the scope of the investigation;
- technical lessons learned, with special attention to any changes in engineering codes or procedures;
- procedural and ethical lessons learned, particularly legal repercussions;
- educational aspects of the case; and
- a detailed reference list, including investigation reports, published papers, and newspaper and journal accounts.

The cases in this book were originally each developed with an individual reference list. References for all cases have now been collected into a single list. Most cases now have a section entitled "Essential Reading," which notes the most valuable and informative references on the topic.

Notes to the Student

I was introduced to the topic of failure case studies as a young graduate student at the Massachusetts Institute of Technology, more than two decades ago. I took a course entitled "Construction Technology," which was made up of graduate students from civil engineering, architecture, and real estate development. In this course, we did a lot of problem solving using the case study method.

Two case studies we investigated were the collapse of the Hotel Vendome during a fire and the construction collapse of 2000 Commonwealth Avenue. Both were in the city of Boston, just across the Charles River.

The cases made a strong impression on me. However, I made the mistake of focusing too intently on the technical aspects of the collapses (such as high bearing stresses or misplaced reinforcing steel) while ignoring the communications and procedural problems that are at the root of many failures. As a result, I got a B on that assignment.

One mistake I often made early in analyzing these case studies was to home in too quickly on the "correct" solution. H. L. Mencken (1949), the Sage of Baltimore, said, "There is always an easy solution to every human problem—neat, plausible, and wrong." Failures are complex problems; the clear, simple, and obvious answer may be incomplete or incorrect.

A year later, I took a course taught by Bill LeMessurier entitled Structural Design of Buildings. In the course, he related his side of the story concerning the Citicorp Tower case in midtown Manhattan, which could have been a catastrophic collapse but wasn't. It was fascinating to hear the story firsthand. LeMessurier's story is provided in Chapter 10, along with an opposing viewpoint.

A decade later, I resumed my graduate studies at the University of Texas at Austin. There, I was fortunate enough to take a course in forensic engineering under David Fowler. I became fascinated by the problems of failure analysis.

Few of you reading this book will become forensic engineers. Instead, I would like to instill a sense of failure literacy in you. Poets and authors are expected to have intimate familiarity with the work that has gone on before: Shakespeare's sonnets, Hemingway's short stories, and so forth. In the same way, engineers analyzing and designing structures and systems need to know how similar facilities have performed in the past and when and how they have failed.

A few of these cases are from my own consulting practice. They are not as spectacular as the others, but they still demonstrate some important engineering principles.

Sources for Case Materials

An extensive reference list is provided. There are many other sources for case studies. These sources include books, technical papers and magazine articles, videos and television programs, and PowerPoint presentations. Unfortunately, although there are many relevant websites, they change too often to provide a useful list.

Books

Three excellent texts are Kaminetzsky (1991), Levy and Salvadori (1992), and Feld and Carper (1997). McKaig (1962) is also good. Ross (1984) contains cases reprinted from *Engineering News Record*, which is a weekly publication covering the construction industry that often contains examples of recent failures. Shepherd and Frost (1995) contains short summaries of a wide variety of cases. Four excellent recent sources of case studies are the proceedings of the 1st, 2nd, 3rd, and 4th ASCE Congresses on Forensic Engineering (Rens 1997, Rens et al. 2000a, Bosela et al. 2003, and Bosela and Delatte 2006).

Some books, such as Levy and Salvadori (1992) and Petroski (1985), do an excellent job of explaining fundamental structural behavior without relying on complex theories or mathematics and are particularly appropriate for undergraduate students.

Papers and Articles

Engineering News Record is a good source of news on recent cases. Another excellent source is the *Journal of Performance of Constructed Facilities*, published by the American Society of Civil Engineers (ASCE). A useful bibliography on failures was assembled in a paper by Nicastro (1996).

Television, Video, and DVD

An excellent video illustrating case studies is "When Engineering Fails," written and presented by Henry Petroski. This videotape closely parallels the book *To Engineer Is Human* (Petroski 1985) and provides dramatic footage of the Kansas City Hyatt Regency walkway collapse, the Tacoma Narrows Bridge in Washington state, and other cases. This program was originally developed for the BBC.

The History Channel, as part of its Modern Marvels series, has a series of programs entitled *Engineering Disasters*. A list of the case studies in the series is provided in Appendix C.

Presentations

PowerPoint presentations have been prepared to accompany many of the case studies in this book. In the past, these presentations have been made available through a series of faculty workshops sponsored by the American Society of Civil Engineers and the National Science Foundation. It is anticipated that these will also be available in the future.

2

Statics and Dynamics

STATICS IS GENERALLY THE SUBJECT OF THE FIRST COURSE students encounter within engineering mechanics, and it forms the basis for mechanics of materials, structural analysis and design, soil mechanics, mechanical engineering, and many other subsequent course topics. Another topic often addressed is static fluid pressure. A statics course is generally followed by a dynamics course, which at some institutions is combined with statics into a single course. Statics and dynamics are part of the mechanics of rigid bodies. Also, the statics course or the combined course is often followed directly by a course on deformable body mechanics or mechanics of materials.

In an introductory statics course, calculations are usually carried out in two and three dimensions. Three-dimensional calculations often use vectors. Ironically, in subsequent courses, analysis is usually conducted purely in two dimensions, and the three-dimensional character of actual engineering problems may be lost.

Structures may be either statically determinate or indeterminate. Indeterminate structures are often also referred to as redundant. Principles of equilibrium (or statics) can solve for unknown forces and reactions acting on statically determinate structures but not on indeterminate structures without additional information. Nevertheless, students taking statics must learn to distinguish between structures that are determinate and indeterminate, as well as those that are not properly restrained. To be stable, structures must have enough supports, and they must be properly arranged, to prevent movement. If not, collapse is possible.

Not all of the topics discussed on this list lend themselves easily to demonstration through failure case studies. However, some representative cases are discussed in this chapter, and others appear in other chapters.

Topics in a dynamics course include particle kinetics and kinematics and rigid-body kinetics and kinematics. *Kinematics* is the study of the geometry of motion, whereas *kinetics* analyzes forces that cause motion. Kinetics problems (both particle and rigid-body) may be analyzed using methods of force and acceleration, work and energy, or impulse and momentum. An important application is impact forces acting on structures and machines, using impulse-momentum methods along with coefficients of restitution.

The study of vibrations may or may not be included in a dynamics course. Vibrations are important for mechanical systems, as well as for structures. Excessive vibration is a serviceability issue that is often related to poor performance.

Hyatt Regency Walkway

Every civil engineer should be familiar with the circumstances of the Hyatt Regency walkway collapse. It is a landmark case, both because of the number of people killed and injured and because of the effect on the engineering profession. Although it has been extensively studied, doubts remain as to whether the key lessons have in fact been learned.

Design and Construction

In July 1980, the Hyatt Regency Crown Center in Kansas City, Missouri, opened to the public after four years of design and construction. A 40-story tower, an atrium, and a function block housing all of the hotel's services, combined to form this impressive building. Three walkways spanned the 37-m (120-ft) distance between the tower and the function block. The front of the building is shown in Fig. 2-1.

The walkways were suspended from the atrium's ceiling by six 32-mm (1¼-in.) diameter hanger rods. The second-floor walkway, directly below the fourth-floor walkway, was suspended from the beams of the fourth-floor walkway, and the third- and fourth-floor walkways hung from the

Figure 2-1. The Kansas City Hyatt Regency exterior.
Courtesy National Bureau of Standards/National Institute of Standards and Technology.

ceiling (Feld and Carper 1997, p. 216). The walkways before the collapse are shown in Fig. 2-2.

The erection of this hotel, however, had not been without incident. During construction, the atrium roof collapsed as a result of inadequate provision for movement in the expansion joint and improper installation of a steel-to-concrete connection. Concerned about the building's structural integrity, the owner hired another engineering firm to investigate the collapse and check the roof design. The consulting structural engineering company also rechecked all of the connections and found nothing to cause alarm. Construction resumed, and the hotel opened a little less than two years later (Roddis 1993, p. 1549). The expansion joint that failed was similar to one in the Pittsburgh Convention Center, which is discussed in Chapter 6.

Collapse

On the evening of July 17, 1981, between 1,500 and 2,000 people were on the atrium floor and on the suspended walkways to see a local

Figure 2-2. The Kansas City Hyatt Regency atrium walkways.
Courtesy National Bureau of Standards/National Institute of Standards and Technology.

radio station's dance competition (Feld and Carper 1997, pp. 214–215). At 7:05 P.M., a loud crack echoed throughout the building, and the second- and fourth-floor walkways crashed to the ground, killing 114 people and injuring more than 200 others. It was the worst structural failure in the history of the United States (Levy and Salvadori 1992, p. 224). The scene of the collapse is shown in Fig. 2-3.

Two weeks later, *Newsweek* reported,

> Flags flew at half mast throughout Kansas City last week, and funeral processions wound through the streets. Outside the Hyatt Regency Hotel, where 111 people died in the collapse of two aerial walkways two weeks ago, "No Trespassing" signs barred the curious and the morbid. Inside, a fine dust covered the floor, and a few balloons clung wanly to the ceiling. But the wreckage was gone, trucked to a nearby warehouse, and the sole remaining "sky bridge" had been dismantled. With investigators arriving daily and lawyers lining up to file suits, the city was beginning to come to grips with a tragedy that may not have

Figure 2-3. Scene of the Kansas City Hyatt Regency collapse.
Photo provided by Wiss, Janney, Elstner Associates, Inc. (WJE).

ended yet: 81 victims still lay in hospitals, 8 of them on the critical list. (McGrath and Foote 1981)

Causes of Failure

Upon investigation, the National Bureau of Standards (then the NBS, now the National Institute of Standards and Technology or NIST) discovered that the technical cause of this collapse was quite simple: The hanger rod pulled through the box beam, causing the connection supporting the fourth-floor walkway to fail. If the structural system had been redundant, with alternate load paths, it would have been possible for the other hanger rods to hold the walkways up. However, the other rods could not handle the increased load once the adjacent rod failed. Because of this lack of redundancy, this failure caused the collapse of both walkways. The NBS report was careful about not assigning blame, to preserve cooperation with the various litigating parties (Ross 1984, p. 402).

Originally, the second- and fourth-floor walkways were to be suspended from the same rod (Fig. 2-4a) and held in place by nuts. The preliminary design sketches contained a note specifying a strength of 413 MPa

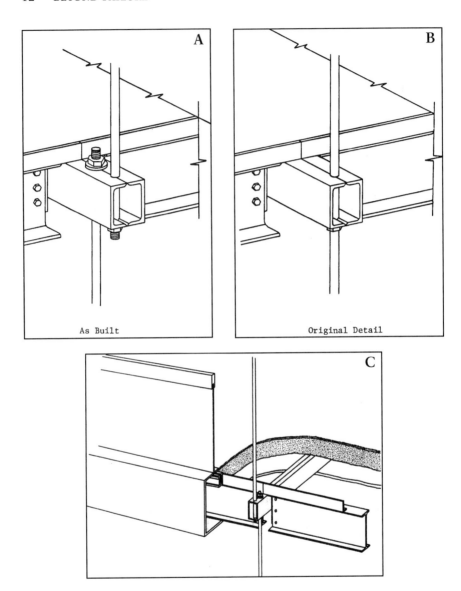

Figure 2-4. Original and as-built hanger details a, b, c.
Courtesy National Bureau of Standards/National Institute of Standards and Technology.

(60 kip/in.^{2}) for the hanger rods, which was omitted on the final structural drawings. Following the general notes in the absence of a specification on the drawing, the contractor used hanger rods with only 248 MPa (36 kip/in.^{2}) of strength.

This original design, however, was impractical because it called for a nut 20 ft (6.1 m) up the hanger rod and did not use sleeve nuts. The contractor modified this detail to use two hanger rods instead of one (Fig. 2-4b), and the engineer approved the design change without checking it. This design change doubled the stress exerted on the nut under the fourth-floor beam. Now this nut supported the weight of two walkways instead of just one (Roddis 1993, p. 1548). Figure 2-4c shows how the connection was integrated with the rest of the walkway. It was concealed and could not be easily inspected.

Moncarz and Taylor (2000) discussed the structural system in detail and analyzed the demand-to-capacity ratios of the components. Their firm, Failure Analysis Associates, had been retained by the project architect. Each walkway was made of lightweight concrete on metal decking, supported by longitudinal I-beam W 16 × 26 stringers. The designation "W 16 × 26" means an I-shaped wide flange section, 16 in. (406 mm) deep and weighing 26 lb per linear foot (mass 38.7 kg per linear meter). The transverse beams were each made of two MC 8 × 8.5 channels, welded toe to toe to form a box. A channel shape designated as MC 8 × 8.5 is 8 in. (203 mm) deep and weighs 8.5 lb per linear foot (mass 12.7 kg per linear meter). Holes were drilled through the welds to hold the support rods. The bearing area for the hanger rod washer was flattened by grinding, further reducing capacity (Moncarz and Taylor 2000, p. 47). The failed fourth-floor beam is shown in Fig. 2-5, and the hanger rod is shown in Fig. 2-6.

Failure Analysis Associates carried out sophisticated computer modeling (finite element) analysis of the box beam connection. The connection was near failure with dead load only, and an additional live load of 7 people on the upper bridge and 56 on the lower bridge proved to be enough to trigger failure. Plastic deformation of the box beam was estimated at 3–6 mm (0.12–0.24 in.), which was hidden by the finish and fireproofing materials (Moncarz and Taylor 2000, p. 49). The third-floor beam, which did not fail, is shown in Fig. 2-7. This box beam showed substantial permanent deformation.

Analysis of these two details revealed that the original design of the rod hanger connection would have supported 90 kN (20,000 lb), only 60% of the 151 kN (34,000 lb) required by the Kansas City building code. Even if the details had not been modified, the rod hanger connection would have violated building standards. As built, however, the connection only supported 30% of the minimum load, which explains why the walkways collapsed at well below maximum load (Feld and Carper 1997, pp. 218–222). The NBS built and tested a full-scale mock-up to simulate the failure and found that the static effects were much more significant than the dynamic effects (Kaminetzky 1991, p. 219).

Figure 2-5. Failed fourth-floor beam.
Courtesy Lee Lowery, Texas A&M University.

Figure 2-6. Hanger rod.
Courtesy Lee Lowery, Texas A&M University.

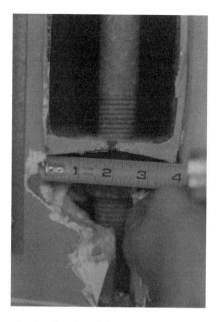

Figure 2-7. Third-floor beam showing deformation.
Courtesy Lee Lowery, Texas A&M University.

Events Leading Up to the Collapse

Luth (2000) details the steps in the design and construction process. He makes the important point that the critical connection was never designed and that the view represented by Fig. 2-4 was never drawn until after the failure. Luth's illustration of the various stages in the history of the connection is shown in Fig. 2-8.

At the time of the collapse, Luth was a recent graduate working at the firm that performed the Hyatt Regency's structural design. Luth's figures and tables illustrating the sequence of events are of particular interest. He notes that

> The project design was performed under the "fast track" method of delivery that came into vogue in the latter part of the 1970s. As with many projects delivered by this method, construction preceded design, structural design preceded architectural design, and both the design and construction phases were plagued by a lack of time and quality control. Thrown into the mix were multiple changes in personnel on both the construction and design sides. (Luth 2000, p. 51)

16 BEYOND FAILURE

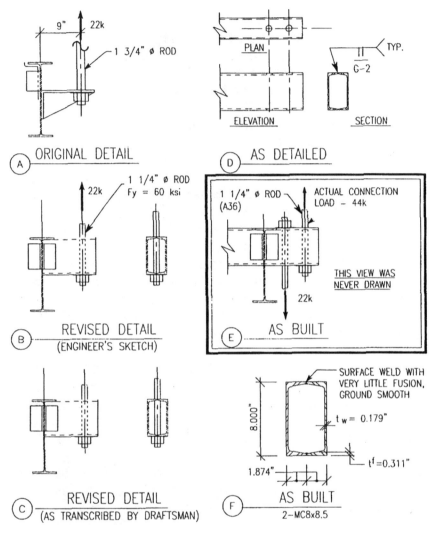

Figure 2-8. The evolution of the failed connection.
Source: Luth (2000).

The following sequence of key events leading up to the collapse is modified from Luth (2000, pp. 51–57):

- Early 1976–August 1978: The project moved from the owner's master planning to the evolution of the design, with major revisions

to the sunscreen framing near the walkways. Design was carried out using fast-track methods.
- June 1978: Both the project engineer and the senior design engineer on the project, each of whom had extensive knowledge of the product history, left the firm. A gap in continuity of this magnitude, particularly on a fast-track project, provides many opportunities for details to fall through the cracks.
- Mid-1978: The first walkway detail showed eccentric angles to support the hanger rods, not box beams. At roughly the same time, the architect requested a change from 44-mm (1¾-in.) to 32-mm (1¼-in.) diameter rods to enhance appearance. Because the rods would later be covered by fireproofing, there would actually be no visible difference.
- At roughly the same time, mid-1978: The original W 8 × 10 purlin (wide flange 8 in. or 203 mm deep, weighing 10 lb per linear foot, or mass 14.9 kg per linear meter) was changed to the box beam with 2 MC 8 × 8.5 channels. The engineer's revised sketch showed a single rod, a notation for a yield strength (F_y) of 414 MPa (60 kip/in.^{2}) for high-strength steel, and an axial load of 98 kN (22 kip). The weld between the channels was not shown. The intent of the load notation was to indicate to the fabricator that the connection still needed to be designed.
- August 1978: The draftsman transcribed the detail to the final drawings, but the rod yield strength and axial load were missing. The plans and specifications were issued for construction.
- December 1978: The fabricator started in-house work on the shop drawings, including heavy truss connections and all beam connections for which the forces were shown on the drawings. The $390,000 contract was not considered to be particularly large for the fabricator.
- January 1979: The fabricator called the structural engineer's project manager requesting a change from the continuous rod to two rods, offset, as shown in Figs. 2-4a and b. The project manager checked the moment and shear in the box beam and responded that the change was acceptable. He asked the fabricator to submit the request through channels for approval, but the fabricator never did this.
- January 12, 1979: The fabricator landed a large contract and transferred the shop drawings for the Hyatt Regency to an outside engineering firm. The drawing showed that the box beam had been started but not completed. The new firm assumed that the connection design had been completed and added the weld for the box beam.

- February 7, 1979: The detailer checked the shop drawings for internal consistency and completeness and did not find any problem with the hanger connection.
- February 16, 1979: The structural engineer received the shop drawings for approval. The contractor requested expedited approval, so the engineer assigned the checking to a senior technician. The technician noted that the hanger rod was not large enough using A36 steel, and the engineer responded from memory that it was high-strength steel. The drawings were returned on February 26.
- Summer 1979: Construction problems arose at an expansion joint because embedded plates had been left out. A repair detail using supporting seat angles (directly under the beam) and expansion bolts was developed, but only 5 of 14 bolts were installed correctly. Earlier, the first testing lab had been fired, and now the second testing lab was fired for poor performance on concrete testing. The project was completed without a testing lab.
- October 14, 1979: The expansion bolt repair detail failed during a fall cold snap, and two bays of the roof collapsed. Fortunately, there were no injuries. A complete design check was performed (by Luth). Again, the strength of the hanger rod was questioned, and the same assurance was provided (without checking the project documents). The design check revealed that W 6 × 16 members in the sunscreen truss had replaced the original W 6 × 15.5 sections, which were no longer available. The correction was completed by November 1979. The general impression of the project team was that disaster had been averted.
- July 1980: Hotel grand opening.
- July 1981: Collapse.

Legal Repercussions

Kansas City did not convict the Hyatt Regency engineers of criminal negligence because of lack of evidence. However, the billions of dollars in damages awarded in civil cases brought by the victims and their families dwarfed the half-million dollar cost of the building (Roddis 1993).

Feld and Carper (1997, p. 215) suggest that this was the most heavily litigated failure in history until that point. Claims under review at one point were up to $3 billion, with a single class action suit settled for $143 million. The 72 rescue workers sued for $150 million for emotional trauma and long-term psychological effects; the claim was settled out of court for $500,000.

Technical Concerns

Neither the original nor the as-built design for the hanger rod satisfied the Kansas City building code, making the connection failure inevitable under service loading conditions. If, however, the building design had provided for redundancy, this failure might not have resulted in the complete collapse of the walkway. The technical issues in this case are not particularly difficult. The procedural concerns are of much greater interest.

Procedural Concerns

The Hyatt Regency walkway collapse highlighted the lack of established procedures for design changes, as well as the confusion over who is responsible for the integrity of shop details (Roddis 1993). The legal repercussions experienced by the Hyatt engineers established the engineer of record's responsibility for the structural integrity of the entire building, including the shop details. It is important for all parties to understand fully and accept their responsibilities in each project (Feld and Carper 1997).

Certain procedural changes have been suggested to help prevent similar collapses (Kaminetzky 1991, p. 220):

- The engineer of record (EOR) should design and detail all nonstandard connections, although perhaps American Institute of Steel Construction (AISC) standard connections do not require similar care.
- All new designs should be thoroughly checked.
- All of the contractor's modifications to design details should require written approval from the EOR.
- Redundancy must be provided to prevent progressive or disproportionate collapse.
- Cross plates or stiffeners must be used for similar box beam-type connections to improve bearing capacity.

Given the history, this type of connection detail is unlikely to see much use in the future.

Ethical Concerns

Pfatteicher (2000, pp. 62–63) notes that this collapse provided a first test of the new ASCE Code of Ethics, officially adopted by the Board of Direction in 1976. The most significant change was the addition of Fundamental Canon 1: "Engineers shall hold paramount the safety, health, and welfare of the public in the performance of their professional duties." This canon has since been revised to encompass sustainable development. The ASCE Code of Ethics is provided in this book as Appendix B.

Once the NBS had completed its investigation (published as NBS 1982), the Missouri licensing board began a quiet inquiry. The public was outraged at the failure and the loss of life, and the local newspapers were filled with calls for justice (Pfatteicher 2000).

The county prosecutor and assistant district attorney announced more than two years after the collapse that they would not file any criminal charges because they did not find enough evidence to support them. The Missouri licensing board continued with its investigation of Jack Gillum and his employee, David Duncan. The board did not contact them during the investigation, nor did it investigate the architects, despite the fact that the board had jurisdiction over architects as well as engineers. At the end of its investigation, the Missouri Board of Architects, Professional Engineers, and Land Surveyors convicted the EOR and the project engineer of gross negligence, misconduct, and unprofessional conduct in the practice of engineering. Both had their Missouri professional engineering licenses revoked. The two combined had licenses in 30 jurisdictions, and most of those licenses were also revoked (Pfatteicher 2000, pp. 64–65).

After the Missouri board's action, the ASCE Committee on Professional Conduct held a confidential hearing on the matter. The committee deliberated for 12 h and recommended that Gillum be expelled for three years. Duncan was not an ASCE member (Pfatteicher 2000, p. 65).

Writing within a few years after the collapse, two attorneys believed that the punishment was appropriate. Their paper appeared in the ASCE *Journal of Performance of Constructed Facilities*.

The attorneys Rubin and Banick concluded:

1. Based on the facts found by the administrative law judge regarding the events that led up to the Hyatt collapse, license revocation was more than warranted. The engineers' conduct cannot be justified under any standard of professional practice. They were callously indifferent to life and safety after questions relating to the particular connection that failed were repeatedly brought to their attention. They were not hapless victims of the system in any sense.
2. It is paradoxical that while tragedies such as the Hyatt failure provide an incentive to change practices, the Hyatt failure is a poor example on which to base recommendations for change, because essentially no change in practices would likely have averted the Hyatt tragedy.

3. Examination of the facts discloses an ironic twist. With all of the alleged deficiencies in current practices, oddly enough, Hyatt proves that the system really does work. The right people asked . . . the right questions before the collapse occurred . . .

> The need to police professions (law and engineers alike) and to continually punish professional misconduct must be recognized. It is healthy; it is necessary. It instills public confidence—it removes from practice those who may cause loss of life. . . . *Most* important, however, is its prophylactic effect on the profession. It is an effective weapon against complacency. (1987, pp. 165–166)

Undoubtedly, some engineers continue to hold this harsh view, but the opinions of others have mellowed somewhat with time and with understanding of the complexity of the case.

The Human Factor

Jack Gillum, the EOR, was well respected. He published an excellent paper discussing the failure, his actions before and after, and the responsibilities of the EOR. His paper addressed two fundamental issues: the role and responsibility of the EOR and whether design responsibility can be delegated (Gillum 2000).

He presented his paper at ASCE's Second Forensic Congress, held in San Juan, Puerto Rico, in May 2000 (Rens et al. 2000). His presentation followed those of Moncarz, Luth, and Pfatteicher and closed a special plenary session for the congress. At the end of his presentation, the audience of engineers gave him a standing ovation. I was present and deeply moved, and I believe that the ovation was for his courage in presenting his story to his fellow engineers. Surely, the temptation to turn his back on the case and avoid discussing it must have been great. He has spoken on this topic to many groups across the country, including ASCE student chapters.

Gillum's paper begins with the phone call that engineers dread.

> It was a Friday evening at about 7:45 P.M. when my wife and I returned home to a ringing telephone. The call was from Herb Duncan, one of the principal architects with the Kansas City firm of Patty, Berkebile, Nelson, Duncan, Monroe, Lefebvre (PBNDML), the firm with whom we had worked on the Kansas City Hyatt. His first words to me—"There has been a collapse at the Hyatt"—shattered me to the core. Herb told me that one of the walkways had collapsed, and upon questioning he

indicated, "Several may have been killed and many injured." I asked him what had happened and he had no answer. He had called to inform me of the collapse and asked me what the weight of an individual walkway unit was, as the rescue workers had to determine the type of equipment needed to remove the debris. (Gillum 2000, p. 67)

Gillum continued to narrate the events of the hours and days after the phone call. He picked up Dan Duncan and chartered a plane, arriving by 11:15 P.M. that evening. The rescue effort was well underway. Arriving at the site, they quickly identified the problem with the connection, and quick calculations verified that the as-built capacity was grossly inadequate. Over the weekend, they met with the architects, their company personnel, and their attorneys. Gillum immediately took steps to clarify his firm's procedures for responsibility and accountability of design work (Gillum 2000, pp. 67–68).

The collapse was of great interest to the engineering profession. Many letters to the editor of *Engineering News Record* (*ENR*) discussed the problems with the original connection. An initial round of letters claimed that the connection could not be built, and a second round suggested several possible solutions, such as sleeve nuts and other details (Ross 1984, pp. 398–402). All of these writers knew, in hindsight, that the connection was critical and that it had failed.

Many articles written referred to the connection as a "designed connection," and many alternate, satisfactory designs were presented. Two are shown in Fig. 2-9, based on Kaminetzky (1991, pp. 220–221).

The NBS report recommended that concentrated loads never be applied to flanges of steel sections. Load-distributing plates (shown in Fig. 2-9) should be used (Feld and Carper 1997, pp. 222–223).

Of course, this analysis misses the point. It was not a poor connection design; it was a critical connection that somehow made it through the entire project *without* being designed. A proper design of this connection would have been easy if it had been flagged at any point during the process. The review of the facts presented in the Gillum paper closely follows Luth's analysis.

Gillum closes his paper by saying,

There is hardly a day that goes by that I don't think about the Hyatt collapse, the lives that were lost or marred forever, the relatives that lost their loved ones, and the effect it has had on Kansas City, the construction industry, and everyone connected with the project. My hope is that we, as a profession, can and will continue to learn, practice, dis-

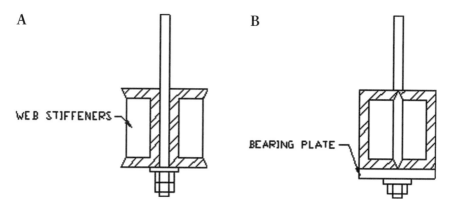

Figure 2-9a, b. Proposed alternative connection designs.

seminate, change, and adopt procedures and policies that will prevent a tragedy like this from occurring again. (2000, p. 70)

Educational Aspects

The free-body diagram is the basic equilibrium analysis tool to determine forces acting on a body. If the diagram is not drawn correctly, the forces cannot be calculated accurately, and the design may be unsafe. The importance of a correct free-body diagram may be shown through analysis of the Kansas City Hyatt Regency walkway collapse. Isolating the force of the box beam bearing against the nut shows that the force transfer changes between the configurations shown in Figs. 2-4a and 2-4b. The load on the connection was doubled, and it failed.

However, the issues of communication and responsibility in engineering and construction are of even more interest than the technical issue of the overloaded connection. This is not really a case study about free-body diagrams. This is about the design and construction process, the pressures for speed and economy inherent in any engineering endeavor, and the care and failure literacy necessary to protect the public.

This case was revisited in four papers published in a special issue of the ASCE *Journal of Performance of Constructed Facilities* (Gillum 2000, Luth 2000, Moncarz and Taylor 2000, and Pfatteicher 2000). In addition, authors of all four papers published and presented abbreviated versions of the papers at ASCE's Second Forensic Congress (Rens et al. 2000a). In his editor's note to the issue in which the four papers appear, the editor of the journal, Ken Carper, asked how much the business of designing and constructing buildings had truly changed.

Lessons Learned

According to Luth (2000, p. 59), some lessons have been learned from this failure:

- Procedures must ensure that every connection is designed. It must be possible to verify the capacity of every connection on the job without referring to the piece drawings.
- A formal peer review must be performed on every detail in structural drawings. Spot-checking is not sufficient.
- When questions come up, such as the strength of the hanger rods, they must be answered by referring to the project documents, in case the engineer's memory does not reflect what is actually on those documents.
- Changes in personnel require careful management to ensure that the project is handed off without errors creeping in.
- Any changes in concept must be handled by a formal review process. Changes should not be approved over the phone.

Gillum (2000, pp. 69–70) cites additional procedural changes that resulted from the Hyatt Regency collapse:

- Florida and Connecticut have mandated special inspection procedures for "threshold buildings," where some components are designed by registered engineers working for suppliers and manufacturers.
- New York State has adopted rules that state, in essence, that each engineer is responsible for his or her own work.
- ASCE published *Quality in the Constructed Project* in 1990, which was updated a decade later (ASCE 2000). This publication assigns responsibility for connections, as well as for other design elements.

Lessons Not Learned

According to Luth, the lessons that were not learned were the following:

1. The implications of structural failures, even though they are relatively rare, are far too serious for the scope of services to be defined through a "low bid" process.
2. City building departments do not—and cannot—provide adequate checking on major projects. A formal peer-review process on such

projects should be mandatory. Peer reviewers should be excluded from liability by law and should be held to a higher standard of qualifications than ordinary engineers.
3. Structural engineers cannot continue to allow the legal profession to define the duties, obligations, and specific actions that constitute "good engineering practice" on a case-by-case basis after the fact. (Luth 2000, p. 59)

Conclusions

According to Moncarz and Taylor (2000, p. 46), the collapse was due to "the doubling of the load on the connection resulting from an ill-considered change of an ill-defined structural detail." They blame the design process control and note that most investigation efforts concentrated on the design procedures, but not the process.

Petroski has commented that,

> Just as no one who knows of the Tacoma Narrows Bridge is likely to ignore the effect of wind on a suspension bridge, so no one who remembers the Hyatt Regency skywalks is likely to let another rod-beam connection escape close scrutiny. Thus the tragedy no doubt made a lot of inexperienced detailers suddenly much more experienced. And it is precisely to keep these lessons in the minds of young engineers that failures should be a permanent part of the engineering literature. (1985, p. 91)

When Petroski wrote those words, the disaster was still fresh. However, this collapse occurred before most current undergraduate students were born, and today's students are thus not likely to be familiar with the story unless it is discussed in the classroom.

Essential Reading

Essential reading for this case is the four papers published in the May 2000 special issue of the ASCE *Journal of Performance of Constructed Facilities* (Gillum 2000, Luth 2000, Moncarz and Taylor 2000, and Pfatteicher 2000), along with the editor's note by Carper. An excellent discussion of this case, with emphasis on ethical issues, is provided by Roddis (1993). This case study is featured on the History Channel's Modern Marvels series *Engineering Disasters 11* videotape and DVD.

Tacoma Narrows Bridge

On July 1, 1940, the Tacoma Narrows Bridge, connecting Seattle and Tacoma with the nearby Puget Sound Navy Yard, opened to the public after two years of design and construction. Its 853-m (2,800-ft) main span connected two 128-m (420-ft) towers from which cables were draped (Levy and Salvadori 1992).

Even though it then was the third longest span in the world, the Tacoma Narrows Bridge was much narrower, lighter, and more flexible than any other bridge of its time. It accommodated two lanes of traffic quite comfortably, while maintaining a sleek appearance. This appearance was so important to the bridge's designer, Leon Moisseiff, that he designed it without the use of stiffening trusses, replacing them with shallower plate girders of half the weight. This modification left the Tacoma Narrows Bridge with one-third the stiffness of the Golden Gate and George Washington bridges. These unique characteristics, coupled with its low damping ability, caused large vertical oscillations in even the most moderate of winds. This design soon earned it the nickname "Galloping Gertie" and attracted thrill seekers from all over (Feld and Carper 1997).

Oscillations had been noted during construction. A movie camera was placed atop the toll collector's building to film the motion. Vertical waves of up to 1.2 m (4 ft) double amplitude at up to 25 cycles per min were recorded. Winds as low as 13–17 km/h (8–10.5 mi/h) could generate motions of up to 760 mm (2.5 ft) double amplitude (Scott 2001, p. 47).

Although these undulations could be quite unnerving to motorists, no one questioned the structural integrity of the bridge. Leon Moisseiff was a highly qualified and well-respected engineer. Not only had he been the consulting engineer for the Golden Gate, Bronx–Whitestone, and San Francisco–Oakland Bay bridges, but he had also developed the methods used to calculate forces acting on suspension bridges (Levy and Salvadori 1992).

Moisseiff's development of the deflection theory substantially improved the efficiency of suspension bridges under gravity loads, making longer spans more economical. The deflection theory was first used in the design of the Manhattan Bridge. It replaced the much more conservative elastic theory. Moisseiff contributed in some way to almost every suspension bridge built during the first four decades of the 20th century (Scott 2001, pp. 15–17).

This was a period of increasing span lengths, coupled with increasing slenderness under the economic constraints of the period. One key element was the replacement of the stiffening trusses with plate girders, as with the Tacoma Narrows Bridge. This innovation emerged in Germany and provided a much shallower profile (Scott 2001, p. 29).

The Tacoma Narrows Bridge had a 1:350 span-to-depth ratio, about one-tenth of the 1:40 ratio of the Williamsburg Bridge. The 1:72 width-to-span ratio was much less than the 1:47 of the recently completed Golden Gate Bridge. One engineer who reviewed the plans for the Toll Bridge Authority, Charles Andrew, expressed concern about the design's narrowness. He anticipated that it would possibly lead to motorist discomfort. Another engineer named T. L. Condon also reviewed the design and believed that it was excessively narrow (Scott 2001, pp. 42–44). The narrowness and slenderness represented a dramatic leap beyond previous practice, and these engineers were concerned that this leap was not safe.

Even though the Tacoma Narrows Bridge adhered to all the safety standards and its oscillations were not considered a threat, F. B. Farquharson of the University of Washington began researching ways to reduce its motion. By studying how different winds affected a highly accurate model of the Tacoma Narrows Bridge and by testing new devices on it, Farquharson was able to propose helpful modifications to the bridge (Ross 1984).

Hold-down cables were installed and anchored to concrete blocks buried in the slopes. They were intended as interim measures, but one of the ties snapped a week after installation. Nevertheless, it was reinstalled in a matter of days. In addition to these hold-down cables, center stays and inclined cables, which connected the main cables to the stiffening girder, were installed. Wind tunnel testing by Farquharson suggested that aerodynamic measures, such as deflector vanes or fairings, could eliminate lift and reduce motions. The bridge collapsed before these remedies could be applied (Scott 2001, pp. 48–49).

Finally, an untuned dynamic damper, similar to the one that had proved quite successful in curtailing the motions of the Bronx–Whitestone Bridge, failed immediately after its installation on the Tacoma Narrows Bridge. It was discovered that the leather used in this device was destroyed during the sandblasting of the steel girders before they were painted, rendering it useless (Levy and Salvadori 1992).

The Collapse

A storm blew in on the night of November 6. At 7:30 A.M. on November 7, 1940, Kenneth Arkin, the chairman of the Washington State Toll Bridge Authority, arrived at the Tacoma Narrows Bridge. Although the wind was not extraordinary, the bridge was undulating noticeably and the stays on the west side of the bridge, which had broken loose, were flapping in the wind. Just before 10:00 A.M., after measuring the wind speed at 68 km/h

(42 mi/h), Arkin closed the bridge to all traffic because of its alarming movement. The bridge was twisting at 38 oscillations per min with an amplitude of about 1 m (3 ft) (Levy and Salvadori 1992). Suddenly, the north center stay broke, and the bridge began twisting violently in two parts. The bridge rotated more than 45°, causing the edges of the deck to have vertical movements of 8.5 m (28 ft) and the motions at times to exceed the acceleration of gravity (Ross 1984).

The onlookers had not been concerned up to this point. Although the wind was the strongest so far since the bridge had been built, the motions were in line with what had been observed earlier. However, the sudden change in motion was alarming. Without any intermediate stage, violent torsional movement started. The movement changed from nine or ten smaller waves to the two dominant twisting waves. Farquharson was surprised because he had not observed this phenomenon in his wind tunnel tests (Scott 2001, p. 50). Figure 2-10 shows the torsional motions of the bridge.

Figure 2-10. Tacoma Narrows Bridge torsional motion.
Courtesy University of Washington Special Collections, photo UW21413.

Two cars were on the bridge when this wild movement began: one with Leonard Coatsworth, a newspaper reporter, and his cocker spaniel, and the other with Arthur Hagen and Judy Jacox. All three people crawled to safety (Levy and Salvadori 1992, Scott 2001, p. 51).

The concrete sidewalks and curbs were beginning to crumble, and lampposts were coming loose. A couple of minutes later, the stiffening girders in the middle of the bridge buckled, initiating the collapse. Then the suspender cables broke, and large sections of the main span dropped progressively, from the center outward, into the river below. Loud bangs were heard, similar to gunfire. The weight of the sagging side spans pulled the towers 3.7 m (12 ft) toward the shore, and the ruined bridge finally came to rest. The side spans, no longer balanced by the main span, dropped 18 m (60 ft) before coming back to a final sag of about 9 m (30 ft) (Feld and Carper 1997, Scott 2001, pp. 51–53). The bridge's only fatality was Coatsworth's cocker spaniel. Figure 2-11 shows the bridge collapsing.

Figure 2-11. Tacoma Narrows Bridge collapse.
Courtesy University of Washington Special Collections, photo UW21422.

Because Farquharson was present that day studying the bridge, its collapse is well documented, photographed, and recorded on film (Levy and Salvadori 1992). A detailed moment-by-moment account of the collapse is provided by Scott (2001, pp. 49–53).

Causes of Failure

The questions before an incredulous engineering profession were troubling: how could a span designed to withstand 161 km/hr [100 mi/h] winds and a static horizontal wind pressure of 146 kg/m^2 [30 lb/ft^2] succumb under a wind of less than half that velocity imposing a static force one-sixth the design limit? (Scott 2001, p. 53)

Clearly, the new deflection theory was not by itself enough to safely design suspension bridges. Somehow, the dynamic effects of wind on flexible suspension bridges had to be accounted for.

The Public Works Administration (then PWA, later the Federal Works Agency or FWA) assembled a board of engineers to investigate. This board was known as the Carmody Board, after the PWA head, John M. Carmody. The board had three members: Othmar H. Ammann, Theodore von Kármán, and Glenn B. Woodruff. Ammann had designed many of the great bridges around New York City. Woodruff was the design engineer for the Transbay (San Francisco–Oakland Bay) Bridge. The third member, von Kármán, was the director of the Guggenheim Aeronautical Laboratory at the California Institute of Technology (Scott 2001, p. 53).

The PWA/FWA report found the following:

- The bridge was well designed and well built. Although it could safely resist all static forces, the wind caused extreme undulations, leading to the bridge's failure.
- Efforts were made to control the amplitude of the bridge's oscillation.
- No one realized that the Tacoma Narrows Bridge's exceptional flexibility, coupled with its inability to absorb dynamic forces, would make the wild oscillations that destroyed it possible.
- Vertical oscillations were caused by the force of the wind and caused no structural damage.
- The failure of the cable band on the north end, which was connected to the center ties, probably started the twisting motion of the bridge. The twisting motion caused high stresses throughout the bridge, which led to the failure of the suspenders and the collapse of the main span.

- Rigidity against static forces and rigidity against dynamic forces cannot be determined using the same methods.
- Subsequent studies and experiments were needed to determine the aerodynamic forces that act on suspension bridges.

In other words, the FWA concluded that because of the Tacoma Narrows Bridge's extreme flexibility, narrowness, and lightness, the random force of the wind that day caused the torsional oscillations that destroyed the bridge. The FWA believed that wind-induced oscillations approached the natural frequencies of the structure, causing resonance (the process by which the frequency of an object matches its natural frequency, causing a dramatic increase in amplitude). This theory explained why the relatively low-speed wind 68 km/h (42 mi/h) caused the spectacular oscillations and destruction of the Tacoma Narrows Bridge (*ENR* 1941).

The FWA's theory, however, is not the only explanation. Many people believe that this explanation overlooks the important question as to how wind, random in nature, could produce a periodic impulse. One explanation proposed by von Kármán, an aeronautical engineer, attributed the motion of the bridge to the periodic shedding of air vortices, which created a wake known as a von Kármán's street. This wake reinforced the structural oscillations, eventually causing the collapse of the bridge. The problem with this theory is that the calculated frequency of a vortex caused by a 68 km/h (42 mi/h) wind is 1 Hz, whereas the frequency of the torsional oscillations of the bridge measured by Farquharson was 0.2 Hz (Petroski 1991).

Another explanation, proposed by Billah and Scanlan, admits that vortices associated with the von Kármán's street were shed but suggests that they did not affect the motion of the bridge. Another kind of vortex, one associated with the structural oscillation itself, was created, with the same frequency as the bridge. The resonance between the bridge and these vortices caused excessive motion, destroying the bridge (Billah and Scanlan 1991). Although these three theories differ as to what exactly caused the torsional oscillations of the bridge, they all agree that the extreme flexibility, slenderness, and lightness of the Tacoma Narrows Bridge allowed these oscillations to grow until they destroyed it.

A contributing factor may have been slippage of a band that retained the cables.

> For months the motions, while disturbing, had been symmetrical, and roadway had remained flat. The lampposts on the sidewalks stayed in the vertical plane of the suspension cables even as they rose, fell, and twisted. But on November 7 a cable band slipped out of place at

mid-span, and the motions became asymmetrical, like an airplane banking in different directions. The twisting caused metal fatigue, and the hangers broke like paper clips that had been bent too often. (Freiman and Schlager 1995b, p. 226)

Technical Concerns

The Tacoma Narrows Bridge collapse showed engineers and the world the importance of damping, vertical rigidity, and torsional resistance in suspension bridges (Ross 1984). Once the threat of twisting was realized, there are many ways that the disaster of the Tacoma Narrows Bridge could have been averted. Making any one of the following adjustments could have prevented the collapse:

- using open stiffening trusses, which would have allowed the wind free passage through the bridge;
- increasing the width-to-span ratio;
- increasing the weight of the bridge;
- dampening the bridge;
- using an untuned dynamic damper to limit the motions of the bridge;
- increasing the stiffness and depth of the trusses or girders; or
- streamlining the deck of the bridge (Levy and Salvadori 1992).

Earlier Problems with Suspension Bridges

The Tacoma Narrows Bridge collapse highlighted the importance of failure literacy, a knowledge of historical failures and their causes. Although the mid-20th century engineering profession was surprised by the incident, many similar stories were buried in the engineering history of the previous century. Two early collapses delayed the progress of suspension bridges in Europe: the Dryburgh Abbey Bridge in England in 1818 and an 1824 failure of a German span. Descriptions of the oscillations before the collapse of the Brighton Chain Pier, also in England, in 1823 provided clear evidence of resonance and were similar to those of the Tacoma Narrows Bridge. Thomas Telford's Menai Straits Bridge in Wales almost collapsed in a gale barely a week after opening, with 4.9-m (16-ft) oscillations of the deck. Again, the torsional motions were similar. The repaired and strengthened bridge failed 10 years later. These early bridges were suspended using chains, rather than the later wire cables (Scott 2001, pp. 3–4, 55–56).

Between 1818 and 1889, the wind destroyed or seriously damaged 10 suspension bridges (Petroski 1994). Most of these bridges, like the Tacoma

Narrows Bridge, had small width-to-span ratios, ranging anywhere from 1:72 to 1:59. They also experienced severe twisting right before collapse, as the Tacoma Narrows Bridge did. In 1854, the bridge over the Ohio River at Wheeling, West Virginia, also collapsed because of wind. Many other bridges suffered a similar fate (Levy and Salvadori 1992). Detailed descriptions of eight suspension bridges destroyed by wind during this time are provided by Scott (2001, pp. 55–58).

In fact, it was not until the success of John Roebling that suspension bridges became widely accepted. Through his understanding of the importance of deck stiffness and knowledge of past failures, Roebling was able to make suspension bridges accepted as strong railway bridges (Feld and Carper 1997). Roebling relied on a combination of weight, trusses, and stays to stiffen his bridges, which were able to avoid the stability problems of earlier designs (Scott 2001, p. 9).

Soon, however, the success of suspension bridges led engineers to forget the failures of the previous century. Once again, suspension bridges evolved toward the longer, sleeker designs, as engineers forgot the cornerstone of their success, wind resistance (Petroski 1994).

Engineer D. B. Steinman designed the first two-plate girder suspension bridges in North America, the Thousand Islands Bridge across the St. Lawrence River between the United States and Canada (1938) and Deer Isle Bridge in Maine (1939). The projects were designed and built under severe economic constraints. Both bridges exhibited substantial oscillations in the wind and required stiffening with cable ties. Vertical and torsional motions were both observed. The cable ties substantially reduced, but did not eliminate, the motions, and were thought to be an adequate repair. Engineer Othmar Ammann's slender Bronx–Whitestone Bridge also opened in 1939. It was longer and heavier than Steinman's two bridges and was also prone to oscillation during construction and after opening. Both midspan cable ties and friction brakes were installed (Scott 2001, pp. 33–40).

Problems were also observed in Norway, which had built 22 thin, flexible, rolled I-beam suspension bridges between 1927 and 1937. Several spans in the 70–230-m (230–755-ft) range were prone to either mild or severe oscillations. The worst was Norway's longest span, the 230-m (755-ft) Fykesesund Bridge, which had oscillations up to 1.6 m (5.2 ft) (Scott 2001, p. 58).

Later Problems with Suspension Bridges

The Tacoma Narrows Bridge collapse focused attention on the performance of similar structures. Several of the plate girder suspension bridges built at about the time of the Tacoma Narrows Bridge remain susceptible to aerodynamic oscillations.

The Deer Isle Bridge has continued to have problems, despite an elaborate retrofit system of stays, cross-stays, and braces. The bridge had to be closed temporarily in May 1978. The structure had been modified four times without success; the fifth attempt added streamlining fascias to the plate girders. The Fykesesund Bridge had a shallow truss added. The Bronx–Whitestone Bridge had tower stays and a truss added in the 1940s, but those retrofits did not succeed in eliminating significant motion. There has also been concern about long-term fatigue from vibration. In the 1980s, a passive tuned mass damper was added. As of 2001, a more comprehensive and radical retrofit was proposed to streamline and further reinforce the bridge (Scott 2001, pp. 354–356). Overall, it seems that nothing has been able to compensate for the inherent flexibility of plate girder suspension bridges.

Other great suspension bridges built leading up to the Tacoma Narrows Bridge have been susceptible to oscillation, but not to the same degree. The George Washington Bridge in New York City has shown vertical movements, but it is much heavier in relation to its length than the Tacoma Narrows Bridge was. The Golden Gate Bridge has oscillated despite its stiffening truss (Scott 2001, p. 59).

Ethical Concerns

In the face of new technology, how do we balance public welfare and progress? If Moisseiff had designed a bridge similar to the ones that had already proven their stability, the Tacoma Narrows Bridge would never have collapsed. It would also have been significantly more expensive. On the other hand, if engineers never tried innovative techniques, suspension bridges might never have been built at all.

Moisseiff tried to create a longer, sleeker, less expensive bridge by pushing the limits of technology. Every time engineers push the limits of technology, they risk a similar loss, sometimes even a loss of life. How much is too much? When is a possible advance worth a risk to public safety? What can the engineering profession do to make the implementation of new technology safer?

Evolution of Bridge Aerodynamics

From the dramatic photographs and film footage of the Tacoma Narrows Bridge collapse, it was obvious that the bridge was twisting in two alternating waves just before it came apart. What remained to be understood was how the wind forces and the bridge had interacted to produce the waves. It is not intuitively obvious how lateral pressure would produce such oscillations.

Possible explanations include flutter, buffeting, turbulence, galloping, and vortex shedding. The Tacoma Narrows Bridge collapse, followed by Farquharson's pioneering work, launched a new discipline of bridge aerodynamics. Because subsequent suspension bridges generally used stiffening trusses or streamlined box girders, the behavior of plate girders in long suspension bridges has mostly been of academic interest for new bridges (Scott 2001, pp. 345–351).

At present, one commonly accepted explanation for the torsional oscillations of the Tacoma Narrows Bridge is the shedding of vortices from the upwind plate girder, causing alternating high and low pressure moving across the deck. Scott (2001) said,

> The explanation is as follows . . . the bluff windward girder creates a leading edge vortex, say, on the top side of the deck, and a previously created vortex is positioned beneath the deck centerline. The newly generated upper vortex . . . creates . . . a positive suction force as it begins to move across . . . at the same time, the bottom vortex, rotating with an opposite spin, moves across the leeward half of the cross section. (p. 352)

These forces alternate and reinforce each other once the critical velocity has been reached.

Billah and Scanlan (1991) reviewed several dozen undergraduate physics textbooks, as well as high school science texts and other similar works, and decided that the way the Tacoma Narrows Bridge case is often portrayed is oversimplified and misleading. It is often used to illustrate the concept of resonance, which is when a system capable of oscillation is acted on by a periodic series of impulses at or near one of the natural frequencies of the system, causing oscillations with increasing amplitude. This explanation implies that the wind had a periodic, oscillating nature close to one of the natural frequencies of the bridge. The term "wind gusts" has been used to suggest a periodic forcing function.

Periodic vortex shedding has also been suggested, and the earlier vertical oscillations of the bridge occurred because of this shedding. This situation is, however, much more complex than a simple case of resonance. Periodic vortex shedding is also a self-limiting mechanism, and as the amplitude increases, so do compensating forces (Billah and Scanlan 1991, p. 120).

However, the final, torsional destructive mechanism was different. At a wind speed of 68 km/h (42 mi/h), the frequency of natural vortex shedding would be close to 1 Hz. However, Farquharson observed that the torsional motions occurred at 12 cycles per min, or 0.2 Hz, which would rule out natural vortex shedding as the final destructive mechanism. In fact, it took about three decades of research and model testing to determine that the actual cause was single-degree-of-freedom torsional flutter (Billah and Scanlan 1991, pp. 120–121). The explanation of the correct mechanism relies heavily on Scanlan's work in the early 1970s with my friend and colleague, John Tomko, a member of the faculty at Cleveland State University for many years.

Billah and Scanlan (1991) said, "Instead of characterizing the force that excited the single-degree oscillator as a purely external function of time, it was characterized as an aerodynamic *self-excitation* effect that was able to impart a net negative damping characteristic to the system" (p. 121). Because damping of a resonant system causes the amplitude of motion to decay, negative damping increases the amplitude of motion until the structure becomes unstable and is destroyed.

Controversy persisted for some time as to whether the vortices caused the motion of the bridge or the motion caused the vortices. The work of von Kármán is often interpreted to imply the former. Billah and Scanlan (1991, p. 122) argue that it was the other way around: that the motion of the bridge produced a flutter wake, which led to the failure.

> Could this be a resonant phenomenon? It would appear not to contradict the qualitative definition of resonance quoted earlier, if we now identify the source of the periodic impulses as *self-induced*, the wind supplying the power, and the motion supplying the power-tapping mechanism. If one wishes to argue, however, that this was a case of *externally forced linear resonance*, the mathematical distinction ... is quite clear, self-exciting systems differing strongly enough from ordinary linear resonant ones. The texts we have consulted have not gone this far in explanation. (Billah and Scanlan 1991, p. 122)

Billah and Scanlan (1991, p. 122) note that further confusions often encountered in the literature include conflating the bridge flutter of a bluff body with the flutter of a streamlined aircraft wing. They are in fact quite different because flow around highly streamlined airfoils is different from flow around the sharp corners of bridge girders.

Educational Aspects

A suspension bridge's resistance to motion is provided by mass in addition to stiffness. The resistance to rotation is provided by the mass moment of inertia I, defined as

$$I = \int_m r^2 dm \qquad (2\text{-}1)$$

The redesigned bridge had four lanes rather than two, and 10-m (30-ft) deep stiffening trusses rather than the 2.4-m (8-ft) deep plate girders. Because the mass m of the bridge doubled, and the moment arm r also doubled, the new bridge had 8 times the resistance to rotational motion of the original, destroyed bridge, even ignoring the contribution of the stiffening trusses.

Strength of structures is important, but stiffness—resistance to deflection and vibration—is also important. Optimized structures, using minimum amounts of material, can be subject to instability.

Conclusions

The Carmody Board, and von Kármán in particular, made a significant effort after the investigation to advance bridge aerodynamics. Until that point, designers' concerns had primarily been focused on treatment of wind as a static pressure. The post-Tacoma era ushered in a fertile field of wind tunnel testing for bridges (Scott 2001, pp. 67–70).

The actual failure mechanism of the Tacoma Narrows Bridge is difficult to understand and is still not clear to many, even after decades of research. On the other hand, the remedies are relatively straightforward and were known to John Roebling: mass of the deck, stiffening trusses, and stays. Subsequent suspension bridge designs have benefited from the lessons of the Tacoma Narrows Bridge.

Essential Reading

A thorough review of the technical aspects of the collapse and of suspension bridge aerodynamics in general is provided in *In the Wake of Tacoma: Suspension Bridges and the Quest for Aerodynamic Stability* (Scott 2001). The case is also discussed in detail in Section 11 of Chapter V of *Engineers of Dreams: Great Bridge Builders and the Spanning of America* by Petroski (1995, pp. 294–308). Other, briefer discussions of the Tacoma Narrows Bridge case are provided by Levy and Salvadori (1992) and Wearne (2000, pp. 37–45). Billah and Scanlan (1991) provide an important discussion of the destruction mechanism.

An illustrated account of the collapse is provided on the University of Washington Libraries Special Collections website at http://www.lib.washington.edu/specialcoll/exhibits/tnb/default.html.

Aircraft Impacts

Buildings are designed for different loads and load cases, such as wind, earthquake, or snow. One case that is rarely considered is vehicle impact, although this type of event occurs. On a small scale, drivers (impaired or not) sometimes mistake a small building for a continuation of the roadway.

Aircraft impacts are more serious because of the large size and speed of the vehicles. Historically, some structures, such as nuclear reactor containment vessels, have been designed against aircraft impacts. This design case is analyzed because, even though such an event used to be considered highly unlikely, the consequences of a release of radioactive material are considered catastrophic.

Four buildings in Manhattan, the Empire State Building, 40 Wall Street, and the two World Trade Center (WTC) towers, were hit on three occasions by aircraft. The first two crashes happened in the 1940s, and the third happened more than a half century later. The crashes into the Empire State Building and 40 Wall Street were accidents, whereas the WTC attacks were deliberate. The performance of the WTC towers during and after the attack is discussed in Chapter 6.

On July 28, 1945, a 10,000-kg (10-ton or 22,000/32.2 = 683 slug) B-25 bomber crashed into the Empire State Building at an estimated 400 km/h (250 mi/h) into the north face of the 79th floor. The flight ceiling was less than about 300 m (1,000 ft), and visibility was poor. The aircraft was evidently weaving between the skyscrapers of Manhattan. One of the plane's two engines passed entirely through the building and into a building across the street, starting a fire. The other engine fell down the Empire State Building's elevator shaft. Although several people were killed by the burning gasoline from the plane, the building remained standing. The building was a heavy, robust, riveted structure, and the aircraft impact missed a critical column. Almost a year later, another twin-engine plane lost in the fog hit the 40 Wall Street building, which also survived (Levy and Salvadori, 1992, pp. 25–30).

There is an ironic foreshadowing of the September 11, 2001, WTC attacks in the B-25 crash. The deaths in the building were caused by the burning aircraft fuel and not by the impact. Even for the day, the B-25 medium bomber was a relatively small aircraft, and the amount of fuel carried was small. In contrast, the massive fuel loads of the much larger airliners that hit the WTC towers would prove enough to destroy both buildings.

To estimate the force applied to the building, we must estimate either the deceleration of the bomber as it crashed into the building (and use the force-acceleration method), estimate the *distance* it took the bomber to come to rest (and use the work-energy method), or estimate the *time* it took the bomber to come to rest (and use the impulse-momentum method). The bomber did not pass through the building, so a distance of 10–20 m (33–66 ft) for the plane to come to rest could be used for calculations. If the bomber came to rest in 20 m (66 ft) with a constant deceleration, the force would be about 4,440 kN (1 million lb) exerted over about 0.25 s. The kinematics of impact are shown in Fig. 2-12.

All three of these methods are derived from $F = ma$. The acceleration is assumed to be constant and is integrated once for velocity and twice for distance traveled. With an initial velocity of 400 km/h or 111 m/s (250 mi/h and 364 ft/s), a final velocity of 0, an initial position of 0, and a final position of 20 m (66 ft), the velocity equation becomes

$$v(t) = at + 111 \qquad (2\text{-}2)$$

and the position equation becomes

$$s(t) = \frac{at^2}{2} + 111t \qquad (2\text{-}3)$$

Solving these two equations simultaneously gives a t of about 0.36 s and an acceleration of 308.6 m/s² or (1,010 ft/s²). This change is actually a deceleration, which would affect the direction of the force.

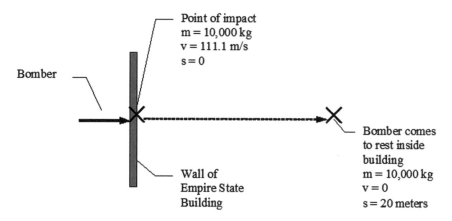

Figure 2-12. Bomber impact into Empire State Building.
Source: Delatte (1997).

Force-Acceleration Method

From $F = ma$,

$$F = 10{,}000 \text{ kg} \times 308.6 \text{ m/s}^2 = 3{,}086 \text{ kN} \tag{2-4}$$

or
$$F = 683 \text{ slug} \times 1{,}010 \text{ ft/s}^2 = 690{,}000 \text{ lb}$$

Work-Energy Method

Initial kinetic energy plus work done equals final kinetic energy:

$$T_1 + \sum U_{1-2} = T_2 \tag{2-5}$$

where T_1 = initial kinetic energy,
U_{1-2} = work done between position 1 and position 2 (initial and final), and
T_2 = final kinetic energy.

Initial kinetic energy is

$$T_1 = \frac{1}{2} mv^2 \tag{2-6}$$

where m = mass (10,000 kg or 683 slug, given above) and v is initial velocity of 111.1 m/s (364 ft/s). Thus, the initial kinetic energy is 61,700,000 newtons meter (N·m) or 45,500,000 ft-lb.

$$U_{1-2} = F \times d \tag{2-7}$$

where F is the unknown force and d is the distance of 20 m (66 ft). The final kinetic energy is 0 because the final velocity is 0 after the bomber comes to rest. Finally, to get the force:

$$F = \frac{U_{1-2}}{d} = \frac{61{,}700{,}000 \text{ N} \cdot \text{m}}{20 \text{ m } (1000)} = 3{,}086 \text{ kN} \tag{2-8}$$

or
$$F = \frac{U_{1-2}}{d} = \frac{45{,}500{,}000 \text{ ft-lb}}{66 \text{ ft}} = 690{,}000 \text{ lb}$$

which is exactly the same result as before.

Impulse-Momentum Method

The initial momentum plus the impulse equals the final momentum, or

$$m_1 v_1 + \int_{t_1}^{t_2} F\, dt = m_2 v_2 \tag{2-9}$$

For a constant force, this becomes

$$m_1 v_1 + F(t_2 - t_1) = m_2 v_2 \tag{2-10}$$

Substituting previous values of 10,000 kg (683 slug) for m_1 and m_2, 111.1 m/s (364 ft/s) for v_1, 0 for v_2, and 0.36 s for the time elapsed ($t_2 - t_1$), we once more calculate:

$$F = \frac{m_1 v_1}{(t_2 - t_1)} = \frac{10{,}000 \times 111.1}{0.36} = 3{,}086 \text{ kN} \tag{2-11}$$

or

$$F = \frac{m_1 v_1}{(t_2 - t_1)} = \frac{683 \times 364}{0.36} = 690{,}000 \text{ lb}$$

This problem is useful for demonstrating the equivalence of the three methods. A bomber crashing into a building is an extremely unlikely occurrence. However, the consequences of the collapse of a large building are grave. How should the engineering profession guard against rare but severe events? In a large and complicated project, is there an obligation to go beyond building code requirements? For example, for nuclear reactor containment vessels, what extreme events should engineers consider? What about other buildings subject to terrorist attacks?

This case study was originally written in 1997 (Delatte 1997), well before the terrorist attacks of September 11, 2001. The same methods may be applied to calculate the impact forces applied by any aircraft of known weight and speed, with an estimate of the average distance before the aircraft (or its parts) came to rest. The characteristics of the aircraft used to attack the two WTC towers are provided in Chapter 6.

Essential Reading

This case study is discussed in Levy and Salvadori (1992).

42 BEYOND FAILURE

Other Cases

Many other cases presented in this book are relevant to a statics, dynamics, or combined course.

Mianus River Bridge

The failure of the Mianus River Bridge in Connecticut (Chapter 6) illustrates the principles of three-dimensional equilibrium, stability, and redundancy. The bridge span that fell was 30 m (100 ft) long and three travel lanes wide, with a 53° skew. Three travel lanes plus the sidewalk and shoulder were about 12 m (40 ft) wide. The bridge span was a parallelogram in plan view.

The hanger system used to support the bridge span was roughly analogous to a link (two-force member) support at each corner of the parallelogram. In theory, then, the structure with four cable supports has one redundant support and is stable if one is removed.

However, the remaining three cable supports do not constrain the free-body against movement under all loading conditions, as shown in Fig. 2-13. The removed cable is shown as dashed.

Without external load, the center of gravity of the bridge section lies on the straight line between cables B and C, and there is no tension in cable A. If a vehicle drives onto the bridge between point A and a straight line

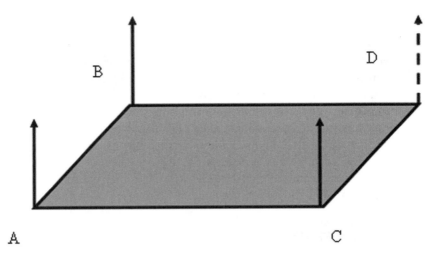

Figure 2-13. Stability of a skewed simple span bridge.

between points B and C, then cable A supports the load. If, however, the vehicle drives past the line, then the bridge section topples over. Therefore, all four supports must be intact to keep the bridge safe, and the bridge is conditionally unstable if any one support is removed.

Agricultural Product Warehouse Failures

The agricultural product warehouse failures case from Chapter 4 may be used to discuss force resultants and support conditions. This case study discusses two cottonseed warehouse buildings investigated by the author. Although the two buildings did not collapse, the distress caused a considerable economic loss to the owner. Lateral pressure of cottonseed, or soil, is similar to static water pressure. The differences are in the unit weight of the material and in pressure coefficients that depend on cohesion and internal friction.

In the case of the cottonseed, we have a triangular pressure distribution. The resultant of the distributed load is the area of a triangle (½ times the base times the height), and acts outward ⅓ of the height from the base. Of course, the main problem was that evidently the cottonseed pressure was never taken into account in the design.

The fixed support at the base of each column in the warehouse implies two force resultants and one moment resultant (in two dimensions). In the design of the foundation, the force resultants were accounted for but not the moment. As a result, the column base plates twisted and cracked the concrete foundation, as shown in Fig. 4-3.

I noted that the connection was assumed to be pinned but was built as fixed. I was asked which is correct. Either one could be correct, but the assumption and the design and detailing have to match. If it is built as fixed, then some prying moment must be assumed and the foundation must be reinforced to resist the tensile forces. It could be built with a pin, although that would probably require a complicated joint detail. Assuming a pin and yet building a fixed support is a case of the worst of both worlds, and it almost guarantees a cracked foundation.

Quebec Bridge

The Quebec Bridge (Chapter 3) was a cantilever truss bridge, and tension and compression members may easily be identified using statics. There are two identical cantilever spans on either side of the bridge, with a simple span suspended between them. The suspended span has compression in the top members and tension in the bottom members, and the four cantilever arms have tension in the top members and compression in the bottom.

As the two arms of the bridge were built out from the pier, the moments on the truss arms increased, and the compressive stresses in the bottom chords of each arm also increased. Both the method of joints and the method of sections, traditionally taught in statics courses, may be used to analyze the compressive strut forces at the different stages of bridge construction.

Austin Concrete Dam Failure

The case of the 1911 Austin, Pennsylvania, dam failure is provided in Chapter 9. This case illustrates the importance of dry friction analysis of gravity dams. Dry friction in the case of a gravity dam is somewhat of a misnomer if there is water seepage under the dam, but it does illustrate the fundamental principle of dam stability. Stability of gravity dams may be easily analyzed using statics.

A gravity dam may fail by either overturning or sliding. Because a gravity dam generally has a relatively constant cross section across the stream, it may be analyzed in two dimensions. Consider the free-body diagram of a simplified dam cross section as shown in Fig. 2-14.

The gravity dam shown has a height H_d and a base width B and is shown as a triangle of concrete, excluding any parapet at the top. The height of the water behind the dam is H_w, which is kept below H_d to prevent an overtopping failure of a dam. Generally, a spillway is used to make sure that the water cannot rise above H_w.

The magnitude and location of the weight of the dam, W_D, and the water force behind the dam, F_w, may now be calculated, per unit length of the dam in either meters or feet:

$$W_D = \frac{1}{2} H_d B \gamma_c \qquad (2\text{-}12)$$

and
$$F_w = \frac{1}{2} H_w^2 \gamma_w \qquad (2\text{-}13)$$

where γ_c and γ_w are the unit weight of concrete and water, respectively. These weights are typically assumed to be 2,300 kg/m^3 (22.6 kN/m^3; 145 lb/ft^3) and 1,000 kg/m^3 (9.81 kN/m^3; 62.4 lb/ft^3). The resultant of the dam weight is ⅔ B from the downstream toe of the dam, and the water force resultant is ⅓ H_w from the base of the dam.

The other two forces acting on the free-body diagram shown in Fig. 2.14 are the normal force N and the friction force F_f acting on the base of the dam. The location of N is unknown and is designated by x. The

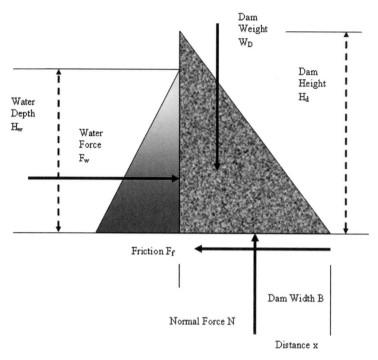

Figure 2-14. Free-body diagram of a concrete gravity dam (Austin, Pennsylvania).

maximum value of the friction force F_{max} is N times the static coefficient of friction μ.

$$F_f \leq F_{max} = \mu N \qquad (2\text{-}14)$$

It is now possible to calculate the factors of safety against overturning and sliding for the dam. The dam overturns when the overturning moment caused by F_w is greater than the resisting moment provided by W_D. When the dam tips, $x = 0$. Therefore, the factor of safety against overturning is

$$FS_{overturning} = \frac{\Sigma M_{resisting}}{\Sigma M_{overturning}} = \frac{W_D\left(\frac{2B}{3}\right)}{F_w\left(\frac{H_w}{3}\right)} = \frac{\frac{2}{3}\left(\frac{1}{2}H_d B^2 \gamma_c\right)}{\frac{1}{3}\left(\frac{1}{2}H_w^3 \gamma_w\right)}$$
$$= \frac{2H_d B^2 \gamma_c}{H_w^3 \gamma_w} = \frac{4.65 H_d B^2}{H_w^3} \qquad (2\text{-}15)$$

For the Austin Dam, the height H_d of the dam was 15 m (49 ft), the width of the dam base B was 9.8 m (32 ft), and the height of the spillway H_w was 13 m (42 ft). The crest, which is ignored in this calculation, was only 0.76 m (2.5 ft) thick. Therefore, the factor of safety against overturning was 3.15. Note that this factor of safety is purely a function of the geometry of a concrete dam.

The factor of safety against sliding is determined by comparing the sliding force of the water to the resisting force of friction against the base of the dam in the x direction. First, from summing forces in the y direction, the normal force N is equal to the weight of the dam, W_D. The factor of safety against sliding is

$$FS_{sliding} = \frac{\Sigma F_{resisting}}{\Sigma F_{sliding}} = \frac{F_{max}}{F_w} = \frac{\mu W_D}{F_w} = \frac{\mu \frac{1}{2} H_d B \gamma_c}{\frac{1}{2} H_w^2 \gamma_w} = \frac{2.32 \mu H_d B}{H_w^2} \quad (2\text{-}16)$$

For the Austin Dam, the factor of safety against sliding was 2.06μ. Therefore, the dam will slide if $FS_{sliding} \leq 1$ or $\mu \leq 0.48$.

These calculations assume that there is no water seeping under the dam. If, however, there is seepage, then the water pressure provides an uplift force against the base of the dam. The Austin Dam had clear evidence of seepage. The revised free-body diagram of the dam, including seepage, is provided as Fig. 2-15.

The uplift force is at a maximum on the upstream side of the base of the dam and is 0 at the toe, so there is a triangular pressure distribution acting at ⅔ B from the toe of the dam. Essentially, the uplift force reduces the effect of the dam weight, W_D. Inclusion of the uplift force F_u modifies the two factor-of-safety equations:

$$FS_{overturning} = \frac{\Sigma M_{resisting}}{\Sigma M_{overturning}} = \frac{(W_D - F_u)\left(\frac{2B}{3}\right)}{F_w\left(\frac{H_w}{3}\right)} \quad (2\text{-}17)$$

$$FS_{sliding} = \frac{\Sigma F_{resisting}}{\Sigma F_{sliding}} = \frac{F_{max}}{F_w} = \frac{\mu(W_D - F_u)}{F_w} \quad (2\text{-}18)$$

Clearly, the uplift force reduces the factor of safety in each case. Determining the uplift force requires knowledge of the properties and permeability of the soils under the dam and is a fairly complicated process.

Two techniques are used to improve the safety of concrete gravity dams. One technique is cutoff walls or trenches to block water flow and reduce the

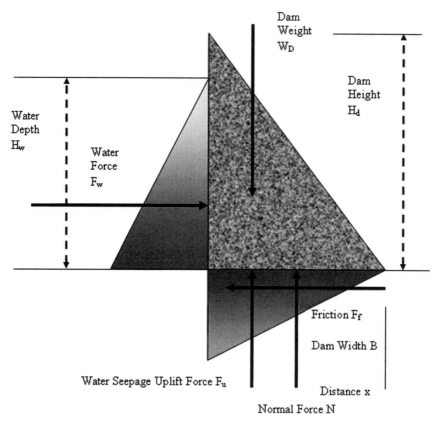

Figure 2-15. Revised free-body diagram of dam, considering seepage.

hydrostatic pressures. The other is the use of keyways under the dam, which resist sliding. Soil resistance against a buried toe of the dam, or against a keyway, improves the factor of safety against sliding. The revised equation becomes

$$FS_{sliding} = \frac{\Sigma F_{resisting}}{\Sigma F_{sliding}} = \frac{F_{max} + F_{soil}}{F_w} = \frac{\mu(W_D - F_u) + F_{soil}}{F_w} \quad (2\text{-}19)$$

The Austin Dam's foundation was 1.2 m (4 ft) below grade, but there was minimal soil resistance to sliding from such a shallow embedment. A shallow block of soil simply slid along with the dam foundation.

This two-dimensional analysis ignores possible additional safety provided by friction and interlock between the dam abutments and the adjacent hillside. However, because the Austin Dam broke into pieces when the concrete cracked before failure, this resistance was not available.

Some old dams are now being improved by providing downstream buttresses of roller compacted concrete (RCC). Essentially, an RCC dam is built against the downstream face of the existing dam, increasing W_D and B, which improves both factors of safety.

Vaiont Dam Reservoir Slope Stability Failure

A landslide or slope stability failure in soil mechanics represents another sort of "dry" friction problem. One failure of this type was the Vaiont Dam Reservoir in Italy, discussed in Chapter 7.

There are some significant differences between gravity dam analysis and slope stability analysis. For one, overturning a slope can be neglected because it doesn't happen. Another important difference is that the location of the failure surface isn't known in advance for a slope stability failure. Typically, multiple potential failure surfaces must be tried, and the failure surface with the lowest factor of safety provides the overall slope factor of safety.

That being said, the effects of water on slope stability are important. Saturated soil increases the weight and driving force of the soil mass. At the same time, water pressure within soil reduces the frictional resistance (μ) along the failure plane. The numerator in the $FS_{sliding}$ equation decreases while the denominator decreases, and if the equation equals 1, there is a landslide or slope stability failure.

As a result, landslides tend to happen during or just after periods of heavy precipitation. The Vaiont Dam slide represents a classic example.

L'Ambiance Plaza Collapse

One of the many theories proposed for the construction failure of L'Ambiance Plaza in Bridgeport, Connecticut, discussed in Chapter 4, was the instability of the wedges temporarily holding the lift slabs in place. Wedges may be analyzed using dry friction.

At L'Ambiance Plaza, the failure theory assumes that the wedges rolled out sideways because the load and the support were not in line. The wedges appeared stable in two dimensions, but not in three. This example illustrates the universal principle of statics and of structural engineering that although we may analyze in two dimensions, actual structures and machines are three-dimensional and behave accordingly.

Teton Dam

The Teton Dam was an earth fill dam in eastern Idaho. The story of the Teton Dam failure is recounted in Chapter 7. Unlike concrete gravity dams, overturning and sliding are not issues with earth fill dams. These dams, however, are vulnerable to erosion from water within the dam or its foundation or flowing over the dam. As a result, water pressures and erosion of soils become important.

Static water pressure is a function of the dam height. When complete, the Teton Dam earth embankment had a maximum height of 93 m (305 ft) above the riverbed and formed a reservoir of 356 million m³ (288,000 acre-ft) when filled to the top. The static water pressure P_w is given by

$$P_w = H_w \gamma_w \tag{2-20}$$

Because γ_w = 1,000 kg/m³ (9.81 kN/m³; 62.4 lb/ft³), the pressure at the base of the dam at full pool becomes 912 kPa (132 lb/in.²).

The water stored behind a dam represents a significant amount of potential energy, which is why dams are so valuable for generating hydroelectric power. Of course, if the dam fails, then the potential energy becomes kinetic energy, often with widespread destruction and loss of life. Gravitational potential energy is

$$U_{1-2} = W(\Delta z) \tag{2-21}$$

where W is weight and Δz is the change in height, in this case the depth of the water held back behind the reservoir. The initial and final kinetic energy, T_1 and T_2, are given by

$$T = \frac{1}{2}mv^2 \tag{2-22}$$

Initial velocity before the dam break, of course, is 0. Therefore, the work-energy equation is

$$T_1 + \sum U_{1-2} = T_2 = \frac{1}{2}mv_2^2$$

and

$$0 + W(\Delta y) = T_2 = \frac{1}{2}mv_2^2 \tag{2-23}$$

Solving for final velocity v_2,

$$v_2^2 = \sqrt{\frac{2W(\Delta z)}{m}} = \sqrt{\frac{2mg(\Delta z)}{m}} = \sqrt{2g(\Delta z)} \tag{2-24}$$

Assuming that the dam is almost full, Δz would be about 91 m (300 ft), and g is 9.81 m/s^2 (32.2 ft/s^2), so v_2 is 42.2 m/s (139 ft/s). This relation may also be derived by using the Bernoulli equation from elementary fluid mechanics as illustrated in Chapter 8.

As the flood proceeded down the riverbed, there was an increase in velocity due to the increase in Δy from the slope, and the theoretical velocity at any point downstream may be estimated from the elevation difference. On the other hand, this example neglects the energy losses due to friction and turbulence and the energy dissipated from smashing houses, structures, and other sources of energy loss. Also, dam failures occur over some period of time, however short, not suddenly. The initial velocity due to a small cut in the dam would be low, would increase as the cut widened, and would decrease as the head behind what was left of the dam dropped.

Similar calculations may be made for the failures of the South Fork Dam that caused the Johnstown Flood in Pennsylvania and the Malpasset Dam in France (both in Chapter 8) and the Austin concrete dam failure in Pennsylvania (Chapter 9). The potential energy stored behind a dam is considerable, and it represents substantial risk if the dam fails. The impacts of dam failures are discussed in much more detail in Chapters 7 and 8.

3

Mechanics of Materials

MECHANICS OF MATERIALS, SOMETIMES CALLED STRENGTH OF materials or mechanics of solids, builds directly on statics. It introduces more powerful but complex tools for solving engineering mechanics problems. The tools, combined with equilibrium (statics), may be used to solve statically indeterminate problems.

Design and performance of connections is a topic that may be illustrated through a wide variety of failure case studies. Connections are vulnerable points that often fail before other structural elements.

Fatigue failures are also well represented in case studies, particularly where stress concentrations are involved. These problems also occur at connections.

Quebec Bridge

The story of the Quebec Bridge spans two decades, from the founding of the Quebec Bridge Company in 1887 to the bridge's collapse in 1907. A cantilever bridge was proposed as the most feasible design to bridge the harsh, icy waters of the St. Lawrence River. The bridge collapsed during construction on August 29, 1907, killing 75 workers. Only 11 of the workers on the span were saved. Some bodies were never found. A second attempt to bridge the St. Lawrence River was then made, with a completely new design. However, it also suffered a partial collapse when the middle span fell into the river. Thirteen workers were lost in the second collapse. The bridge was finally completed in 1917, and it stands today. A review of the history of the bridge and the two collapses was written by Middleton (2001).

Conception, Design, and Construction

The St. Lawrence River was the main channel of trade for Quebec during the summer. During the winter, it filled with ice, and trade was completely cut off until the river iced over and travel was possible again across a dangerous ice bridge. The desire to bridge the St. Lawrence River was fueled by Quebec's need to be competitive in trade. Montreal already had the Grand Trunk railway system, which passes through it from the west connecting it to Toronto. Quebec was left even farther behind when Montreal began construction in 1854 of the Victoria Bridge, which was completed in 1859, connecting it to western ports. This development quickly established Montreal as Canada's leading eastern port. Although the need was great, the job of bridging the St. Lawrence would prove to be no easy task (Middleton 2001, pp. 7–8).

The St. Lawrence River was approximately 3.2 km (2 mi) wide at its narrowest section. Its waters were about 58 m (190 ft) deep at its middle. The velocity of the river reached 13 or 14 km/h (8 or 9 mi/h) at times, and the tides ranged as high as 5 m (18 ft). During the winter, ice became stuck in the narrowest part of the channel and piled up as high as 15 m (50 ft) (Middleton 2001, p. 3).

Interest in building the Quebec Bridge arose as early as 1850. However, the project did not gather momentum until 1887, when a group of businessmen and political leaders came together and formed the Quebec Bridge Committee. Because of the high level of interest in the project, the Canadian Parliament passed an act that incorporated the committee into the Quebec Bridge Company, with a million-dollar capital and the power to issue bonds (Middleton 2001, p. 27).

The company now faced the problem of financing the great bridge. Government funding was requested. However, no money could be awarded for the project until the bridge site was selected. With some financial help from the local Quebec legislature, preliminary surveys were made. In 1898, after years of debate, the Chaudiere site was selected from the three recommended sites to be the location of the Quebec Bridge. With a site selected, bridge design proposals poured in (Middleton 2001, p. 26).

On June 16, 1897, the chief engineer of the Quebec Bridge Company, Edward Hoare, wrote to a friend, who was the president of the Phoenix Bridge Company (Holgate et al. 1908). In response, the Phoenix Bridge Company sent its chief engineer to meet with the Quebec Bridge Company's chief engineer at an American Society of Civil Engineers (ASCE) meeting in Quebec in 1897. The Phoenix Bridge Company offered to prepare plans for the bridge free of charge. In return, the Quebec Bridge Company would then be obligated to give the contract for construction of the bridge to the Phoenix Bridge Company. Theodore Cooper, who learned of the Quebec Bridge project at the ASCE meeting, offered his consulting services to the Quebec Bridge Company (Middleton 2001, pp. 32–33).

Edward Hoare had never before worked on a bridge longer than about 90 m (300 ft). The company decided to hire a consulting engineer, and Theodore Cooper was selected from a list of six prominent engineers for the project (Holgate et al. 1908, Middleton 2001, p. 36). Cooper was an independent consultant operating out of New York City. He was one of the foremost American bridge builders of his day. To Cooper, this project would be the crowning achievement to his life's work.

Petroski (1995) notes Cooper's strong qualifications for this project. In his long career, he had written an award-winning paper pioneering the use of steel for railway bridges and had prepared general specifications for iron and steel bridges. His method of accounting for railroad loads on bridge structures became widely used (Middleton 2001, p. 37).

Proposals were called for on September 6, 1898, and received until March 1, 1899 (Holgate et al. 1908). They were then reviewed by Cooper. The specifications called for a cantilever structure. However, suspension bridge designs were allowed, providing they came with their own set of specifications. Earlier, noted French engineer Gustave Eiffel had considered the problem and observed that a cantilever design would be superior to either a suspension or an arch bridge for the Quebec site (Middleton 2001, pp. 29–30).

A cantilever bridge has balanced cantilever arms that extend on either side of the two towers at the end of the clear span. Two anchor arms extend to the shore, and two other arms extend across the gap. A small simple span bridge is suspended between cantilever arms to close the gap.

Six proposals were submitted for the superstructure and two for the substructure. After review, Theodore Cooper stated, "I hereby conclude and report that the cantilever superstructure plan of the Phoenix Bridge Company is the 'best and cheapest' plan and proposal submitted to me for examination and report" (Holgate et al. 1908, p. 15, Middleton 2001, p. 34).

The Phoenix Company had been in correspondence with Cooper throughout this process (Holgate et al. 1908). In addition, the Quebec

Bridge Company was in favor of the Phoenix Bridge Company to win the contract (Tarkov 1986). This resolution provided at least an impression that the process was not fair and open, even though Holgate et al. (1908) concluded that it was.

Two months later, the company awarded contracts to the Phoenix Bridge Company for construction of the superstructure and to the Davis Firm for construction of the substructure. However, the Phoenix Bridge Company refused to sign a contract with the Quebec Bridge Company because of the financial provisions, which left the bridge company open to considerable risk. Financial matters were finally resolved in 1903, when additional funds became available from a government grant. On June 19, 1903, a final contract was entered into between the two companies, and the name of the Quebec Bridge Company was changed to the Quebec Bridge and Railway Company (Middleton 2001, pp. 45–47).

The proposed Quebec Bridge would bridge the St. Lawrence River approximately 14 km (9 mi) north of Quebec, connecting into the Grand Trunk rail line. The cantilever arms would reach a distance of 171.5 m (562.5 ft). They were to support a suspended span with a length of 205.7 m (675 ft). It would stand 45.7 m (150 ft) above the river. The initial design's clear span length was 487.7 m (1,600 ft).

However, in May 1900, this span was increased to 548.6 m (1,800 ft) by Theodore Cooper. He stated that this would eliminate the uncertainty of constructing piers in such deep water, lessen the effects of ice, and shorten the time of construction of the piers. Although there were sound engineering reasons for this change, it was also true that the lengthening of the span would also make Cooper the chief engineer for the longest cantilever span in the world (Petroski 1995, Middleton 2001, p. 46).

Construction of the bridge officially began on October 2, 1900, after a grand ceremony. The Quebec Bridge Company had enough funds to begin erecting the substructure. The completed piers would stand approximately 8 m (26.5 ft) above the highest water level. The piers were made of huge granite facing stones with concrete backing. The top 5.8 m (19 ft) of each pier was made of solid granite. The piers were tapered 1 in 12 (1 in. per foot) until they reached the dimensions of 9.1 m (30 ft) by 40.5 m (133 ft) at the top. Each pier rested on a concrete-filled caisson that was 14.9 m (49 ft) wide, 7.6 m (25 ft) high, and 45.7 m (150 ft) long, weighing 16.2 MN (1,600 tons) (Middleton 2001, pp. 48–50).

Because of the unprecedented size of this structure, innovative construction methods proved necessary. These methods were well documented in frequent reports with extensive illustrations in *Engineering Record,* the forerunner of the present-day *Engineering News Record.*

Under a separate contract with the Quebec Bridge Company, the Phoenix Bridge Company began the construction of the approach spans in 1902 and completed them in 1903. Erection of the superstructure portion of the bridge did not begin until July 22, 1905. The Phoenix Bridge Company agreed to have the structure completed by the last day of 1908. Otherwise, the company would pay $5,000 per month after this deadline to the Quebec Bridge Company until the project was finished (Holgate et al. 1908, Middleton 2001).

Events Leading Up to the Collapse

As the bridge was erected, workers and supervisors found noticeable midpoint deflections in some of the chords. When the workers tried to rivet the joints between these chords, the predrilled holes did not line up. In addition, bends (deflections) were observed in some of the most heavily loaded compression members. Over time, some of the member deflections increased.

The last photograph taken of the bridge before the collapse is shown in Fig. 3-1. The panels were numbered from 1, at the outer ends of the cantilever arms, through 10, at the piers. The anchor arm panels added

Figure 3-1. The Quebec Bridge just before the collapse.
Courtesy Archives Canada.

the notation "A". The A9L notation, therefore, refers to the chord located in the anchor arm, within the ninth panel, and on the left or west side of the bridge. Some of the major chords with their corresponding deflections, along with the dates of measurement, are presented in Table 3-1.

Deflections were first noticed as early as mid-June and were reported to Cooper by his on-site inspector, Norman McLure. Compression members had been cambered, so that under load the joints would line up and could be riveted together. However, some of the joints failed to close. The lack of fit was puzzling because the truss members had been carefully designed so that the rivet holes would line up during construction. Both men presumed that the relatively small deflections had occurred because of some unknown preexisting condition. They were not alarmed (Middleton 2001, p. 72).

Subsequent inspections turned up more deflecting chords in August. Again, these findings were reported to Cooper on the same day that they were discovered. Cooper wired a message back referring to chords 7L and 8L, asking, "How did bend occur in both chords?" The chief engineer of the Phoenix Company replied, saying that he did not know (Middleton 2001, pp. 72–73).

The chief design engineer for the Phoenix Company, Peter Szlapka, was certain that the bend was put in the chord ribs at the shop. He later admitted that he never actually saw the chords in question. However, Norman McLure wrote, "One thing I am reasonably sure of, and that is that the bend has occurred since the chord has been under stress, and was not present when the chords were placed." While this dispute of how the bend occurred

TABLE 3-1. *Bridge Member Deflections*

Date of observation	Member	Amount of deflection	
		mm	inches
June 15		1.5–6.5	1/16–1/4
June	A3R & A4R	1.5–6.5	1/16–1/4
June	A7R & A8R	1.5–6.5	1/16–1/4
June	A8R & A9R	1.5–6.5	1/16–1/4
June	A8L & A9L	19	3/4
August 6	7L & 8L	19	3/4
August	8L & 9L	8	5/16
August 20	8R	Bent	Bent
August	9R & 10R	—	—
August 23	5R & 6R	13	1/2
August 27	A9L	57	2 1/4

in chords 7L and 8L was going on, McLure reported to Cooper another similar bend in chords 8L and 9L (Middleton 2001, pp. 73–74). The members with these deflections were the lower chords of the truss on either side of the pier, the members with the highest compressive loads under the negative moment across the pier.

A disturbing pattern was emerging. The members under the highest compressive loads were gradually buckling. These members were built up with latticing, and as they deflected, higher stresses were placed on the latticing as well as on the rivets attaching the lattices to the main compression members.

Being dissatisfied with the theories offered by the engineers on site, Cooper developed his own theory.

> None of the explanations for the bent chord stand the test of logic. I have evolved another theory, which is a possible if not the probable one. These chords have been hit by those suspended beams used during the erection, while they were being put in place or taken down. Examine if you cannot find evidence of the blow, and make inquiries of the men in charge.

McLure did as he was instructed, and reported back to Cooper that there was no evidence of such an incident (Middleton 2001, p. 74).

Some of the engineers were unconcerned about the problem, believing that it was nothing serious. Others were still insisting that the bends were the result of a preexisting condition. The manufacturer guaranteed that all the members had been perfectly straight when they left the yard. Another incident had occurred during the 1905 construction season, when chord A9L was dropped and bent while being handled in the storage yard. It was repaired and placed into the structure. Although at the time the repair was thought to be satisfactory, this member was later found to be the triggering cause of the collapse.

Cooper, although the most experienced, seemed to be the most confused by the problem. He was 60 years old at the time he accepted the position of consulting engineer for the Quebec Bridge project. He also accepted the responsibility of shop inspector of the steel fabrication and erection. His health was poor, and because of this, he never visited the site once construction began on the superstructure. His consulting services were based on the information that was reported to him by others. Cooper's official eyes and ears on the construction site belonged to Norman McLure, a young civil engineer who had been appointed by Cooper himself. Cooper was also poorly compensated for his work (Petroski 1995, Middleton 2001).

McLure continued to argue that the bends in the members had occurred after they were installed. Some of the workers had observed the deflecting chords and were concerned enough to not report to work for a few days. However, when McLure and Cooper disagreed on the cause of the deflections, McLure did not have the confidence to contradict Cooper. Work continued on the bridge. There had already been a three-day strike over working conditions, and the workers who had not agreed with the new terms had left. This walkout greatly reduced the number of workers on the project, and there was concern that a temporary stoppage would cause more workers to leave and delay the project.

After another routine inspection, chord A9L was placed under observation when its initial deflection of 19 mm (¾ in.) had increased to 57 mm (2¼ in.) in less than two weeks. The opposite chord, A9R, was bent in the same direction. There was growing concern about the deflections. One of the construction foremen decided to halt work on the bridge until matters could be resolved. On August 27, the same day that construction was halted, McLure sent a message to Cooper informing him that construction would not resume until he reviewed the matter. The next day, McLure went to New York to seek advice from Cooper (Middleton 2001, pp. 78–79).

The erection foreman who had ordered the work to stop changed his mind, and with reassurance from Edward Hoare, the chief engineer of the Quebec Bridge Company, resumed work again that day. The only reason given for this decision was in a note from Hoare to Cooper stating that "the moral effect of holding up the work would be very bad on all concerned and might also stop the work for this season on account of losing the men." Two days later, news of the matter reached the Phoenixville office, and the project superiors there met and discussed the problem. They relayed a message back by telephone, saying that it was safe to resume work on the bridge. They had somehow reached the conclusion that the bends in the chords had occurred before they left the yard. The Phoenix Company's chief engineer had stated that the chord members were carrying "much less than maximum load" (Middleton 2001, pp. 78–80).

In the meantime, McLure was meeting with Cooper in New York. Neither of the men was aware that construction had resumed on the bridge. After a brief discussion between the two men on August 29, Cooper wired the Phoenixville office saying, "Add no more load to the bridge until after due consideration of facts. McLure will be over at five o'clock." Cooper's reasoning for informing the Phoenixville office, rather than directly relaying it to the site, was that he felt that action would be taken faster if the information went to the site through Phoenixville. McLure had assured Cooper that he would wire the information to the site on his way to the Phoenixville office. In his haste to get to his destination, he neglected to send the information (Middleton 2001, p. 80).

The message from Cooper reached the Phoenixville office at 1:15 P.M. It was ignored in the absence of the chief engineer. At around 3:00 P.M., Phoenixville's chief engineer returned to his office. After seeing the message, he arranged for a group meeting as soon as McLure arrived. McLure arrived at roughly 5:15 P.M., and the men discussed the circumstances briefly before deciding to wait until the next morning to decide a course of action (Middleton 2001, pp. 78–80).

Collapse

Meanwhile, back at the construction site, at about the same time the decision makers in Phoenixville were ending their meeting, the Quebec Bridge collapsed at 5:30 P.M. The thunderous roar of the collapse was heard 10 km (6 mi) away in Quebec (Middleton 2001, p. 80). The entire south half of the bridge, approximately 189 MN (19,000 tons) of steel, fell into the waters of the St. Lawrence River within 15 s. Of the 86 workers on the bridge at the time, only 11 survived.

The A9L bottom compression chord, which was already bent, gave way under the increasing weight of the bridge. The load transferred to the opposite A9R chord, which also buckled. The piers were the only part of the structure that survived. The wreckage is shown in Fig. 3-2, looking from the south bank toward the pier. Of 38 Caughnawaga Mohawk ironworkers who had left their village to work on the bridge, 33 were killed and two were injured (Middleton 2001, p. 84).

The Royal Commission Report

The Governor General of Canada formed a Royal Commission composed of three civil engineers, whose task was to investigate the cause of the collapse. They were Henry Holgate of Montreal, John George Gale Kerry of Campbellford, Ontario, and John Galbraith of Toronto. Their completed report consisted of more than 200 pages plus 21 appendices. As stated by Middleton (2001, p. 91), ". . . the thoroughness and objectivity of their inquiry and report stand even today as models of their kind."

The immediate cause of failure was found to be the buckling of compression chords A9L and A9R. The official report attributed the collapse to a number of reasons. Listed below are some of the major findings (Holgate et al. 1908, pp. 9–10):

1. The collapse of the Quebec Bridge resulted from the failure of the lower chords in the anchor arm near the main pier. The failure of these chords was due to their defective design.

Figure 3-2a, b. The collapsed Quebec Bridge. Courtesy Archives Canada.

2. We do not consider that the specifications for the work were satisfactory or sufficient, the unit stresses in particular being higher than any established by past practice. The specifications were accepted without protest by all interested.
3. A grave error was made in assuming the dead load for the calculations at too low a value and not afterwards revising this assumption. This error was of sufficient magnitude to have required the condemnation of the bridge, even if the details of the lower chords had been of sufficient strength, because, if the bridge had been completed as designed, the actual stresses would have been considerably greater than those permitted by the specifications. This erroneous assumption was made by Mr. Szlapka and accepted by Mr. Cooper, and tended to hasten the disaster.
4. The loss of life on August 29, 1907, might have been prevented by the exercise of better judgement on the part of those in responsible charge of the work for the Quebec Bridge and Railway Company and for the Phoenix Bridge Company.
5. The failure on the part of the Quebec Bridge and Railway Company to appoint an experienced bridge engineer to the position of chief engineer was a mistake. This resulted in a loose and inefficient supervision of all parts of the work on the part of the Quebec Bridge and Railway Company.
6. The work done by the Phoenix Bridge Company in making the detail drawings and in planning and carrying out the erection, and by the Phoenix Iron Company in fabricating the material was good, and the steel used was of good quality. The serious defects were fundamental errors in design.
7. The professional knowledge of the present day concerning the action of steel columns under load is not sufficient to enable engineers to economically design such structures as the Quebec bridge. A bridge of the adopted span that will unquestionably be safe can be built, but in the present state of professional knowledge a considerably larger amount of metal would have to be used than might be required if our knowledge were more exact.

Causes of Failure

The fall of this massive bridge can be traced back to several technical factors. The top and bottom chords for the anchor and cantilever arms of a bridge were typically designed as straight members. This common practice made the fabrication of these members easier. The bottom chords for the anchor and cantilever arms in the Quebec Bridge were slightly curved, as shown in Fig. 3-1, for aesthetic reasons. This design added difficulty to the fabrication of such unusually large members. The curvature also increased the secondary stresses on the members, reducing their buckling capacity. Secondary stresses are much more dangerous in compression than in tension members.

Another concern during the erection of the bridge was the joints. The ends of all the chords were shaped to allow for the small deflections that were expected to occur when the chords came under their full dead load. These butt splices were bolted to allow for movement. The splices initially touched only at one end, and would not fully transfer their load until they had deflected enough for full bearing at the splices. At this point, they were to be permanently riveted in place. The result was to be a rigid joint that transferred loads uniformly across its area to ensure only axial loading. Great care had to be taken while working around these joints until they were completely riveted (Middleton 2001, pp. 70–72).

Adding to the design problems, Cooper increased the original allowable stresses for the bridge. He allowed 145 MPa (21 kip/in.^{2}) for normal loading and 165 MPa (24 kip/in.^{2}) under extreme loading conditions. These allowances were questioned by the bridge engineer for the Department of Railways and Canals as being unusually high. The new unit stresses were accepted based solely on Cooper's reputation (Holgate et al. 1908).

Cooper developed an allowable compressive stress formula (in MPa and lb/in.^{2}) based on the slenderness ratio (l/r) of the member:

$$\text{allowable compressive stress} = 165 - 0.69 \, (l/r) \text{ MPa} \\ = 24{,}000 - 100 \, (l/r) \text{ lb/in.}^2 \quad (3\text{-}1)$$

where l = length of compression member,
r = radius of gyration = $\sqrt{(I/A)}$
I = moment of inertia, and
A = cross-sectional area.

Cooper's formula is compared to contemporary allowable stresses from the American Institute of Steel Construction (AISC 1998), as well as the 96.5 MPa (14 kip/in.^{2}) compressive allowable stress adopted for the second bridge (Middleton 2001, p. 107) in Fig. 3-3. AISC curves are shown for both 230 MPa (33 kip/in.^{2}) and 250 MPa (36 kip/in.^{2}) steel. Cooper's allowable stresses are higher than those in use today by 3.3–8.7% over a range of slenderness ratios from 10 to 100. Given the lower and uncertain quality of the materials available to Cooper, along with the less developed state of

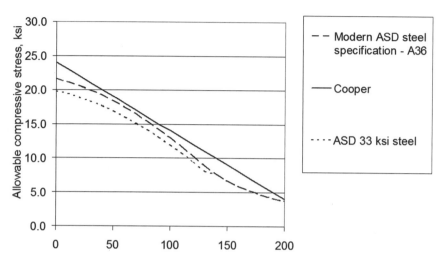

Figure 3-3. Comparison of allowable compressive stresses.

knowledge of compression members at that time and the lack of testing of compression members, Cooper's formula represents an unsafe practice.

The Quebec Bridge was an enormous structure, and little was known about how it would behave. The Quebec Bridge Company lacked funding to test adequately. Cooper had required extensive tests on the eyebars, which formed major tension members in the top chords. He did not require the compression members to be tested. Later, Cooper would state the reason for this: "There is no machine or method existing by which any such test could be made" (Middleton 2001, p. 56). However, when the company did finally secure funds for testing, Cooper rejected the idea, stating that too much time had been wasted already.

Another oversight was that the stresses were not recalculated once Cooper increased the span from 487.7 m (1,600 ft) to 548.6 m (1,800 ft). The stress calculations were based on the original span dimension. Once this error had been discovered and brought to Cooper's attention, he immediately made an estimate of the new stresses. He found that they would be approximately 7% higher. Weights were then recalculated from the new information. They were found to be as much as 10% higher than those previously calculated (Middleton 2001, p. 65). The initial design weight for the bridge was expected to be 276 MN (62 million lb). The real weight of the bridge was estimated at 325 MN (73 million lb), an increase of 18% (Tarkov 1986).

In the rush after the final financial arrangements of 1903, the necessity of revising the assumed weights was overlooked by the engineers of the Phoenix Bridge Company and by those of the Quebec Bridge Company alike, with the result that the bridge members would have been considerably overstressed after completion.

Table 3-2, based on the Royal Commission Report, compares the actual and assumed dead loads.

> The difference between these two sets of concentrations indicates a fundamental error in the calculations for the bridge. In a properly computed bridge the assumed dead load concentrations upon which the make-up of the members is based should agree closely with the weight computed from the dimensions in the finished design and with actual weights. (Holgate et al. 1908, p. 57)

At the time this error was discovered, a large portion of the fabrication had been completed and a considerable amount of bridge erection was finished. Cooper accepted these heavier loads and stresses, in addition to the already high stresses set for the bridge, as being within acceptable limits. His only other alternatives were to start over, strengthen the bridge in place, or abandon the project.

Procedural and Professional Aspects

Cooper insisted on retaining full control of the project, even at a considerable distance. Schreiber recommended that the governmental agency of the Department of Railways and Canals hire a consultant on their behalf. This engineer would, in a sense, be double-checking Cooper's work and ultimately have the final authority. After finding this out, Cooper, the Quebec Bridge Company, and the Phoenix Bridge Company immediately objected. In a letter to Edward Hoare, Cooper wrote, "This puts me in the position of

TABLE 3-2. *Comparison of Assumed and Actual Dead Load*

Element	Assumed dead load		Actual dead load		Percent difference
	kN	lb	kN	lb	
Half suspended span	21,538	4,842,000	25,328	5,694,000	17.6
Cantilever arm	58,740	13,205,200	70,300	15,804,000	19.7
Anchor arm	59,240	13,317,600	77,034	17,318,000	30.0

a subordinate, which I cannot accept" (Middleton 2001, p. 52). Cooper met with Schreiber. After this meeting, Schreiber revised his recommendation to eliminate the need to hire an additional project consultant. The amended order in council to the Department of Railways and Canals failed to define clearly how much authority Cooper would have over the project.

According to Middleton (2001, p. 53),

> While there remained a requirement to submit all plans for the approval of the chief engineer of Railways and Canals, it was treated as a perfunctory formality. When later modifications to the specifications appeared desirable, Cooper made them without reference to the government engineers, and there was no evidence that Schreiber ever questioned a decision made by Cooper or interfered in any way with the work.

This opinion is supported by a statement in a letter from John Sterling Deans, Chief Engineer for the Quebec Bridge Company, to Cooper. In it he wrote, "The suggested action by Mr. Schreiber would place the business in a much worse condition than it was originally in." He also wrote, ". . . it is simply being necessary to have Schreiber's signature as a matter of form." To further these implications, in another letter from Deans to Cooper, he wrote,

> I have written him again [Schreiber], and urged him to stop entirely this proposed plan, and explaining that the sole purpose of the order in council was to give you the final authority to settle all details, the government approval being a mere formality, and in this way save time which was so valuable. (Holgate et al. 1908)

No clear chain of responsibility or command existed. It was assumed that the final authority rested with Theodore Cooper. All concerns were directed toward him, even though, because of illness, he was unable to travel to the job site. There was no one on the job site qualified to oversee this type of work or in a position to make a decision, including a decision to stop work if conditions became unsafe. Whenever the need arose, the authorities on site would confer with each other before making any decision. On the few occasions where a decision had been reached, there was hesitation in carrying it out. The authors of the Royal Commission Report wrote,

> It was clear that on that day the greatest bridge in the world was being built without there being a single man within reach who by experience, knowledge and ability was competent to deal with the crisis. (Holgate et al. 1908)

66 BEYOND FAILURE

Capacity of Compression Members

The commission had suspected that the A9L compression chord failed because of improper latticing. Compression tests were performed on one-third scale models of the compression chords in November 1907 and January 1908 to verify this theory. The compression chord members for the Quebec Bridge consisted of four multilayered ribs (Fig. 3-4). They were stiffened by the use of diagonal latticing to make them act as one unit. During testing, the lattice system failed explosively because its rivets sheared, immediately followed by buckling of the chord. This result confirmed the commission's findings that the chords were inadequately designed. In Schneider's opinion,

> These members consist of four separate ribs, not particularly well developed as compression members, and their connections to each other are not of sufficient strength to make them act as a unit. (Middleton 2001, p. 97)

Member Cross-Section

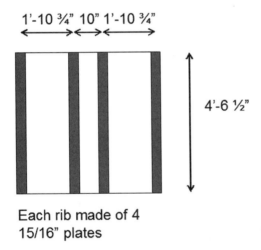

Each rib made of 4
15/16" plates

Plan View of Chord

Figure 3-4. Built-up compression members.

Schneider wrote in his report, which was published as an appendix to the Royal Commission Report,

> If a column is made up of several shapes or parts, they have to be connected in such a manner that they will act as a unit. In an ideal column each part would take its share of the load and no connection would be required. In practice, however, as stated before, bending will occur before the buckling load is reached, causing shearing strains which have to be transferred through the connections, as latticing, tie plates or cover plates. These connection parts have, therefore, to perform the same function as the web of a girder or the web system of a truss. (Holgate et al. 1908, p. 193)

This unprecedented large-scale testing and studying of compression members and their connections led to major advances in the field of engineering. Bridge specifications were improved after this collapse (Shepherd and Frost 1995). Another advance was the formation of two organizations, the American Institute for Steel Construction in 1921 and the American Association of State Highway Officials in 1914. These organizations advanced the field of engineering by providing the means to fund research, which had become too difficult and expensive for fabricating companies to conduct on their own (Roddis 1993).

Ethical Aspects

Several ethical concerns may be illustrated through this case. The major one is that deformations went unheeded for so long. The engineers on site argued among themselves as to the cause but did not stop the work. Although the workers who failed to report to work because of the deformations lacked the technical expertise, they seemed to be the only ones who understood what was really happening to the bridge (Middleton 2001, p. 78). Engineers and others in charge must be open-minded to the ideas of the laborers because many laborers have years of experience.

Another ethical concern was Cooper's rejection of an independent engineer to check his work. His decisions were not questioned, even when they seemed unusual. An independent consultant might not have allowed the higher than normal design stresses. Some of the other errors, such as the underestimated dead loads and the failure to recheck the weight, could have been discovered before the bridge collapsed. In the end, "Cooper's engineering expertise became the sole factor that was relied upon for assuring structural integrity of the bridge" (Roddis 1993).

Aftermath

The lives of those involved with the Quebec Bridge, from the designers to the construction workers, were forever changed after the accident. None, however, was affected as much as the families of the ones who died and Theodore Cooper. Edward Hoare went to work for the National Transcontinental Railway Commission. John Deans continued to serve as chief engineer of the Phoenix Bridge Company. He eventually became vice president of the company. Szlapka continued his duties as chief designer for the Phoenix Company. Cooper withdrew from practice to live out a lonely retirement. He died only two days after the Prince of Wales officially dedicated the completed Quebec Bridge (Middleton 2001).

The Second Bridge

After the collapse, the government took over the design and construction of the new bridge. This takeover also provided the financial support for the project. The second bridge was substantially heavier than the first. Petroski (1995) compared dimensions of the two bridges, showing the dramatic increase in member sizes. The cross-sectional area of the critical compression member for the old bridge had been 543,000 mm^2 (842 in.^{2}), whereas that of the new bridge was 1,250,000 mm^2 (1,941 in.^{2}) (Petroski 1995, p. 113, Middleton 2001, p. 116). The second bridge is shown in Fig. 3-5.

The second attempt to bridge the St. Lawrence also encountered problems. The project suffered a second collapse in 1916, when a casting in the lifting apparatus broke, causing the center suspended span to fall into the water as it was being hoisted into place from a barge. Thirteen workers lost their lives in this accident. The 50-MN (5,000-ton) span sank to the bottom of the river to rest beside the wreckage of the first bridge, which still remains there today. The second bridge was finally completed in 1917 and weighed two and a half times as much as the first one (Tarkov 1986). The construction of the second bridge was well documented by a report of the Government Board of Engineers of the Canada Department of Railways and Canals (Modjeski et al. 1919).

Conclusions

At Quebec, the greatest bridge in the world was under construction in 1907 under severe financial constraints, with inadequate funds provided for either engineering work or the bridge construction itself. These constraints had delayed engineering analysis and led to adoption of unconserva-

Figure 3-5. The second Quebec Bridge.
Courtesy Archives Canada.

tive specifications. When the error in the calculation of the dead load was identified, the measures taken to reanalyze the structure were not adequate. On this project, virtually every conflict between safety and economy was resolved in favor of economy. Most of the poor engineering decisions were made by the prominent consulting engineer, Theodore Cooper.

Cooper's reluctance to travel to the site because of his poor health led to confusion about responsibility and site supervision. When skilled ironworkers observed the growing deflections, indicating a gradual collapse of the structure, a confident site supervisor might have realized the gravity of the situation and halted construction. However, the engineers on the site lacked the confidence and the authority to contradict Cooper's judgments.

The Royal Commission Report, which investigated the collapse and identified the engineering and procedural errors that had led to it, remains an important document in the field of forensic engineering. The lessons learned from the case had many important effects on the engineering profession, particularly in Canada and the United States.

Essential Reading

Middleton's *Bridge at Quebec* (2001) is an important and readable reference. The Royal Commission report (Holgate et al. 1908) is also interesting and is available in many university engineering libraries. Another account is provided by Tarkov (1986). The case is discussed at length, along with its place in the evolution of bridge design, in Chapter III of Petroski's 1995 book, *Engineers of Dreams: Great Bridge Builders and the Spanning of America* (pp. 66–121).

Point Pleasant Bridge

The Point Pleasant Bridge, also known as the Silver Bridge, shown in Figs. 3-6 and 3-7, provided crossing for U.S. Route 35 over the Ohio River. It connected Columbus, Ohio, to Charleston, West Virginia. The bridge crossed between Point Pleasant, West Virginia, and Gallipolis, Ohio. The Point Pleasant Bridge not only allowed direct access between the state capitals of Ohio and West Virginia but also provided clearance for the cargo ships that passed through the Ohio River. Figure 3-8 shows the bridge cross section.

Figure 3-6. The Point Pleasant Bridge.
Source: NTSB (1970).

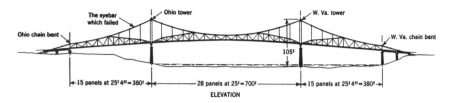

Figure 3-7. The Point Pleasant Bridge.
Source: Scheffey (1971).

On December 15, 1967, the bridge collapsed while handling heavy traffic and killed 46 people. This accident changed the nation's view of bridges across the country. The government ultimately changed the way bridges were inspected to ensure safety and functionality.

Design and Construction

The design criteria for the bridge were established by the J. E. Greiner Engineering Company of Baltimore, Maryland. They prepared plans for a suspension bridge and completed a bid document with specifications for the bridge. This document included calculated stresses for the structure and an acceptable basic design of a suspension bridge (ASCE 1968). The engineering company set the estimated cost of the suspension bridge at $825,000 (Lichtenstein 1993). An incentive was added to the bid document if an alternative bridge could be built under $800,000 with acceptable design stresses. The winning bid was from the American Bridge Company of Pittsburgh, which designed an eyebar suspension bridge consisting of heat-treated eyebars and a rocking base (Steinman 1924). The contractor's proposal emphasized a unique material that, in combination with the new design, would cost less.

D. B. Steinman was one of the great suspension bridge engineers of the 20th century, and his work culminated in the great Mackinac suspension bridge in Michigan. He died in 1960, a few years before the Point Pleasant Bridge collapse. The Point Pleasant Bridge was a Florianopolis truss of a type originally developed in Brazil. Similar bridges were later built in Australia (1936) and Japan (1954). Key features of the Florianopolis suspension bridge type are described by Scott (2001, pp. 18–19):

- It acts as a continuous truss between anchorages.
- The suspending eyebar chain also acts as the top chord of a variable-depth stiffening truss.

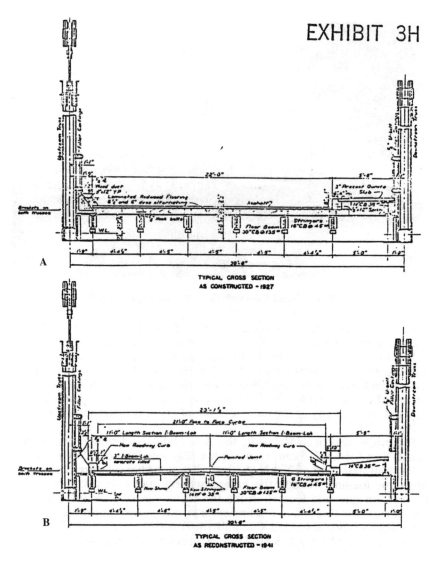

Figure 3-8. The Point Pleasant Bridge cross section.
Source: Lichtenstein (1993).

- It greatly improved bridge stiffness with less steel.
- It uses rocker towers, hinged at the bases, to accommodate thermal deformations.
- It was designed using the deflection theory, which was promoted by Steinman to make suspension-type bridges more efficient.

The bridge was completed in 1928. The General Contracting Company of Pittsburgh, constructed the piers. The American Bridge Company of Pittsburgh built the bridge. It was a unique design because it used high-tension eyebar chains, a unique anchorage system, and rocker towers. The eyebars were two force bar members with rounded eyes at each end. Because bedrock was at a great depth, piles were octagons of reinforced concrete. The anchorage consisted of reinforced concrete troughs that were filled with soil and concrete. This substructure supported the bridge towers, which were joined by dowel rods. This connection allowed for a flexible connection that was designed to shift under different loads on the tower, and as the length of the chains changed because of temperature influences. This type of design was only used in one other bridge in the United States: the St. Mary's Bridge in West Virginia, also spanning the Ohio River. Although not integral to the Florianopolis truss design concept, the Point Pleasant and St. Mary's bridges introduced the use of high-strength, heat-treated carbon steel eyebars to the United States. The eyebar connection is shown in Fig. 3-9.

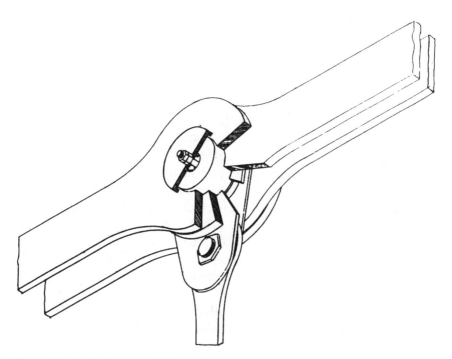

Figure 3-9. Point Pleasant Bridge eyebar connection.
Source: NTSB (1970).

Collapse

On a cold winter evening, the bridge suddenly collapsed. The connection of the eyebar to the first joint on the north side of the bridge failed. This failure caused the Ohio tower to fall and resulted in the collapse of the West Virginia tower. The center span broke and flipped over into the water below, carrying 64 people from the roadway (Lichtenstein 1993).

The immediate response to the accident by local and federal authorities was to block traffic on U.S. Highway 35 while emergency response teams quickly began searching for survivors. To make the recovery more efficient, flood reservoirs were closed to decrease the current. The St. Mary's Bridge was also closed for safety because of its similar design. Once all the survivors had been rescued and vehicles had been recovered, the operations shifted to removing the collapsed structure. This response was important because the Ohio River played a major role in the transportation of freight, and the wreckage prevented any ships from navigating the waterway. This phase included removing wreckage for reassembly and investigation into the cause of the failure (U.S. Army Corps of Engineers 1968).

President Lyndon B. Johnson responded by creating the President's Task Force on Bridge Safety. Their objectives were to identify what caused the bridge to fail, to figure out how to fund a new bridge in an efficient manner, and to define a system of standards that would ensure the safety of the nation's bridges (U.S. Army Corps of Engineers 1968).

Investigation

The experts investigated many possible causes of the collapse. Local citizens spoke of hearing a loud boom before the collapse. Independent engineers at the scene confidently told the media that the collapse happened because of overloading and that "the wreckage not in the river showed no structural fatigue or stress corrosion" (ASCE 1968). It was known that the weight of the trucks passing the bridge had continually increased since the design and that high traffic volumes occasionally loaded the bridge to capacity (Nishanian and Wiles 1970). However, this theory was put forward before any concrete evidence for the collapse had been established.

The U.S. Army Corps of Engineers was able to remove much of the steel debris from the river for the investigation. Because of the need to open the Ohio River again rapidly for navigation, the wreckage was collected and dumped in an 11-hectare (27-acre) field next to the river. Fortunately, each piece was photographed as it was removed from the river (Wearne 2000, pp. 50–51).

The U.S. Army Ordinance Group ruled out explosions and impacts. They first verified with the military that no planes had passed through in the vicinity at the time of the collapse. Vehicles removed from the wreckage were checked for explosives, and structural members were inspected for signs of damage due to explosives. Design features also indicated that the towers were protected from possible contact with a vehicle (U.S. Army Corps of Engineers 1968).

Next, the analysis moved on to the physical attributes of the bridge. The fact that the entire bridge collapsed on failure pointed out that the weakness must have occurred in one of the primary structural members of the bridge (Shermer 1968). It was up to analysts to study the different failure hypotheses, which included the substructure, improper design, failure stress, and corrosion (Scheffey 1971).

One feature considered potentially faulty was the substructure of the bridge. If the piers shifted, the towers would be out of alignment, inducing a sag in the cables. This sag would change the load distribution throughout the structure (ENR 1967). This shift could have been induced by sediment around the piers being washed away (known as *scour*), a deficiency in the foundation that would cause the anchors and piers to move, or an impact to the piers from a water vessel (U.S. Army Corps of Engineers 1968). Scour was eliminated because of the observations of the divers at the scene. They reported no holes around the piers or any signs of damage to the piers. Further verification showed the piers to be in proper orientation.

The design was also evaluated. The bridge did not provide any redundancy in the structural members that connected the suspension chain to the eyebars. Each joint contained two eyebars that connected to the suspension chains. A failure of one eyebar was enough to cause the collapse of the bridge (Lichtenstein 1993). The critical eyebar connection is shown in Fig. 3-9. Photographs of the connection after removal from the river are provided in Figs. 3-10 and 3-11.

The stress calculations that had been prepared for the design were verified for assurance that the bridge met the engineering criteria that were used at the time of construction. This analysis took into consideration static and dynamic loads. The stresses that occurred during the moment of collapse were less than the maximum stresses identified in the design. However, there was no evidence that calculations were made for the correct dynamic loading. This stress increase was considered in terms of an increased percentage of the original live loads (U.S. Army Corps of Engineers 1968, p. 15).

To verify the fact that dynamic effects were not a factor in the collapse, vibration response and dynamic stress amplification tests were performed on the St. Mary's Bridge. The tests sought to establish the natural frequencies

Figure 3-10. Fractured eyebar.
Source: NSTB (1970).

Figure 3-11. The failed joint reassembled.
Source: NTSB (1970).

of the bridge and to determine the excitation created by heavy vehicle use. These tests concluded that the live loads were not excessive and that the loading caused some vibration to occur in the individual members of the structure (Varney 1971, p. 165).

Evidence of Fracture, Corrosion, and Fatigue

As this analysis was underway, the reassembly of the bridge showed evidence of a cleavage fracture in eyebar 330, joint C13N, on the Ohio side of the truss. However, different opinions surfaced for the cause of this fracture. One opinion cited evidence for fatigue stress, whereas a different opinion explained the cause as a combination of stress corrosion and corrosion fatigue.

John Bennett of the U.S. Bureau of Standards noted the fracture pattern on the eyebar. One part of the fracture surface showed deeply encrusted rust, about 3 mm (⅛ in.) long, which contrasted with the light rust from the Ohio River. This flaw was obviously old and might have even gone back to the forging of the eyebar four decades earlier. Cross sections of the eyebar revealed a number of small cracks, about the size of pinholes. One of these, it appeared, had grown into the 3 × 6 mm (⅛ × ¼ in.) flaw that had initiated the fracture (Wearne 2000, pp. 51–53).

The growth of the crack was most likely due to stress corrosion. The two factors were application of tensile stress and the attack of pollutants on the steel. The stress came from traffic loads and thermal effects. The coal-burning factories and locomotives and automotive pollution provided many airborne chemicals that could attack steel. The narrow gap between the eyebar and the connecting pin made the problem worse. It was wide enough to provide an air space for pollutants to attack the steel. However, it was not large enough to allow engineers to inspect the connection (Wearne 2000, p. 53).

The brittleness of the steel connection, and thus its susceptibility to sudden fracture, was enhanced by two factors. The first factor was the use of higher strength steel, which is more brittle than the mild steel typically used for structural steel and reinforced concrete construction. The other factor was the low temperature that day, near freezing. The susceptibility of steel to sudden fracture at low temperatures was dramatically demonstrated by the sinking of several Liberty ship freighters in Arctic waters during World War II while carrying supplies to the Soviet Union (Wearne 2000, pp. 53–54). The Liberty ship fracture case study is discussed in Chapter 9.

The Charpy V-notch test is used to assess the toughness, or resistance to fracture, of various steels. The test measures the energy absorbed during fracture, with more energy required to fracture more ductile materials. The high-strength steel used for the bridge was tested by the Battelle Memorial Institute in Columbus, Ohio. At 74 °C (165 °F), the specimens fractured at 9.2 N·m (6.8 ft-lb) for steel from the interior of the specimen and 11.6 N·m (8.6 ft-lb) for steel from the outside. In contrast, at 0 °C (32 °F), the energy absorbed was 3.0–3.5 N·m (2.2–2.6 ft-lb) respectively, only about a third of that at the higher temperature. The Battelle tests also indicated that at low temperatures a 3-mm (⅛-in.) crack would be enough to initiate a sudden, catastrophic fracture (Wearne 2000, pp. 54–55).

The tiny cracks had most likely been formed during the quenching and tempering process used to produce the high-strength steel. The bars were heated to 816 °C (1,500 °F) and then plunged into water to cool them (quenching) before being tempered. With this process, the outside cools more rapidly than the inside, becoming harder and more brittle. The steel on the inside is softer and more ductile, and it develops tension as it cools more slowly against the tensile restraint provided by the outside of the bar. These processes were not understood when the eyebars were manufactured (Wearne 2000, p. 55).

The evidence supporting failure by fatigue stress was based on changes in the live loads of the structure, which changed the orientation of the eyebars. As the loads redistributed in the eyebar chain, the eyebar heads and the connecting pins would rotate. This adjustment would cause a sliding friction to occur, rather than a rolling friction. Live loads produced by a heavy vehicle would cause the hole in the eyebar to be placed in tension. After the vehicle passed, the connection would respond in compression to return to its original static dead load.

This oscillation caused fatigue stress in the member and could, over time, lead to failure. The eyebar would still be vulnerable to failure even if the resulting stress did not exceed the yield strength of the material (Nishanian and Wiles 1970). It was determined that the stress reversal was unique to a few eyebars because the vehicle lanes were not centered on the roadway. A sidewalk and a vehicle lane on the south side, instead of two vehicle lanes on the north end, caused this lack of symmetry, as shown in Fig. 3-8. This change of symmetry caused uneven loads across the suspension spans. The combination of uneven loading and changes in stress levels caused the pins to shift. This theory implied that the pin cap would become responsible for resisting the lateral forces within the connection. It was possible for these forces to exceed the available resistance. This exceedance would cause the eyebar to detach from the pin, allowing the connection to rotate and thus cause the collapse.

A different theory of the collapse found support in the metallurgical aspects of the eyebar chains and the consequence of both loading and the environment on the steel. Tests of different eyebars showed a consistent composition between the specimens, which indicated that the fractured eyebar was not unique. However, certain characteristics suggested the cause of failure of the material. The hardness was not uniform throughout the thickness of the material. This lack of consistency indicated that the eyebar had a more severe susceptibility to stress corrosion and hydrogen embrittlement in the harder section of the material (Bennett 1973). Further examination of the eyebar showed a crack that had occurred during manufacturing.

It was also found that the eyebar was affected by the atmospheric conditions. These factors made the high-strength steel vulnerable to the propagation of cracks and reduced the time for these cracks to occur (Phelps 1969). Testing showed that as the temperatures in the environment decreased, the material was more susceptible to fracture (Bennett 1973). This finding had significance, considering the extreme temperature changes of the region.

The composition of the atmosphere also played a critical role in the state of the steel. The linkage of the eyebar chains to the suspension cables by a pin created a section of the material that could not be protected by paint. The exposed surface was thus left open to natural elements and created a built-in weakness. Inspection of the wreckage showed corrosion in many sections of the bridge. Studies were conducted to identify the influence of the environment on the material. Tests showed that the eyebar pin connection was susceptible to pitting corrosion due to water, hydrogen sulfide, and salt (NTSB 1970, p. 58). The reaction most prevalent to the cause of the collapse was found in the eyebar chains and pins. Calculations showed a 3% loss of steel due to this degradation of material (U.S. Army Corps of Engineers 1968).

The study of high-strength steel in the environment determined that the combination of stress corrosion (due to tensile stress in a corrosive environment) and fatigue corrosion (induced by the oscillation of loading in the corrosive environment) led to the failure of the eyebar and ultimately caused the collapse of the bridge.

Standards

When the bridge was designed, the loading was based on AASHTO H-15 specifications for a 15 tonne (15-ton) truck with 106 kN (24 kip) on the rear axle and 27 kN (6 kip) on the front axle (Lichtenstein 1993). Since then, the design load has been increased to the heavier AASHTO HS-20.

Lichtenstein (1993) noted that reduced factors of safety, 2 against ultimate strength and 1.5 against yielding, were used in the original bridge design. This limit contrasted with higher factors of safety, 2.75 and 1.75 respectively, for an alternate conventional suspension design. However, the steel used was of unusually high strength, and its properties were not well known. A higher, not lower, factor of safety should have been use for this bridge.

Changes to Bridge Inspection

The National Bridge Inspection Standards were created in response to the collapse. In 1968, the U.S. federal government enacted a procedure for national bridge inspection. Before this law, bridges were not consistently inspected (Lichtenstein 1993). The new procedure provided exact protocols to be followed, and inspectors were required to be evaluated and certified by the Federal Highway Administration. Each bridge was to be inspected every two years.

The system of inspection is broken down into a standard procedure. The format categorizes the superstructure, the substructure, and the deck of the bridge into a set range of values. The importance of this standardization is to allow the accurate rating and comparison of the bridges in the inventory.

Other improvements to bridge management include nondestructive evaluation methods. In 1996, the Federal Highway Administration created the Nondestructive Evaluation Center in McLean, Virginia. The responsibility of this center is to identify inspection methods through the use of technology that includes laser measurement, monitoring systems, and ultrasonic tests.

Educational Aspects

The key lesson from the Point Pleasant Bridge failure is the dangerous interaction between stress concentrations and fatigue. The small, built-in eyebar flaw, subjected to many load cycles over four decades, led to a sudden and rapid crack propagation through the eyebar.

Another lesson is the value of structural redundancy. The eyebar connection was not redundant. Because it was composed only of two pairs of eyebars, the failure of one eyebar allowed the entire connection to rotate and separate. In contrast, when the Tacoma Narrows Bridge was torn apart by wind, 500 of 8,700 wires in the suspension cables fractured, but the cables as a whole survived. At Florianopolis, Steinman had thickened the heads of the eyebars and used four rather than two bars per link. However, these safety features were not used at Point Pleasant (Wearne 2000, pp. 56–58).

Furthermore, the structure as a whole was also not redundant. The loss of any one structural element would lead to a complete loss of the bridge. With the failure of the Tacoma Narrows Bridge, the towers were bent but stayed up, and the side spans of the bridge remained after the center span fell into the water. In contrast, because the Point Pleasant Bridge towers were supported on rockers at the bottom, they toppled over and all three spans were lost.

The alternate chain link suspension design submitted by the American Bridge Company lacked technical justification for the low factors of safety used. The company did not make public its test results or material processing details. Steinman had complained about the company's secrecy and stated that he could not assume responsibility and that American Bridge Company was responsible for the performance of the bars. After the collapse, the company could find no record of any testing of the eyebars. In fact, the company did not own any equipment large enough to test the eyebars. The little testing that was done used smaller samples specifically manufactured for testing (Wearne 2000, pp. 56–58).

Conclusions

The collapse of the Silver Bridge left the nation wondering how such a catastrophe could happen. The public had not previously been greatly concerned about the safety of existing bridges. The design of the bridge had been state of the art. It allowed the combination of new design techniques with new materials to create a bridge that was considered unique and cost-effective. However, the safety of the structure could not be verified by means available at that time, and it did not include sufficient redundancy or structural reliability. Even though the high-strength steel theoretically was stronger, no machinery existed that was capable of testing the large eyebar chains. Proper testing of a new material along with calculations can greatly increase the safety of a unique design.

When the Silver (or Point Pleasant) Bridge was completed, little was known of long-term fatigue combined with crevice corrosion. No one would have considered that a small, exposed surface flaw embedded in the design would cause the failure of the bridge. Nor did anyone consider the effect of the unsymmetric loading on fatigue.

More than 30 years after the collapse, an article in the *Charleston Daily Mail* in Charleston, West Virginia (*Charleston Daily Mail* 1999), noted,

> Point Pleasant was abuzz with Christmastime activity on the afternoon of Friday, Dec. 15, 1967. Busy shoppers shuttling back and forth between the town and its neighbor across the river, Gallipolis, Ohio,

had slowed traffic to a crawl on the Silver Bridge, which connected the two via U.S. 35. When about 72 cars and three large trucks were stopped on the bridge, the end closest to Ohio began to sway wildly and then collapsed, bringing most of the span down behind it.

Witnesses described the collapse by saying that the bridge "slithered like a snake" into the icy Ohio River below. With tragically few exceptions, the occupants of every car on the 1,500-foot span that was thrown from the bridge died. A fast moving current, a massive pile of steel from the bridge and cold water temperatures hampered rescue efforts, leaving emergency crews and local residents with little alternative but to watch helplessly from shore.

In all, 46 people were killed in the collapse—one of the worst such disasters in national history. Almost as soon as the bodies had been gathered, a public outcry rose up in Point Pleasant demanding that the safety of the 38-year-old span be investigated. Many claimed that the bridge was known to be unsafe before the collapse, and the state and federal highway authorities had acted negligently. In a hearing before the National Highway Safety Board in Washington, which included testimony from several esteemed engineers and a re-creation of the Silver Bridge, it was determined that the structural flaw in the bridge was of such a nature that it could not have been detected by inspectors. Even so, in the wake of the disaster, the inspection process for older bridges was made more rigorous.

Essential Reading

The National Transportation Safety Board report (NTSB 1970) provides a thorough discussion of the failure. Lichtenstein (1993) also provides an excellent review. The case is also discussed by Levy and Salvadori (1992) and Wearne (2000, pp. 45–58). This case study is featured on the History Channel's Modern Marvels *Engineering Disasters 4* videotape and DVD.

Comet Jet Aircraft Crashes

One of the first commercial jet aircraft was the British de Havilland Comet, which first flew in 1952. By 1954, seven of the Comets had crashed under unknown circumstances. Because only 21 had been built, the one-third failure rate was catastrophic. The Comet crashes caused the loss of Britain's early leadership in commercial passenger jet aircraft. This summary is modified from a more complete case study developed by Levy and Salvadori (1992, pp. 121–126).

One plane crashed about a half hour after takeoff from Rome, with the wreckage falling into the sea between the islands of Elba and Monte Cristo. The Tyrrhenian Sea was shallow enough for the British Royal Navy to recover the wreckage. While the investigation was continuing, 50 structural modifications were made to the Comets. They were cleared to fly again, but within two weeks another crashed.

Researchers assembled and examined the wreckage. A large number of theories, including sabotage and pilot error, were considered and discarded. The hypothesis that best fit the circumstances was metal fatigue.

Jet aircraft are subjected to stress cycles from cabin pressurization and depressurization, as well as from turbulence and vibration in flight. Each of the last two aircraft to crash had completed about 1,000 cyclic load repetitions. The designer had calculated a fatigue life of 10,000 repetitions, or approximately a 10-year service life.

The Comet had square windows, like earlier unpressurized propeller-driven passenger planes. The square windows had severe stress concentrations at the corners, which increased stresses and reduced fatigue life. Examination of the wreckage showed cracks initiating at the corners of the windows and propagating into the plane's fuselage. Stress concentrations and fatigue loads, as at the Point Pleasant Bridge, make a deadly combination. Today, aircraft windows are round or oval, without sharp corners.

The solutions are fatigue-resistant design, coupled with an extensive program of inspection and maintenance. Bridges, aircraft, and other structures subjected to fatigue are designed to eliminate stress concentrations as much as possible. Smooth transitions are used where lines of stress change direction. Connections and details prone to fatigue are identified, and maintenance schedules are followed to inspect or replace these components periodically.

The Crashes

The crash of the British Overseas Airways Corporation (BOAC) Comet off Elba on January 10, 1954, was particularly puzzling because the pilot's transmission to another BOAC flight was cut off in midsentence. Three previous crashes on takeoff had been attributed to either pilot error (attempting to lift the nose before gaining enough speed) or storms. These crashes, therefore, were not initially thought to be the fault of the aircraft design. Nevertheless, the Comet operated at unprecedented speeds, pressures, and altitudes compared to earlier propeller-driven passenger planes (Freiman and Schlager 1995a, pp. 78–80).

Investigations

The scenarios considered for the crashes included the following:

- sabotage;
- a ruptured turbine blade penetrating a fuel tank and causing an explosion;
- clear air turbulence (vibration) structural failure;
- engine fire in flight, weakening the structure;
- a hydrogen explosion from a leaking battery; or
- a fuel vapor explosion in an empty tank (Freiman and Schlager 1995a, p. 81).

While the investigation was ongoing, the modifications to the aircraft included strengthening fuel lines, installing shields between engines and fuel tanks, and improving smoke and fire detectors. None of these changes, unfortunately, addressed the actual cause of the crashes (Freiman and Schlager 1995a, p. 82).

Once another Comet exploded on April 8, 1954, British Prime Minister Winston Churchill ordered the Royal Aircraft Establishment (RAE) to investigate. About two-thirds of the wreckage was recovered from an ocean depth of 1,070 m (3,500 ft). The RAE chief, Arnold Hall, had the wreckage flown to the United Kingdom on a U.S. Air Force cargo plane so that he would not have to wait for a ship. The RAE suspected metal fatigue almost immediately. A missing engine turbine blade took the investigation down a blind alley until the turbine casing was found intact, indicating that the turbine blade broke off during the crash (Freiman and Schlager 1995a, pp. 82–83).

Autopsy reports from 20 victims of the two crashes indicated death due to "violent movement and explosive decompression," occurring before hitting the sea. This result supported the theory of a structural failure that led to immediate loss of cabin pressure (Freiman and Schlager 1995a, p. 83).

Hall ordered construction of a tank large enough to pressure test a complete Comet fuselage. The tank was filled with water, and jacks were used to flex the wings. The fuselage was pressurized to 56 kPa (8 lb/in.^{2}), simulating the pressurization at 10,700 m (35,000 ft), and then reduced to 0 in 3-min cycles. The test fatigued the airframe at 40 times the normal rate expected in service. In late June 1954, at a simulated 9,000 flight hours, the fuselage cracked. After draining the tank, investigators found a 2.4-m (8-ft) crack along the fuselage starting from the corner of a small window. The window corner showed the telltale metal fatigue signatures (Freiman and Schlager 1995a, pp. 83–84).

The Comet crashes were attributed to a combination of factors:

- The inadequate test program did not capture the long-term effects of the pressurization and depressurization cycles on the structure of the airframe. Full-scale tests were only conducted by the RAE after several crashes had occurred.
- The square windows, with high stress concentrations, combined with the lack of measures to prevent cracks from spreading, left the fuselage highly vulnerable to fatigue.
- The fuselage skin was thin (Freiman and Schlager 1995a, pp. 84–85).

Changes to Aircraft Designs

British authorities had required the Comet's designers to increase the fuselage maximum pressure from the 28 kPa (4 lb/in.^{2}) that had been used to design propeller-driven aircraft to 110 kPa (16 lb/in.^{2}). The designers increased that further to 138 kPa (20 lb/in.^{2}). To minimize aircraft weight, the aluminum skin was only 0.7 mm (0.028 in.). The U.S. Civil Aeronautics Administration (CAA), the forerunner of the FAA, questioned the durability of the square windows and suggested switching them to an oval shape. BOAC's engineers maintained that the windows had been tested up to a pressure of 690 kPa (100 lb/in.^{2}) without any signs of fatigue (Freiman and Schlager 1995a, pp. 80–81).

After the Comet crashes, a number of changes were made to aircraft designs:

- Thicker skin was used.
- Triple-strength rounded windows reduced stresses.
- Metal bracing provided additional fuselage structural strength.
- Small metal tabs or "stoppers" were placed at critical points in the fuselage to arrest the growth of fatigue cracks. Boeing tested a 707 airframe with and without stoppers. The fuselage with stoppers held together, although some pressure escaped. Without the stoppers, the fuselage was split open (Freiman and Schlager 1995a, pp. 85–86).

The U.S. aircraft manufacturers Boeing and Douglas had already adopted these changes before the Comet crashes. The stigma of the 1954 crashes, coupled with the newer and larger U.S.-built jets, essentially killed the market for the Comet (Freiman and Schlager 1995a, p. 85).

Petroski observes,

> Metal fatigue was a new phenomenon in aircraft. Before the introduction of jets, airplanes did not fly so high and thus did not have to be so highly pressurized for passenger comfort. In order to gain the fuel efficiency that gave the new engine part of its advantage, jets had to fly higher than propeller-driven airplanes, and as they did so the structural components of the aircraft were subjected to conditions that were beyond the experience of its designers. Previously, metal fatigue was believed to affect only machine parts that were subjected to cycles numbering in the hundreds of thousands, if not millions. Therefore, the Comet's engineers did not believe that fatigue would affect the plane, which would be subjected to far fewer cycles during its lifetime. But because airplanes must be as light as possible, their structural parts must carry more intense loads than the parts of land-based structures or machines, for which weight is less important. The critical combination of load intensity and flight cycles that would lead to the growth of critical cracks proved to be far lower than expected in the Comet. (1996, pp. 121–122)

Essential Reading

The Comet crashes are covered in one chapter of Levy and Salvadori (1992, pp. 121–126). In addition, a chapter of Freiman and Schlager (1995a, pp. 78–86) covers the incidents. This case study is also featured on the History Channel's Modern Marvels *Engineering Disasters 4* videotape and DVD.

Other Cases

Hyatt Regency Walkway

The collapse of the Kansas City Hyatt Regency walkway connections, discussed in Chapter 2, illustrates a bearing stress failure. The actual failure mechanism was a punch-through bearing failure of the box beam against the supporting nut attached to the hanger rod. The poor quality of the box beam welded connection, without stiffeners, substantially reduced the bearing capacity of the beam.

The connection had an unacceptably low factor of safety. One of the possible design alternatives, suggested after the fact, was the use of a bearing plate to distribute the bearing stress against the box beam. This explanation is, of course, a gross oversimplification of the case study, and the reader is encouraged to review the complete narrative. This critical connection was never designed, so the bearing stress was never considered. The ethical implications of this case are also important.

Liberty Ship Hull Failures

Stress concentrations and fatigue are illustrated through the failures of a number of World War II Liberty ships, as discussed in Chapter 9. Like the Point Pleasant Bridge and the Comet jet aircraft, fatigue cracks started at stress concentration sites. These cracks propagated through the entire hull structure. Some ships that cracked in harbors were saved, whereas others were lost at sea. The problem seemed to be aggravated by the low temperatures the ships encountered in the North Atlantic and on convoy routes into Arctic seas to supply the Soviet Union. Once the ships were reinforced, there were no more failures.

Pittsburgh Convention Center Expansion Joint Failure

Mechanics of materials also addresses thermal deformations, stresses, and strains. Expansion joints are used in buildings and bridges to relieve stresses caused by thermal deformations. If the expansion joints don't work properly, structural damage may result.

On Monday, February 5, 2007, an expansion joint at the Pittsburgh Convention Center failed as a tractor-trailer was parked next to it. The cause of the partial collapse was the combination of a cold snap and a poor expansion joint detail. Greg Luth told the *Pittsburgh Tribune-Review* that he had only seen that type of joint once before in almost three decades of structural engineering practice, and that was in the first atrium joint that had failed at the Kansas City Hyatt Regency (Houser and Ritchie 2007). The case study is covered in detail in Chapter 6. The failed expansion joint is shown in Fig. 6-8, along with the retrofit that was installed to fix the problem.

4

Structural Analysis

STRUCTURAL ANALYSIS COURSES BUILD ON THE ENGINEERing science and design concepts developed in statics and mechanics of materials courses. Structural analysis forms the basis for courses in building and bridge design in reinforced concrete, steel, and other materials.

With the wide variety of topics covered, it is easy for students to get lost in the details and lose an overall sense of structural behavior. Failure case studies show how structures perform, particularly at the ultimate state.

In a 2006 paper, I suggested the 12 failure case studies that every structural engineer should know (Delatte 2006). All these cases are reviewed in this book, although some are in different chapters. In chronological order, they are:

- the Quebec Bridge, 1907 (Chapter 3),
- the Tacoma Narrows Bridge, 1940 (Chapter 2),
- the Point Pleasant Bridge collapse, 1967 (Chapter 3),
- Ronan Point, 1968 (Chapter 4),
- 2000 Commonwealth Avenue, 1971 (Chapter 5),
- Skyline Plaza in Bailey's Crossroads, 1973 (Chapter 5),
- the Hartford Civic Center Stadium collapse, 1978 (Chapter 6),
- the Willow Island Cooling Tower collapse, 1978 (Chapter 9),

- Harbour Cay Condominium, 1981 (Chapter 5),
- the Hyatt Regency walkway, 1981 (Chapter 2),
- the Mianus River Bridge collapse, 1983 (Chapter 6), and
- L'Ambiance Plaza collapse, 1987 (Chapter 4).

Because the case studies already discuss the relevant structural aspects, they are not reviewed further in this chapter.

Agricultural Product Warehouse Failures

Two manufactured metal buildings were purchased and erected for use as agricultural product storage warehouses. The product stored was cottonseed. The cottonseed was separated from the seed in a gin building adjacent to each warehouse and then blown in for storage until it could be sold. The blower loaded the warehouses from one end, with the product removed through a door at the other end. The warehouses were intended to be fully loaded with cottonseed. Once the pile of cottonseed reached the peak of the roof, an auger built into the roof was used to move it toward the door. The warehouse storage capability would allow the owner to store cottonseed until it could be sold at a higher price. The two warehouses were located in towns about 32 km (20 mi) apart.

For two years, the warehouses were only partially loaded. During the third season, the warehouses were loaded to the roof, and workers outside of the building observed outward bulging of the warehouse walls. Once the product was removed, the concrete floor slabs were found to be badly cracked. The structure was also damaged, with evidence of permanent deformation. Much of the outward bulging disappeared when the cottonseed was removed, but the distortion in the metal walls remained.

Inspection

Each warehouse was 24.4 × 42.7 m (80 × 140 ft) in plan. The warehouse side walls were a little more than 7.26 m (23 ft 10 in.) high, with the roof rising at a 45-degree angle to a peak 18.3 m (60 ft) high. One of the warehouses is shown in Fig. 4-1. The auger, intended to move the product from the loading end to the door, was at the peak of the roof.

The structural configuration of both warehouses was identical. The interior framing of one of the warehouses is shown in Fig. 4-2. Six gable frames were spaced at 6.1-m (20-ft) intervals. Between the gable frames were interior soldier columns. The tops of the interior soldier columns were

STRUCTURAL ANALYSIS 91

Figure 4-1. Metal building used to store cottonseed.

restrained by steel angle bracing leading back to the gable frames. Each of the pair of braces at the top of each column had two parts, with bolts attaching them to other parts of the structural secondary framing. The first and last bays had X-bracing for longitudinal stability of the structure. Each end wall had seven columns.

These structures were purchased with an interior liner, so the inner and outer walls of the building were separated by purlins. Flange braces were provided to brace the side and end wall columns against lateral torsional buckling and had been attached to the inner liner panel with self-tapping screws.

The buildings were heavily damaged. The metal skin had been stretched, and flange braces had been torn away along the walls. The concrete foundation had extensive cracking, with one crack 1 in. (25 mm) wide (Fig. 4-3). Details of the damage are provided in Delatte (2002).

The need to keep the seed from piling up against the wall reduced the storage capacity of the warehouses to a small fraction of the designed capacity. Clearly, the buildings had failed and could not be used as intended without risk of further damage and possible collapse.

Design and Construction

The metal building manufacturer's plans, shop drawings, and job file were available for review. The design had been carried out with computer

Figure 4-2. Interior framing.

Figure 4-3. Damage to concrete foundation.

software, following industry practices. The software used the direct stiffness method. The Metal Building Manufacturer's Association publishes a manual prescribing loads for this type of structure, such as a snow load of 240 Pa (5 lb/ft^2) and a wind load of 843 Pa (17.6 lb/ft^2) based on a wind speed of 39.3 m/s (88 mi/h) for the building location (MBMA 1996). These loads were shown in the computer program output in the job file.

However, the MBMA manual does not directly address the loads from agricultural products. These loads were high, as much as 6 to 12 times the wind pressure. Loads of agricultural product were not reflected in the computer program output.

The agricultural product loads were shown on one sheet of the plans supplied to the erector and owner as horizontal reactions to be applied to the foundation slab. This sheet showed only force reactions and not moment reactions. A pinned base column is often assumed for the design of low-rise metal buildings (e.g., Lee et al. 1981). Moment reactions occur at the base of a column when it is subjected to lateral loads (and it acts as a beam-column), unless the column base is detailed to allow rotation.

The steel structure plans were supplied without a foundation plan, except for an anchor bolt layout. In accordance with common industry practice, the plans contained a disclaimer stating that the building designer was not assuming liability for foundation, floor, or slab design or construction. The owner was told to hire a foundation designer and provide him or her with the plans.

The owner hired a soil testing firm (laboratory A) for site testing and compaction recommendations. The testing firm prepared recommendations and provided field density reports during construction.

The owner asked the soil testing firm's licensed engineer to prepare a foundation plan. The drawing prepared by the testing firm engineer provided a layout with a specification for 27.6 MPa (4,000 lb/in.^{2}) concrete containing polypropylene fiber reinforcement (specified by a trade name). The slab was 125 mm (5 in.) thick with an outer turndown beam 406 × 406 mm (16 × 16 in.). A footing 1.07 m (42 in.) square was provided under each column.

Two 15.9-mm (⅝-in.) diameter steel reinforcing bars (designation 16M, US #5) were specified to be continuous around the perimeter of the foundation in the turndown beam, with additional bars of the same size in the footer. However, no reinforcing steel was specified in the interior of the slab to resist the outward horizontal reactions and moments at the column bases.

Once the foundation was prepared, a construction firm was hired to erect the building. Adjustments had to be made during construction because the building was purchased with an interior liner, which was not shown on the plans supplied by the metal building manufacturer.

The original detail for the flange braces showed them bolted to the purlins along the walls. However, with the interior liner in place, they could not be attached directly to the purlins. Therefore, the flange braces were attached with self-tapping screws installed through the liner panel into the purlins. Furthermore, some of the flange braces were field-modified to fit by flame cutting or similar methods. In some cases, little metal was left for attachment.

One of the defendants' attorneys hired a testing laboratory (laboratory B) to investigate the soil and concrete slab at each warehouse. This laboratory had not been previously associated with the project.

At warehouse 1, laboratory B performed seven field density tests. Densities varied from 89% to 98%. Three of the seven tests were less than the 95% density required in laboratory A's recommendations. Four field density tests were performed at warehouse 2. Densities ranged from 83% to 93%, and none met the specification prepared by laboratory A.

At warehouse 2, four concrete cores were removed. One of the cores, taken through the nominally 125-mm (5-in.) part of the slab, was only 111 mm (4⅜ in.) long. The others were all taken around the perimeter beam. The short core suggests an inadequate thickness of the slab for at least one point, but it is not enough to form a conclusion.

Two cores were cut to make a total of five specimens for compression testing. Test results were 22.2, 24.1, 29.0, 29.4, and 31.9 MPa (3,221, 3,497, 4,202, 4,260, and 4,622 lb/in.^{2}). Three Windsor Probe tests were made on the foundation. The compressive strengths predicted using the Windsor Probe were 23.6, 32.8, and 34.3 MPa (3,425, 4,750, and 4,975 lb/in.^{2}).

The bulk density of cottonseed is 400 kg/m^3 mass and 3.92 kN/m^3 weight (25 lb/ft^3), and it has an angle of repose (φ) of 45°. The lateral wall pressure may be estimated using the following formula:

$$WP = k \times D \times H \qquad (4\text{-}1)$$

where WP = lateral wall pressure, k = pressure coefficient (0.20, based on the angle of repose of 45° and rounded up), D = density, and H = height or seed depth (Willcut et al. undated). Therefore, with an eave height of 7.26 m (23 ft 10 in.), the pressure at the base of the wall was 5.75 kPa (120 lb/ft^2). The pressure increases linearly from 0 at the top of the wall to a maximum value at the bottom.

Therefore, the resultant of the pressure distribution is

$$F = \frac{1}{2} \times k \times D \times H^2 \times L \qquad (4\text{-}2)$$

where F = resultant force, L = length of wall between columns (3.05 m, or 10 ft), and other variables are as previously defined. The resultant force acts at $\frac{1}{3} H$ from the bottom of the wall (Fig. 4-4).

Agricultural product may be analyzed as a cohesionless soil to determine lateral pressures. For a cohensionless soil, the active earth pressure coefficient k is

$$k = \tan^2\left(45 - \frac{\varphi}{2}\right) \tag{4-3}$$

where $\varphi = 45°$, as noted earlier. This situation gives an active pressure coefficient of 0.172, close to the value of 0.2 suggested by Willcut et al. (undated). These equations are the commonly used Rankine formulas for active soil pressure, which may be obtained from a number of soil mechanics texts.

Therefore, along the side walls, each column had an outward resultant force of 63.2 kN (14,200 lb) acting 2.42 m (8 ft) from the base of the wall. On the end walls, the outward resultant forces were as much as 303 kN (68,000 lb).

Because the building was loaded from one end, it was possible to have a full load of cottonseed on the far wall, with no load against the inside of the near wall (door end). This load provided a total force of 1,670 kN (375 kip) acting on the rear wall, at a location 4.66 m (15.3 ft) from the base of the wall. In fact, this condition was not only possible but also unavoidable in the course of operating the warehouses as intended.

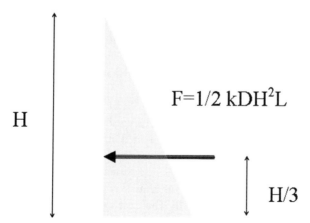

Figure 4-4. Resultant force due to cottonseed pressure.

Failure Hypotheses

Five hypotheses were considered, which included steel structure design error, construction error, foundation design error, low-strength concrete, or differential settlement. The possibility of extreme loading, such as hurricane wind loading, earthquake, or another similar event, was considered but rejected because there was no evidence that such an event had occurred since the buildings had been constructed. As mentioned above, the buildings were some distance apart but exhibited nearly identical distresses. Combinations of causes were considered. A complete discussion is provided in Delatte (2002).

Discussion

The steel structure design and the foundation design were both deficient. Construction errors and inadequate soil compaction seemed to have played little or no role in the failure.

Review of the metal building manufacturer's plan and job file indicated that the cottonseed forces were never accounted for in the design, although they were listed on the table of reactions for foundation design. This was a serious omission because the cottonseed forces were much higher than the forces considered in the design and represented the controlling load case. Under the cottonseed loads, a number of the metal building structural elements were overstressed and damaged, and the structure was in danger of overall collapse. Therefore, the steel structure design was inadequate.

The foundation design was also inadequate and violated the ACI building code (ACI 1995). The registered engineer who designed the foundation slab had not previously designed a foundation. He also missed some of the force reaction notations on the steel structure plans. This engineer was clearly operating outside of his area of expertise. Almost all of his recent experience had been in the preparation of soil test reports and recommendations for laboratory A and its clients.

A contributing factor was the practice of splitting responsibility for the structure and the foundation between two engineers. Neither one took responsibility for the overall project. The metal building manufacturer's engineer disavowed responsibility for the foundation and provided incomplete information on the column base reactions. The foundation engineer misinterpreted some of the notations that were provided. Because of the high outward pressures imposed by the cottonseed, careful coordination of building design and foundation design was needed to ensure satisfactory performance. A single engineer of record could have prevented this failure of communication. The only common contact between the two engineers was the building owner.

Conclusions

Stored agricultural products such as cottonseed or peanuts impose significant outward pressures on warehouse walls. These forces may be predicted using standard soil mechanics equations. Calculating these pressures is straightforward.

Because at least three of these warehouses have failed in the same predictable manner, it is clearly important for designers of future agricultural product warehouses to consider these forces in design. It is particularly important to consider overall structural stability under unbalanced product loads.

It is also necessary to make sure that the designer's assumptions about structural behavior are consistent with the detailing. The column base connections were fixed but were modeled as if hinged. Either a fixed or a hinged connection could have been used, and the structure could have performed properly with either detail so long as the structure and foundation designs took the actual connection behavior into account. Instead, the moment that cracked the foundation slab was not accounted for by either the structural engineer or the foundation engineer.

The practice of splitting responsibility for a structure between a manufacturer's structural engineer (who may know nothing about the site conditions) and a local foundation engineer (who may know nothing about metal building design and behavior) presents a high risk of failure. It is particularly dangerous when loading of this magnitude is present. A single engineer of record is needed to avoid poor performance that happens because of poor communication.

Ronan Point

In the early morning hours of May 16, 1968, the occupant of apartment 90 on the 18th floor of the Ronan Point apartment tower in London lit a match for her stove to brew her morning cup of tea. The resulting gas explosion, caused by a leak, knocked her unconscious.

The pressure of the small gas explosion blew out the walls of her apartment and initiated a partial collapse of the structure that killed 4 people and injured 17. The partially collapsed structure is shown in Fig. 4-5, and the floor plan for apartment 90 is provided in Fig. 4-6.

98 BEYOND FAILURE

Figure 4-5. Ronan Point after collapse.
Source: Pearson and Delatte (2005).

Design and Construction

Many high-rise apartments were constructed in London to replace the housing stock destroyed during the Second World War. The development of prefabricated construction techniques (known as system building) led to the popularity of high-rise apartment buildings. This new style of housing

STRUCTURAL ANALYSIS 99

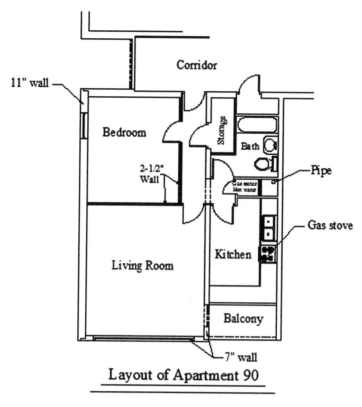

Figure 4-6. Floor plan of Ronan Point apartment.
Source: Pearson and Delatte (2005).

could be constructed quickly, accommodate large numbers of people, and save land and labor.

The Ronan Point Apartment Tower was constructed using the Larsen–Nielsen system. This system was developed in Denmark in 1948. The Larsen–Nielsen system was

> ... composed of factory-built, precast concrete components designed to minimize on-site construction work. Walls, floors and stairways are all precast. All units, installed one-story high, are load bearing. (ENR 1968, p. 54)

This building technique encompassed the patterns for the panels and joints, the method of panel assembly, and the methods of production of the panels.

100 BEYOND FAILURE

Ronan Point was the second of nine identical high-rise, precast, concrete flat plate structures that were erected in London after the war. In this type of structural system, each floor was supported by the load bearing walls directly beneath it. Gravity load transfer occurred only through these load-bearing walls. The wall and floor system panels were fitted together in slots. The joints were then bolted together and filled with dry pack mortar to secure the connections. The connections are shown in Fig. 4-7 and Fig. 4-8.

Ronan Point was 22 stories tall. There were 110 apartment units in the building, grouped five to a floor. The building contained 44 two-bedroom apartments and 66 one-bedroom apartments. Construction took less than two years, and the building was near full occupancy for less than 3 months before a section of it collapsed.

Collapse

The southeast corner of the Ronan Point Tower collapsed on May 16, 1968, at approximately 5:45 A.M. Four people died and seventeen were

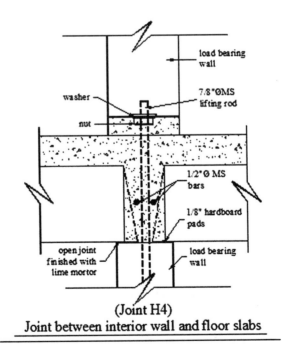

Figure 4-7. Ronan Point connection.
Source: Pearson and Delatte (2005).

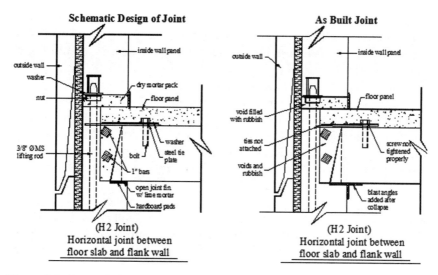

Figure 4-8. Ronan Point connection.
Source: Pearson and Delatte (2005).

injured. The fatality rate could have been considerably higher, given the extent of the structural damage (Feld and Carper 1997). Fortunately, at the time of the disaster, all of the residents but one were sleeping in their bedrooms. The collapse sheared off the living room portion of the apartments, leaving the bedrooms intact, with the exception of floors 17 through 22, where all the fatalities occurred. This corner of the building contained the only three vacant apartments left in the building. The apartment on floor 22 was the only one occupied above floor 18.

The collapse was initiated by a gas-stove leak on the 18th floor in apartment 90. The resident struck a match to light the stove and was knocked unconscious by the resulting explosion. The force of the explosion knocked out the opposite corner walls of the apartment. These walls were the sole support for the walls directly above. This collapse created a chain reaction in which floor 19 collapsed, then floor 20 and so on, propagating upward. The four floors fell onto level 18, which initiated a second phase of progressive collapse. This sudden impact loading on floor 18 caused it to give way, smashing floor 17 and progressing until it reached the ground.

Causes of Failure

The government formed a panel to investigate the collapse. The panel's report was issued later that year (Griffiths et al. 1968). It was quickly determined that the explosion from the gas leak had initiated the collapse of the building. A substandard brass nut had been used to connect the hose to the stove. The nut had a thinner flange than the standard and also had an unusual degree of chamfer. A replicate of this nut was made and tested to determine how much force was required to break it.

It was found that a force of 15.6 kN (3,500 lb) would break the connection. It was also concluded that the hose connecting the stove to the gas would have failed before the nut at a force of 1.6 kN (360 lb). The nut was assumed to have been previously fractured by overtightening during installation, causing it to break and allowing gas to leak into the apartment (Griffiths et al. 1968).

The gas may have accumulated at the ceiling, explaining why the resident did not detect it. The explosion was not large, and the resident's hearing was not damaged. This suggested that the pressure was less than 70 kPa (10 lb/in.^{2}) (Bignell et al. 1977). Items from the kitchen of this apartment were tested. Results indicated that these objects had been exposed to pressures of less than 70 kPa (10 lb/in.^{2}).

The Building Research Station and Imperial College of London performed an extensive battery of tests to discover how much internal force Ronan Point could withstand. The results indicated that the walls could have been displaced by a pressure of only 19.3 kPa (2.8 lb/in.^{2}) (Levy and Salvadori 1992). It was estimated that the kitchen and living room walls were moved at a pressure of only 1.7 kPa (0.25 lb/in.^{2}), and the exterior wall was moved at a gas pressure of 21 kPa (3 lb/in.^{2}) (Griffiths et al. 1968).

Ultimately, the collapse of Ronan Point occurred because of its lack of structural redundancy. It had no fail-safe mechanisms and no alternative load paths for the upper floors should a lower level give way. Without any type of structural frame, the upper floors had no support and fell onto floor 17. The panels forming floor 17 could not support the sudden loading caused by the upper five floors. Consequently, they gave way, and the process continued until it reached the ground level.

The southeast corner of Ronan Point was rebuilt as a separate section of apartments and was then joined to the existing building by means of walkways. Ronan Point was reinforced with blast angles as part of the reconstruction. Gas was also banned from the Ronan Point complex.

Technical Aspects

The overall weakness of the building and the associated deficiencies in building codes became apparent after the public inquiry into the collapse.

The inquiry revealed that strong winds or the effects of a fire in the building could also have caused a progressive collapse. Ronan Point was designed to withstand wind velocities of only 100 km/h (63 mi/h). A wind of 170 km/h (105 mi/h) could be expected to occur at 61 m (200 ft) above the ground every 60 years, within the life expectancy of the building.

The building code used for the design of Ronan Point and its sister buildings was issued in 1952. This set of codes had not been kept up to date. Higher than stated winds were known to occur based on a publication in 1963 by the National Physical Laboratory (Griffiths et al. 1968). It was noted that, "the structure had been designed to comply with fifteen-year-old wind load codes that did not take into account current building heights" (ENR 1970, p. 12). According to the inquiry, "the suction effect of the pressures applied by such winds, in particular the opening of the joints as the tower block bent in the wind, would have similar effect to the explosion" (Wearne 2000, p. 143).

Fire also would have had a similar effect on Ronan Point. The inquiry stated, "it is estimated that fire could so expand and 'arch' the floor slab and bend the wall panel, as to displace or rotate an H-2 joint to a dangerous degree" (Wearne 2000).

Continuing concerns over the building's structural integrity eventually led to its demolition in May 1986. The building had a life expectancy of 60 years but was razed after just 18 years. The building was not demolished in the traditional fashion. Poor workmanship was suspected, and Ronan Point was dismantled floor by floor so that the joints could be studied. The site was an "open site" for anyone interested.

A shocked architect, Sam Webb, commented,

> I knew we were going to find bad workmanship—what surprised me was the sheer scale of it. Not a single joint was correct. Fixing straps were unattached: leveling nuts were not wound down, causing a significant loading to be transmitted via the bolts: panels were placed on bolts instead of mortar. But the biggest shock of all was the crucial H-2 load-bearing joints between floor and wall panels. Some of the joints had less than fifty percent of the mortar specified. (Wearne 2000, p. 154)

Professional and Procedural Aspects

The findings of such magnitude of poor workmanship performed in the construction of Ronan Point led to the demolition of the remaining Larsen–Nielsen system built towers. At the time these buildings were erected, the building codes did not adequately address the system. Large concrete panel construction was the height of innovation at this time, and little was known about how it would perform. The building regulations in effect at the time contained a "catch-all" clause known as the "functional requirement on structure." This clause contained no mention of redundancy or progressive collapse (Bignell et al. 1977).

The collapse of the southeast corner of Ronan Point initiated changes to the codes. Building codes now take into account the possibilities of progressive collapse and of forces from an internal explosion. The codes also require minimum amounts of ductility and redundancy.

One of the outcomes of this inquiry was the development of the "Amendment" to the U.K. building regulations in 1970. According to Allen and Schriever (1972, pp. 39, 45), all buildings of more than five stories were to be designed to resist progressive collapse. This design could be accomplished by analyzing the structure with a critical structural member removed (notional removal), ensuring alternate load paths. It was not necessary to consider notional removal of structural members if the members could resist a specified pressure in any direction. For both alternatives, a safety factor of 1.05 with dead load plus ⅓ live load should be used.

The British also conducted their own research on progressive collapse. The United States followed and also implemented new design criteria (Fuller 1975). The lessons from Ronan Point changed building regulations throughout the world.

The government mandated guidelines for the prevention of progressive collapses. These instructions included the requirement for a fail-safe mechanism in all large-panel system buildings, steel bracing with floor-to-wall connectors, and a minimum tensile strength of 21 MPa (3,000 lb/in.^{2}) across the length and width of the roofs and floors (Feld and Carper 1997, p. 306).

The Portland Cement Association and the Prestressed Concrete Institute also issued their own sets of guidelines. These methods called for "tying building elements together and increasing ductility so that the building elements can better sustain deformations from the failure of a portion of the building's structure" (Ross 1984, p. 273).

The engineering profession was reminded of the need for redundancy in design to prevent a progressive collapse. It is of the utmost importance that building designs contain some measure of continuity (Shepherd and Frost 1995). Extensive research was carried out in the United Kingdom to provide engineers and architects with data for load-bearing walls. When the Larsen–Nielsen system was developed, it was not intended to be used in buildings more than six floors high. However, in the United Kingdom it had been used for taller structures.

It was concluded by the inquiry that the codes governing construction and design methods needed immediate reevaluation. The authors of the inquiry stated in their report,

> . . . we do not consider that in its present form Ronan Point is an acceptable building, and yet it was designed to comply with the statutory standards contained in the Newham by-laws, which are, in all material respects, identical with current Building Regulations. This is so manifestly an unsatisfactory state of affairs that it is necessary to enquire how it came about and to consider remedies for the future. (Griffiths et al. 1968)

The need for quality control in the construction process was reinforced after the dismantling of Ronan Point. Although the design flaw was the primary downfall of Ronan Point, poor construction quality could have led to future problems with the building's structural integrity. It is imperative that quality control be enforced in the construction process to ensure public safety.

> As with all other construction materials, the best designs in precast and prestressed concrete can be ineffective unless the work done in the field is of high quality. If the design is marginal, construction deficiencies can compound the errors increasing the potential for serious problems. . . . Skilled supervisors who understand the design intent and can communicate it clearly to the field workers are needed full-time at the construction site while all prestressed concrete work is erected. (Feld and Carper 1997, p. 302)

Ethical Aspects

Substandard workmanship was detected in the initial inquiry of the collapse. Even though it was not found to be the major factor in the corner collapsing, this information was hidden from the public. By the time the inquiry's findings were published in 1968, many large-panel concrete buildings had been completed. At least six Larsen–Nielsen system buildings had been completed by this time. There was not enough money to strengthen them.

Summary and Conclusions

The investigations found that the Ronan Point apartment tower was deeply flawed in both design and construction. The existing building codes were inadequate for ensuring the safety and integrity of high-rise, precast concrete apartment buildings. In particular, the design wind pressures were too low and did not account for the height of the building. The Larsen–Nielsen building system, intended for buildings with only six stories, had been extended past the point of safety.

The tower consisted of precast panels joined together without a structural frame. The apartment tower lacked alternate load paths to redistribute forces in the event of a partial collapse. When the structure was dismantled, investigators found appallingly poor workmanship of the critical connections between the panels. The already shaky structure had been further weakened by the inadequate construction practices. The result was described by Levy and Salvadori (1992) as a "house of cards." Overall structural integrity is of particular importance in precast concrete building systems.

The relatively low overpressure from the gas explosion should have led to localized damage at most, not a partial progressive collapse and the loss of four lives. The evaluation also found that the building was unusually vulnerable to ordinary wind and fire loading.

Petroski comments,

> Conceptually, the system building was a brilliant alternative to costly on-site construction, and it could be argued that it allowed for the achievement of better quality control in the individual components. However, as the gas explosion revealed, the system had a fundamental flaw in its design, and the loss of one load-bearing wall under the wrong conditions allowed the whole corner of the structure to collapse. The concept of system construction had many advantages over the more traditional schemes, but it clearly had the disadvantage of very little redundancy. . . . The fundamental error of the concept was not revealed in the logic of the design process but in the chance events that led to a gas explosion. When an explosion blew out some upper walls in a more conventionally constructed apartment building in New York's Harlem in 1991, the overall building withstood the structural trauma, as it was expected to. (1994, pp. 26–28)

Essential Reading

The key reference is the report by Griffiths et al. (1968), entitled *Report of the Inquiry into the Collapse of Flats at Ronan Point, Canning Town*. This case study is discussed in Chapter 5 of Levy and Salvadori (1992), as well as Chapter 7 of Wearne (2000). Two other references are Pearson and Delatte (2003, 2005).

L'Ambiance Plaza Collapse

The death of 28 workers in the April 23, 1987, construction collapse of the L'Ambiance Plaza building in Bridgeport, Connecticut, triggered a massive rescue effort and several investigations. Unfortunately, to this day the true cause of the collapse remains in dispute because a settlement ended all investigations. This was a lift-slab project.

In his textbook *Prestressed Concrete: A Fundamental Approach*, Nawy discusses lift-slab construction. He points out the need to keep the slabs level during lifting operations and notes that "the construction technique in lift slabs and the absence of the expertise required for such construction can create hazardous conditions which may result in loss of stability and structural collapse" (Nawy 2006, p. 556).

Design and Construction

L'Ambiance Plaza was planned to be a 16-story building with 13 apartment levels topping 3 parking levels. It consisted of two offset rectangular towers, 19.2 × 34.1 m (63 × 112 ft) each, connected by an elevator (Figs. 4-9 and 4-10). Post-tensioned concrete slabs 178 mm (7 in.) thick and steel columns made up its structural frame (Cuoco et al. 1992).

Post-tensioning overcomes the tensile weakness of concrete slabs by placing high-strength steel wires along their length or width before the concrete is poured. After the concrete hardens, hydraulic jacks pull and anchor the wires or strand, compressing the concrete (Levy and Salvadori 1992).

Using the lift-slab method, the floor slabs for all 16 levels were constructed on the ground, one on top of the other, with bond breakers between them (Fig. 4-9a). Then packages of two or three slabs were lifted into temporary position by a hydraulic lifting apparatus and held in place by steel wedges. The lifting apparatus consisted of a hydraulic jack on top of each column, with a pair of lifting rods extending down to lifting collars cast in the slab (Figs. 4-9b and 4-9c).

Once the slabs were positioned, they were permanently attached to the steel columns. Two shear walls in each tower were meant to provide the lateral resistance for the completed building on all but the top two floors. These two floors depended on the rigid joints between the steel columns and the concrete slabs for their stability. Because the shear wall played such an indispensable role in the lateral stability of the building, the structural drawings specified that during construction the shear walls should be within three floors of the lifted slabs (Heger 1991).

108 BEYOND FAILURE

Figure 4-9. L'Ambiance Plaza Lift Slab Construction.
Courtesy National Institute of Standards and Technology.

STRUCTURAL ANALYSIS   109

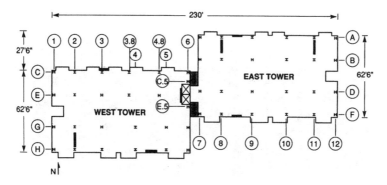

Figure 4-10. Floor plan of L'Ambiance Plaza.
Source: Moncarz et al. (1992).

Details of the lift-slab system, the competing lift-plate system, and other similar systems are provided by Zallen and Peraza (2004, pp. 7–21). The systems are proprietary.

Collapse

At the time of the collapse, the building was a little more than halfway completed. In the west tower, the 9th, 10th, and 11th floor slab package was parked in stage IV directly under the 12th floor and roof package (Fig. 4-11). The shear walls were about five levels below the lifted slabs (Cuoco et al. 1992).

The workers were tack-welding wedges under the 9th-to-11th floor package to temporarily hold them in position, when a loud metallic sound followed by rumbling was heard. Kenneth Shepard, an ironworker who was installing wedges at the time, looked up to see the slab over him "cracking like ice breaking." Suddenly, the slab fell onto the slab below it, which was unable to support this added weight and fell in turn. The entire structure collapsed, first the west tower and then the east tower, in 5 s, only 2.5 s longer than it would have taken an object to free fall from that height. Ten days of frantic rescue operations revealed that 28 construction workers had died in the collapse, making it the worst lift-slab construction accident ever. Kenneth Shepard was the only one on his crew to survive (Levy and Salvadori 1992). The collapsed structure is shown in Fig. 4-12.

Causes of Failure

All of the parties involved in the design and construction of the building hired forensic engineering firms to investigate possible causes of the

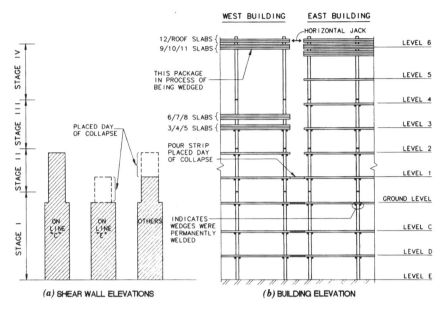

Figure 4-11. Elevation of L'Ambiance Plaza just before collapse. Source: Cuoco et al. (1992).

failure. However, a prompt legal settlement prematurely ended all investigations of the collapse. Consequently, the exact cause of the collapse has never been established. The building had a number of deficiencies, any one of which could have triggered the collapse. The question, however, remains which one of these problems was in fact the triggering mechanism. There are six competing theories. Kaminetzky lists, but does not discuss, a seventh theory: "failure resulting from lateral soil pressure acting on the foundation walls" (Kaminetzky 1991, p. 82).

- Theory 1, National Bureau of Standards (NBS), now the National Institute of Standards and Technology (NIST): An overloaded steel angle welded to a shear head arm-channel deformed, causing the jack rod and lifting nut to slip out and the collapse to begin (Korman 1987).
- Theory 2, Thornton-Tomasetti Engineers (T-T): The instability of the wedges holding the 12th floor–roof package caused the collapse (Cuoco et al. 1992).
- Theory 3, Schupack Suarez Engineers, Inc. (SSE): The improper design of the post-tensioning tendons caused the collapse (Poston et al. 1991).

STRUCTURAL ANALYSIS *111*

Figure 4-12a, b. The collapsed structure of L'Ambiance Plaza.
Courtesy National Institute of Standards and Technology.

112 BEYOND FAILURE

- Theory 4, Occupational Safety and Health Administration (OSHA): Questionable weld details and substandard welds caused the collapse (McGuire 1992).
- Theory 5, Failure Analysis Associates, Inc. (FaAA): The sensitivity of L'Ambiance Plaza to lateral displacement caused its collapse through global instability (Moncarz et al. 1992).
- Theory 6, Oswald Rendon-Herrero: Rapid slump of a column footing precipitated the collapse (Rendon-Herrero 1994).

Theory 1 — Overloaded Steel Angle

The NBS investigation concluded that the failure occurred at the building's most heavily loaded column, E4.8, or the adjacent column, E3.8, as a result of a lifting assembly failure (Fig. 4-13). The shear head reinforced the concrete slab at each column, transferred vertical loads from the slabs to the columns, and provided a place of attachment for the lifting assembly. It consisted of steel channels cast in the concrete slab, leaving a space for the lifting angle. The lifting angle had holes to pass the lifting rods through. These rods were raised by the hydraulic jacks on the columns above them (Levy and Salvadori 1992).

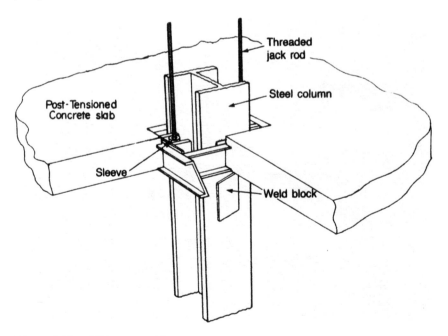

Figure 4-13a. Lifting assembly.
Source: Poston et al. (1991).

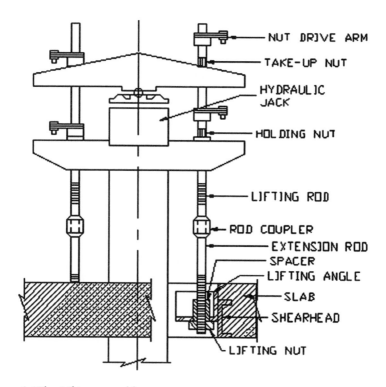

Figure 4-13b. Lifting assembly.
Source: Martin and Delatte (2000).

Shortly before the collapse, the workers lifted the 9th-to-11th floor package to its final position and began tack-welding the steel wedges into place. They used a jack on top of the column E4.8 or E3.8 to slightly adjust the position of the slab, overloading the lifting angles. When the shear heads and lifting angles had lifted the package of three 3.13 MN (320-ton) slabs, they were dangerously close to their maximum capacity, so adding even the smallest of loads could exceed that maximum.

The lifting capacities of the two types of jacks used were too small for the 9.38 MN (960-ton) package. The regular jacks have a maximum load of 869 kN (89 tons), whereas the super jacks have a maximum load of 1.47 MN (150 tons). NBS also tested the shear head and lifting angle and found that the angles tended to twist as the loads approached 781 kN (80 tons), because although the angles had enough strength, they did not have enough stiffness. The force deformed the lifting angle, allowing the jack rod and lifting nut to slip out of the lifting angle and hit the column with 333 kN

(75,000 lb) of force. This load accounted for the loud noise that Kenneth Shepard heard and the indentation found in that column. After this initial slip, the jack rods and lifting nuts in the entire E line progressively slipped, causing the 9th floor slab to collapse, initiating the collapse of the entire building (Korman 1987).

This theory was published before evidence pointed out errors in the alleged facts (Zallen and Peraza 2004). However, the proponents of this theory believe that it is still supported by the available evidence (Culver 2002).

Theory 2 — Unstable Wedges

Thornton-Tomasetti Engineers (T-T) concluded that the instability of the wedges at column 3E caused the 12th floor–roof package to fall, initiating the collapse. They disagreed with the NBS investigation, finding that all the wedges supporting the 9th-to-11th-floor package were mounted before the collapse and that that column had no indentations on it. They, however, did find abnormal tack welds on the wedges that supported the 12th floor–roof package, a large deformation on the top edge of the west wedge of this set, and indentations on the underside of the level 9 shear head. The shallowness of the indentations indicated that, although both lifting nuts slipped out, they were not heavily loaded at the time.

Their investigation also found that the shear head gaps on columns 3E and 3.8E (16 mm, 0.628 in.) were much larger than the gaps on the rest of the building (5.92–8.31 mm, 0.233–0.327 in.) and other buildings built with the lift-slab technique (6.35–9.53 mm, 0.250–0.375 in.). In addition to these abnormally large gaps, the shear heads used on these two columns did not have cutouts in their lifting angles to restrict shifting, and they were installed eccentrically. Finally, until a wedge was completely welded into place, it depended on friction to hold it. Normally, friction is sufficient. The large shear head gaps on columns 3E and 3.8E and the presence of hydraulic fluid on these wedges, however, would have demanded an extremely high friction coefficient to hold the wedges in place.

On the day of collapse, the lateral load from the hydraulic jack was exerted on the heavily loaded wedges, causing the west wedge to roll. Then the local adjustments to slab elevations caused the remaining wedge to roll out, initiating the collapse of the 12th floor–roof package and the west tower (Fig. 4-14). Forces transmitted through the pour strips or the horizontal jack, or the impact of the debris from the west tower, triggered the east tower's collapse (Cuoco et al. 1992). Zallen and Peraza (2004, pp. 28–29) refer to this as "theory 3—wedges falling out."

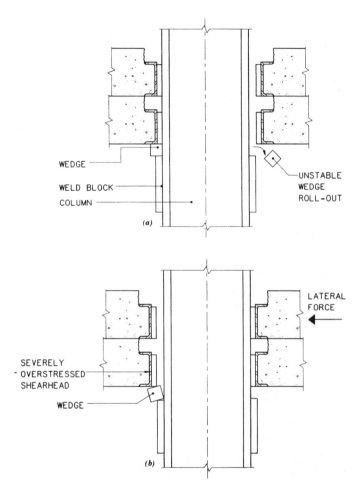

Figure 4-14. Wedges and wedge roll-out mechanism.
Source: Cuoco et al. (1992).

Theory 3 — Improper Design of Post-Tensioning Tendons

SSE analyzed the structural behavior of a typical west tower floor slab with respect to the unusual layout of the post-tensioning tendons (Fig. 4-15). The tendons in the east tower followed a typical two-way banded post-tensioning tendon layout. In this layout, the vertical tendons distributed the weight of the slab to the east–west column lines, which in turn distributed

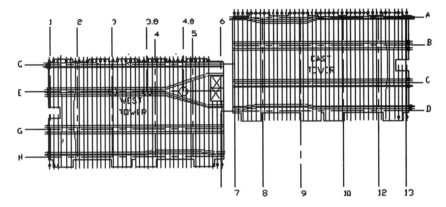

Figure 4-15. Post-tensioning tendon layout.

the weight to the columns. The west tower, however, deviated from this pattern. At column 4.8E, the tendons split in two, both diverging from the column line. In the west tower, the vertical tendons still distributed the slab's weight to the column line. Along line E, however, there are no tendons to carry this weight. This setup violated the American Concrete Institute Building Code (ACI 1983). Kaminetsky points out that the code stipulates "a minimum of two tendons shall be provided in each direction through the critical shear section over columns" (1991, p. 84).

Furthermore, the design details of the post-tensioned floor slabs do not show the location of the shear walls or the openings for the walls at columns 11A, 8A, and 2H. The design did not take these openings into account. Detailed finite-element analysis showed that tensile stresses along column line E, east of column 4.8E, exceeded the cracking strength of the concrete. Therefore, once a crack began, it would immediately spread to column 4.8E. In addition, under ideal lifting conditions, the analysis demonstrated that column 2H would have high compressive and punching shear stresses (Poston et al. 1991). Zallen and Peraza (2004, p. 29) refer to this as "theory 5—improper post-tensioning design."

Theory 4—Poor Weld Details and Welds

OSHA found that the header bar-to-channel welds on one side of the 9th floor shear head at column E3.8 had failed. The use of one-sided square-groove welds for the header bar-to-channel connection was criticized because they were not prequalified joints, according to American Welding Society standards. Because the amount of weld penetration was not known, their strength could not be determined. OSHA hired Neal S. Moreton and

Associates to examine 30 welds around the shear heads at column E3.8 at the 7th, 8th, and 10th floors. They found only 13 of the 30 welds acceptable; the other 17 were substandard. The questionable weld details and substandard welding, coupled with drawings that indicated that the welds would undoubtedly experience forces that they could not resist, all point to weld failure as the trigger of the collapse (McGuire 1992).

Zallen and Peraza (2004, p. 28) refer to this as "theory 2—failure of welds in lifting collar." After weld failure, load would be transferred to the lifting angles, which would then fail in turn.

Theory 5 — Global Instability

The FaAA studied the towers' torsional stability and response to lateral loading to understand their collapse. When the concrete slabs were temporarily resting on the wedges, the connection was rotationally stiff, but as soon as the slab was lifted off one of the wedges into its final position, it could rotate freely from the column. Once the wedges were fully welded into their final position, the connection became rigid again. In the absence of lateral loading, the towers were completely stable.

Lateral loading and displacement, however, could cause the slab to lift off one of its wedges, causing the structure to become laterally flexible. The FaAA used 3D computer modeling and nonlinear stability modeling to study this phenomenon. Their investigation and analysis led them to the conclusion that the towers' sensitivity to lateral displacement caused their collapse. Whereas the FaAA acknowledges that another mechanism could have triggered the lateral displacement, they believed that lateral jacking provided sufficient displacement to initiate the collapse (Moncarz et al. 1992). Zallen and Peraza (2004, p. 29) refer to this as "theory 4—instability."

Theory 6 — Foundation Failure

In a discussion replying to Cuoco et al.'s 1992 paper, Rendon-Herrero suggests that "a closer look warrants reconsideration of the role played by the foundation in the collapse" (Rendon-Herrero 1994). He notes that the NBS report found disintegrated rock, bedrock, and fill materials of varying quality, with some questions as to whether testing of in-place density was performed and as to the rationale for the assumption of the allowable bearing pressure. He concludes that

> The writer feels that descriptions like "mica," "micaceous schist," "highly fractured," "cracks," "disintegrated rock," "fill," "compaction with backhoe," "highly weathered," "thinly laminated," and "very steep dip (nearly vertical)" are red flags that indicate the need for

caution and special attention in the design of a foundation. Punching or local shear is likely when subgrade conditions include loose granular soils (i.e., inadequate compaction); micaceous soils; micaceous schists; and highly fractured, steeply dipping bedrock. (Rendon-Herrero 1994)

Legal Repercussions

All of these theories are plausible, but what triggered the collapse? The answer may never be known. A two-judge panel mediated a universal settlement among 100 parties, closing the L'Ambiance Plaza case. Twenty or more separate parties were found guilty of "widespread negligence, carelessness, sloppy practices, and complacency." They all contributed, in varying amounts, to the $41 million settlement fund. Those injured and the families of those killed in the collapse received $30 million. Another $7.6 million was set aside to pay for all of the claims and counterclaims among the designers and contractors of L'Ambiance Plaza.

Although this settlement kept hundreds of cases out of court and provided rapid closure to a colossal collapse, it also ended all investigations prematurely, leaving the cause of collapse undetermined (Korman 1988). Fortunately, many of the investigators subsequently published their findings (Feld and Carper 1997).

Technical Aspects

Although buildings constructed by the lift-slab method are stable once they are completed, if great care is not taken during construction they can be dangerous. Feld and Carper (1997) reviewed a number of previous lift-slab construction failures and near-failures. The following measures can be taken to ensure lateral stability and safety during construction:

- During all stages of construction, temporary lateral bracing should be provided, unless the lateral stability of the structure is provided through another mechanism.
- Concrete punching shear resistance and connection redundancies should be provided in the structure (Kaminetzky 1991).
- Sway bracing (cables that keep the stack of floors from shifting sideways) should be used. This bracing was required but not used in L'Ambiance Plaza (Levy and Salvadori 1992).

Because of the terms of the settlement, many of the technical lessons that could have been learned from this incident may have been lost.

Professional and Procedural Aspects

The L'Ambiance Plaza collapse highlighted several procedural deficiencies. Responsibility for design was fragmented among so many subcontractors that several design deficiencies went undetected. If the engineer of record had taken responsibility for the overall design of the building or a second engineer had reviewed the design plans, these defects probably would have been detected (Heger 1991). Also, standardized step-by-step procedures for lift-slab construction should be established to ensure the safety of the construction workers. A licensed professional engineer should be present during construction to ensure that these guidelines are followed (Kaminetzky 1991).

According to Zallen and Peraza, three structural engineers should be involved in the design and construction of a lift-slab building. These are the structural engineer of record, the lift-slab engineer, and the post-tensioning engineer. The structural engineer of record is responsible for the integrity of the building in its completed state. The lift-slab engineer, hired by the lift-slab contractor, designs the lift-slab process, including structural stability during lifting operations. The post-tensioning engineer details the tendons and related details and must coordinate carefully with the lift-slab engineer. All three engineers must coordinate their work carefully (Zallen and Peraza, 2004, pp. 62–63).

Ethical Aspects

Although the L'Ambiance Plaza building was designed to be safe once it was completed, during construction it did not have an adequate level of stability. This situation is all too common in the construction industry today (Heger 1991). Canon 1 of the American Society of Civil Engineers (ASCE) Code of Ethics states, "Engineers shall hold paramount the safety, health and welfare of the public and shall strive to comply with the principles of sustainable development in the performance of their professional duties" (ASCE 2006). This safety includes the safety of construction workers. Building regulations do not sufficiently consider structural safety during construction; they should be changed to require a high standard of safety during construction, as well as after a building's completion. In the absence of such regulations, however, an ethical engineer must always consider the safety of the workers (Heger 1991). The ASCE Code of Ethics is provided in Appendix B.

The Human Factor

Many of the organizations involved in the L'Ambiance Plaza project went out of business. One exception was the structural engineer, James

O'Kon. In October 1991, he described his experiences to his colleagues at an annual ASCE conference.

> A federal judge has just told James O'Kon that he wanted $2 million from him—$1 million in insurance money and $1 million from O'Kon's personal funds. ... The judge, Robert C. Zampano, was trying to collect enough money to settle all the lawsuits from the Bridgeport collapse and to provide victims and their families with enough money to live comfortably and educate the victims' children. But O'Kon couldn't see why he should have to help make the families "wealthy." He had been exonerated of wrongdoing and, in his opinion, the workers themselves caused the accident. He also told the judge that his daughter had recently been paralyzed in an accident and he had spent hundreds of thousands of dollars in medical expenses. Zampano eased his demands and asked only for the insurance money, thus saving O'Kon from financial devastation. (Houston 1991)

O'Kon believed that workers' errors in operating the lifting jack triggered the collapse. He immediately contacted his clients to assure them that the accident was not his fault and spoke with the media as little as possible. He had, unfortunately, let his Connecticut license lapse, and had to fight for two years to regain it.

In the unpublished remarks presented at the ASCE meeting, more than four years after the event, O'Kon detailed the pressure the judge put on him and others. He was told that if he didn't pay the requested amount, he would be excluded from the settlement group and financially destroyed in litigation. The hot dog man who provided coffee and sandwiches for the laborers paid $75,000, and the drywall metal framing installation company and drywall installer paid $450,000 and $150,000, respectively, although their work had not been scheduled to start for almost six months. The general contractor and developer paid roughly the same amounts as O'Kon, and both went out of business. The supervising architect paid only $7,500 because he did not have insurance.

Other Lift-Slab Cases

There were other problems in the early days of lift-slab construction, just as there often are with innovative construction technologies. The 1962 system patented by Stubbs used a system with grooves spaced 150 mm (6 in.) apart, with jacks that could not be removed if they jammed between strokes. A Canadian wedge system building under construction in Marion, Indiana, collapsed in 1962. The Canadian wedge connection relied on frictional resistance between the column and wedges. Other buildings failed in sidesway mode because of global instability. These buildings included the Junipero Serra High School Roof in San Mateo, California, in 1954 and the Pigeonhole Parking Garage in Cleveland, Ohio, in 1956. The Pigeonhole Parking Garage swayed 2.1 m (7 ft) in winds gusting from 56 to 97 km/h (35 to 60 mi/h), but the contractor was able to bring it back to plumb (Zallen and Peraza 2004, pp. 22–25). This near-collapse is discussed in the next section. Unlike L'Ambiance Plaza, none of these failures led to a loss of life. These incidents highlighted possible problems during jacking operations, including uneven lifting of slabs, as well as the importance of overall stability.

Conclusions

The L'Ambiance Plaza collapse killed 28 workers and had serious ramifications for all the people involved with the project, as well as for the civil engineering profession as a whole. All of the theories discussed above are plausible, but it seems unlikely that the triggering mechanism of the collapse can ever be determined. This failure severely reduced the use of the lift-slab system and almost eliminated it from use (Moncarz and Taylor 2000, p. 46). The ASCE Technical Council on Forensic Engineering (Task Committee on Lift-Slab Construction) published *Engineering Considerations for Lift-Slab Construction* to benefit future designers and builders of lift-slab projects (Zallen and Peraza 2004).

Essential Reading

Zallen and Peraza (2004) provide an important review of the technical aspects of lift-slab construction. An account is provided by Levy and Salvadori (1992). This case study is featured on the History Channel's Modern Marvels *Engineering Disasters 4* videotape and DVD.

Cleveland Lift-Slab Parking Garage

Although the final cause of the L'Ambiance Plaza collapse has never been determined, the reason of the 1956 near-collapse of the Pigeonhole Parking Garage in Cleveland, Ohio, appears obvious. This was also a lift slab structure.

The building consisted of eight-story twin towers, 28 × 6.4 m (91 × 21 ft) each in plan. The slabs were 200 mm (8 in.) thick and weighed 800 kN

(180,000 lb). The slabs had been jacked into place on 18.5-m (61-ft) tall columns spaced 6.7 m (22 ft) on center. All of the slabs of the west tower were in place, secured by temporary steel wedges. The structure before the incident is shown in Fig. 4-16.

On the evening of April 6, winds gusted 56–80 km/h (35–50 mi/h). The structure shifted 2.1 m (7 ft) out of plumb in the long direction, with the fifth floor close to an adjacent building. The east tower, with only the second and third floors in place, was not affected by the wind. The sidesway displacement of the Cleveland parking garage is shown in Fig. 4-17.

The building was secured with guys and temporary shoring on orders of the Cleveland building commissioner, who remarked, "if they had welded at each floor as they went along the building would have been braced and this would not have happened." The west tower was eventually jacked back into place (Feld 1964, pp. 93–94). The work to restore the Cleveland parking garage to plumb is shown in Fig. 4-18.

Figure 4-16. Cleveland parking garage under construction.
Courtesy *Cleveland Press* Collection, Cleveland State University Library.

STRUCTURAL ANALYSIS 123

Figure 4-17. Sidesway displacement of Cleveland parking garage.
Courtesy *Cleveland Press* Collection, Cleveland State University Library.

Figure 4-18. Restoring the Cleveland parking garage to plumb.
Courtesy *Cleveland Press* Collection, Cleveland State University Library.

The sequence of events is illustrated in photos published in the *Cleveland Press* newspaper, shown here as Figs. 4-16 through 4-18. The newspaper is no longer published, but fortunately its archives were passed on to the Cleveland State University Library's special collections.

Kemper Arena

Stiffness of a flat roof is important for preventing ponding. *Ponding* is the tendency for a flat roof to deflect under rainwater into a "pond," attracting and holding progressively more water. If the roof is stiff enough, it can stand up until the water has a chance to drain away. If it is too flexible, however, deflections continue until the roof becomes unstable and collapses. Ponding combined with wind is the most likely explanation for the collapse of the Kemper Arena in Kansas City, Missouri, on June 4, 1979.

The Structure and Roof

The Kemper Arena opened in 1973 as the new home of the Kansas City Kings basketball team. Because the roof was suspended from above, it featured uninterrupted sight lines. The innovative design earned the architect an Honor Award from the American Institute of Architects (AIA), and the AIA held its 1979 annual convention in Kansas City. Word of the collapse came to the AIA meeting's opening banquet (Wearne 2000, pp. 26–27).

The Kemper Arena's large flat roof, 97 × 108 m (324 × 360 ft) was suspended on hangers from three large space frame cantilever trusses. The three trusses, each 16.5 m (54 ft) wide, were spaced 30 m (99 ft) apart and were made from pipe sections as large as 1.2 m (4 ft) in diameter (Levy and Salvadori 1992, pp. 57–67).

Each of the 42 hangers supporting the roof carried 622 kN (140 kip) in tension. The hangers used ASTM A490 high-strength bolts, which are not recommended for variable or fatigue loads.

To reduce the storm water runoff into the city storm sewers, the roof was designed to hold water as a temporary reservoir. The roof only had eight 130-mm (5-in.) diameter drains. The local code actually required eight times as many. Once the water depth exceeded 50 mm (2 in.), water could pour out over scuppers. This feature could obviously aggravate ponding.

Collapse

At 6:45 P.M., a storm was dumping 108 mm (4¼ in.) of rain per hour, along with wind gusting to 112 km/h (70 mi/h). One arena employee present heard strange noises, followed by explosive bangs. He barely had time to flee. A portion of the roof approximately 60 × 65 m (200 × 215 ft) collapsed into the arena. The pressure wave from the falling roof segment blew out some of the arena walls (Levy and Salvadori 1992, pp. 57–67).

The storm was substantial, but over the 6 years since it had been built the arena had withstood stronger winds and rains. It wasn't a hundred-year storm, just a typical Great Plains thunderstorm (Wearne 2000, pp. 27–28).

Investigation Results

A report prepared by Weidlinger Associates, working on behalf of a subcontractor, found the following:

- The hangers had probably been weakened by fatigue cycles over the six years the arena had been open.
- The roof was susceptible to ponding, and the wind pushed the water to pile up near the point of failure.
- It was necessary to analyze the roof in three dimensions, not just two, to determine the actual flexibility and ponding susceptibility of the roof.
- One hanger fractured from the combined action of wind and water ponding weight.
- Once one hanger failed, the roof had no redundancy. The other hangers could not carry the additional load, and several more hangers failed.

Roger McCarthy of Failure Analysis Associates (FaAA), investigated on behalf of the steel manufacturer KC Structural Steel. He observed that the arena was a modern, optimized, light flexible structure. Although that made the arena highly economical, it also made it susceptible to vibration, ponding, and similar problems. The flexibility had been accentuated by the looseness of the roof bolts, which had never been tightened properly. Because of the roof's overall flexibility, the wind drove the water to the weakest part of the structure. Over the six years, the ponding and flexing during storms had severely weakened the bolts supporting the roof (Wearne 2000, pp. 30–34).

James L. Stratta investigated on behalf of Kansas City. He attributed the bolt failure to weakening of A490 high-strength bolts over time through fatigue. He believed that they failed at ¼ to ⅕ of their static design strength. McCarthy disagreed with this element of the report, noting that the static loads caused more than enough prying force to fail the bolts (Wearne 2000, pp. 30–35).

The roof was rebuilt, with a significant upgrade to the hangers supporting the roof, including ductile welded steel bars. The center of the roof was raised 760 mm (30 in.), and 14 more drains were installed. The rebuilt roof has not had any structural problems (Wearne 2000, pp. 35–36).

Educational Aspects

Failures often occur because of a combination of circumstances, plus a trigger. The roof had previously survived wind and rain events. The circumstances were the fatigue-weakened hanger assemblies and the roof's susceptibility to ponding. The trigger was the accumulation of enough water to fail the hanger.

Carper (2001) notes that several contributing causes were found, but the primary cause was "fatigue failure in the A490 bolds used in the connection."

This case study also illustrates the principle that both strength and stiffness requirements must be satisfied for a structure. The Kemper Arena was originally strong enough, but its flexibility combined with the wind and rain to develop a failure mechanism.

Essential Reading

The Kemper Arena collapse is discussed by Levy and Salvadori (1992, pp. 57–67) and also by Wearne (2000, pp. 25–36). This case study is featured on the History Channel's Modern Marvels *More Engineering Disasters* videotape and DVD.

Other Cases

A number of the other cases in this book are directly relevant to structural engineering. The Sampoong Superstore case, reviewed in Chapter 10, represents an example of a structural collapse of a reinforced concrete building attributed in large part to corruption.

Citicorp Tower

Although the Citicorp Tower in midtown Manhattan did not fail, it was found to have a critical structural problem that could have caused it to collapse under wind loading. The full case study is provided in Chapter 10. Examination of the wind bracing system points out the need to analyze structural behavior in three dimensions, and not merely in two. Analysis for quartering winds illustrates how the force in the critical diagonal members was multiplied by 1.4.

Terrorist Attacks and Building Performance Studies

Unfortunately, terrorist attacks against structures are likely to change the ways we design and analyze structures. The case of Ronan Point illustrated the importance of design against progressive, or disproportionate, collapse. Subsequently, attacks on the Oklahoma City Murrah Federal Building in 1995 (discussed in Chapter 5) and the Pentagon (discussed in Chapter 5) and World Trade Center twin towers on September 11, 2001 (discussed in Chapter 6), reinforced the lesson.

5

Reinforced Concrete Structures

INTRODUCTORY COURSES IN DESIGN OF REINFORCED CONcrete structures in the United States typically follow the American Concrete Institute (ACI) 318 Building Code and Commentary (ACI 2008). Advanced topics, such as torsion, slender columns, yield line analysis, strut and tie models, and shear walls, may be covered but are more often found in a more advanced reinforced concrete course. The ACI code differentiates between ductile failure modes, such as flexure of beams and slabs, and brittle failure modes, such as shear failure, punching shear, and compression (columns and beam columns). Brittle failure modes are assigned a lower resistance factor, and thus a higher safety factor.

One topic that is typically not covered is the design of formwork for freshly cast concrete. This omission is ironic, because failures associated with formwork problems are probably much more common than other issues in concrete construction.

Strength gain of concrete is related to formwork. At some point, shores supporting forms need to be removed to allow construction to continue. If the concrete has not achieved enough strength to hold itself up at this point, a partial or total collapse may occur. This failure is more likely in cold weather because cold retards the strength gain of concrete. If formwork and shoring removal are based on a schedule, as opposed to the actual in-place properties of the concrete, the risk of collapse increases as construction moves into cold weather.

In 1989, ACI published Seminar Course Manual 19 (SCM-19), *Avoiding Failures in Concrete Construction* (ACI 1989). Unfortunately, this document is no longer available from ACI, although some libraries have a copy. However, much of the manual consisted of reprints of articles from ACI's *Concrete International* magazine, particularly the April 1985 issue, which can be found in most university libraries. The articles in SCM-19 cover several case studies, including the Berlin Congress Hall collapse (Buchhardt et al. 1984), the Skyline Plaza in Bailey's Crossroads, Virginia collapse (Carino et al. 1983), the Harbour Cay Condominium collapse in Cocoa Beach, Florida (Lew et al. 1982), and the collapse of Ramp C, Cline Avenue Extension Expressway in East Chicago, Indiana (Russell and Rowe 1985).

Air Force Warehouse Shear Failures

Two warehouse roofs at Air Force bases in Ohio and Georgia cracked and collapsed under combined load, shrinkage, and thermal effects in 1955 and 1956. In each case, 122-m (400-ft) lengths of reinforced concrete roof girders functioned as single units because of defective expansion joints. Other warehouses, built to the same plans, survived because separation between adjacent 61-m (200-ft) bays was maintained by functioning joints. These failures led to more stringent shear reinforcing steel requirements in subsequent editions of the ACI Building Code. In the warehouse structures, the concrete alone, with no stirrups, was expected to carry the shear forces, and the members had no shear capacity once they cracked (McKaig 1962, Feld and Carper 1997).

Wilkins Air Force Depot

At the Wilkins Air Force Base depot in Shelby, Ohio, about 370 m^2 (4,000 ft^2) of the roof collapsed suddenly on August 17, 1955. At the time of collapse, there were no loads other than the self-weight of the roof (Feld 1964, p. 25).

The Air Materiel Command (AMC) built warehouses to a common design at many Air Force bases and depots. The original design was developed in April 1952, with a modification to reinforcement made in March 1954. The Ohio warehouse had been built to the original 1952 design. It was a six-span rigid-frame building, 122 m (400 ft) wide and 610 m (2,000 ft) long. The haunched rigid frames each had six 20-m (67-ft) spans and were spaced approximately 10 m (33 ft) on center. The concrete for each frame was placed continuously in a single working day. Vertical steel plate construction joints were set at the center of each span before concrete placement, but they may not have been effective (Feld 1964, p. 25).

Severe cracking had been observed two weeks before the collapse, so the girder had been supported by temporary shoring. The cracks occurred about 0.45 m (1½ ft) past the end of the cutoff of the top negative reinforcement over the columns (Feld 1964, pp. 26–27). A typical AMC warehouse frame is shown in Fig. 5-1.

Robins Air Force Base

A second warehouse roof collapse took place at Robins Air Force Base near Macon, Georgia, early on the morning of September 5, 1956. This warehouse had been built to the revised design. The revision added top bars and nominal stirrups, at a volume of about 0.06%, for the length of the frames. This collapse included two adjacent girders and about 560 m² (6,000 ft²) of the roof. Before the collapse occurred, cracks in the concrete girders that reached 13 mm (½ in.) in width had been observed. Feld (1964, p. 25) suggests, "It seems that the extent of shrinkage and resulting axial tensions may be somewhat related to the speed of concreting or to the extent of each separate placement."

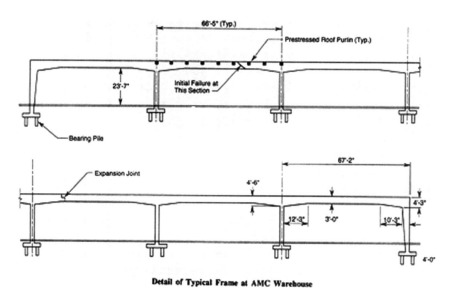

Figure 5-1. AMC Warehouse Frame (1 in. = 25.4 mm, 1 ft = 0.305 m).
Source: Shepherd and Frost (1995).

In both cases, the design, materials, and workmanship were up to the codes and standards of the day. However, the failures had still occurred. Feld believed that

> Failure took place by a combination of diagonal tension (shear) due to dead load and axial tension due to shrinkage and temperature change. Circumstantial evidence suggested that high friction forces were developed in the expansion joint consisting of one steel plate sliding on another; some plates showed no indication of relative displacement since their installation. (1964, p. 27)

In other words, the expansion joints locked and did not function to relieve stress.

Lessons Learned

After these failures, construction of warehouse structures of this type was halted until a new design was developed. Existing warehouse frames were strengthened by adding external shear reinforcement consisting of tensioned steel strapping and steel angles at the lower corners of the girders. The ACI Building Code shear provisions were also revised (Feld 1964, pp. 27–28).

Because these warehouse designs had been built in many locations, questions arose as to why these two warehouses had collapsed and not others. At the time of collapse, these two structures were subject to wide temperature variations and were about a year to a year and a half old. The temperature stresses, unrelieved by the locked expansion joints, combined with shrinkage and shear effects to cause high tensile stress. At the Wilkins Air Force Base (Ohio) structure, the failure plane did not cross any shear stirrups. The rapid, monolithic casting of the frames was thought to exacerbate shrinkage and to contribute to the problem (Feld and Carper 1997, pp. 255–257).

Although it was not required by the code, many designers at that time used at least two continuous top bars and stirrups at a maximum spacing of 300 mm (12 in.) along rectangular continuous beams. After the two failures, tests were carried out at the Portland Cement Association laboratories in Skokie, Illinois. Beams were tested with flexure only and with flexure plus axial tension. In the latter test, the beams under tension and flexure failed in the same manner as the two warehouse frames (Feld and Carper 1997, pp. 258–259).

These two cases illustrate the importance of providing at least some minimum amount of reinforcement, called *temperature steel,* to resist tension forces caused by thermal, shrinkage, and other effects. Real structures do not behave in the same way as our simplified models, and they develop forces and stresses where our analyses suggest that there should be none. The shear resistance of unreinforced, cracked concrete is virtually nil. Feld (1964, p. 29) notes a similar warehouse slab collapse in which the crack through the slab did not cross any reinforcement.

2000 Commonwealth Avenue

Four workers died when about two-thirds of a 16-story apartment building in Boston under construction collapsed on January 25, 1971. The next day's *Boston Globe* newspaper featured dramatic photographs of the remains of the collapsed structure. Rescue operations were delayed because of concerns about the stability of the remaining structure (Blake 1971). Almost 8,000 tonnes (8,000 tons) of debris were removed before the bodies of the workers could be recovered (Granger et al. 1971). A building that had been in development for more than six years collapsed in a few minutes. Fortunately, the collapse occurred slowly enough that most of the people working on the site were able to escape.

Punching shear was determined to have triggered the collapse. An investigation called for by the mayor of Boston found that there had been many errors and omissions associated with the apartment building. Over the long period of development, there had been many changes in the building's owners and designers, leading to considerable confusion (Granger et al. 1971). It was difficult to trace the project ownership and to determine who was responsible for the safety and structural integrity of the project. A summary of the events leading up the incident is provided by King and Delatte (2004).

Design and Construction

The building was cast-in-place reinforced concrete flat-slab construction with a central elevator shaft core. The floor plan is shown in Fig. 5-2. This style of construction is popular for multistory buildings because it reduces the slab thickness and the overall height of the building (Feld and Carper 1997). The flat slabs were 190 mm (7½ in.) thick, except for some bays near the elevator core and at stairwells, which were 230 mm (9 in.) thick. This design made possible a story height of 2.7 m (9 ft) for most of the floors.

The building at 2000 Commonwealth Avenue was designed to be 16 stories high, with a mechanical room above a 1.5-m (5-ft) crawl space on the roof. The building was 55.1 × 20.9 m (180 ft 10 in. × 68 ft 6 in.) in plan. The structure also had two levels of underground parking. A swimming

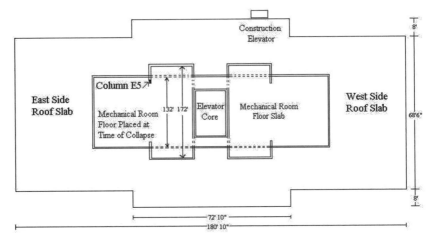

Figure 5-2. Floor plan of 2000 Commonwealth Avenue.
Source: King and Delatte (2004).

pool, ancillary spaces, and one apartment were located on the first floor, and 132 apartments were on the 2nd through 16th floors. Originally, these apartments were to be rented, but the owners later decided to market them as condominiums (Granger et al. 1971).

Construction began on the site late in fall 1969. Almost all of the work was subcontracted. Only one representative from the general contractor was on site during construction.

At the time of collapse, construction was nearing completion. Brickwork was completed up to the 16th floor, and the building was mostly enclosed from the 2nd to the 15th floors. Plumbing, heating, and ventilating systems were being installed throughout various parts of the building. Work on interior apartment walls had also started on the lower floors. A temporary construction elevator was located at the south edge of the building to aid in transporting equipment to the different floors. It is estimated that 100 people were working in or around the building at the time of the failure (Granger et al. 1971).

Collapse

After interviewing many eyewitnesses, the mayor's investigating commission concluded that the failure took place in three phases. These phases were punching shear failure in the main roof at column E5, collapse of the roof slab, and, finally, the progressive and general collapse of most of the structure (Granger et al. 1971).

Phase 1: Punching Shear Failure in the Main Roof at Column E5

At about 10:00 in the morning on the 25th of January, 1971, concrete was being placed in the mechanical room floor slab, wall, wall beams, and brackets. Placement started at the west edge and proceeded east. Later in the afternoon, at about 3:00, most of the workers went down to the south side of the roof for a coffee break. Only two concrete finishers remained on the pouring level.

Shortly after the coffee break, the two men felt a drop in the mechanical room floor of about 25 mm (1 in.) at first and then another of 50–75 mm (2–3 in.) a few seconds later. The labor foreman was directing the crane carrying the next bucket of concrete. He instructed the operator to "hold the bucket" and went down to the 16th floor by way of a ladder in the east stairway. That is when he noticed the punching shear around column E5. He stated, "I can't believe my eyes. I see this slab coming down around the column" (Granger et al. 1971, p. 13).

The carpenter foreman was also in the area and immediately yelled a warning to the men working on the 16th floor and roof of a possible roof collapse. The slab had dropped 125–150 mm (5–6 in.) around the column, and there was a crack in the bottom of the slab extending from column E5 toward column D8. Column E5 is located directly below where the concrete was being placed for the mechanical room floor slab on the east side of the building, as shown in Fig. 5-2 (Granger et al. 1971).

Phase 2: Collapse of the Roof Slab

After hearing the carpenter foreman's warning, most of the workers near column E5 managed to run to an east balcony and stay there until after the roof slab collapse. Eyewitness testimony concluded that the collapse happened fairly quickly. The roof slab began to sag in the shape of a belly, and reinforcing steel began popping out from the mechanical room floor slab. The structure started to shake, and the east half of the roof slab collapsed onto the 16th floor. Then it stopped, giving the workers a chance to run down the stairs to the ground.

At the time of failure, the structural subcontractor was placing reinforcing steel for the stairs on the 14th and 15th floors on the east side of the building. When the workers were making their way from the roof and floors above, most of them crossed over to the west side of the building when they reached the 15th floor (Granger et al. 1971). Thus, the portion of the building that remained standing provided an escape for many of the workers who survived.

Phase 3: Progressive and General Collapse

After the roof collapsed, it settled, and most of the stranded workers could be rescued using the crane and construction elevator. However, about 10–20 min after the roof failed, the east side of the structure began to collapse. A resident of 1959 Commonwealth Avenue described the collapse as a domino effect (or progressive collapse). The weight of the collapsed roof caused the 16th floor to collapse onto the 15th floor, which then collapsed on the 14th floor, and so on to the ground (Litle 1972).

At first, the different floors were distinguishable, but later dust and debris made it difficult to tell them apart. When the dust finally settled, two-thirds of the building had collapsed. The east side and areas on either side of the elevator shaft were gone. Four workers were killed during the collapse, and 30 workers suffered injuries (Granger et al. 1971). The extent of damage is shown in Fig. 5-3. The elevator core probably prevented the collapse from propagating to the other half of the structure.

The Commission Investigation

A commission of inquiry was appointed by the mayor of Boston and convened a week after the collapse. The Associated General Contractors of Massachusetts, the Boston Society of Architects, the Boston Society of Civil Engineers, and the Boston Building and Construction Trades Council had

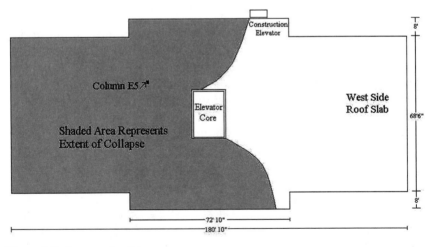

Figure 5-3. Extent of collapse.
Source: King and Delatte (2004).

representatives on the commission. Professor William A. Litle of the Massachusetts Institute of Technology helped draft the commission report and later reported on the failure at an ASCE conference in Cleveland in April 1972 (Litle 1972).

The commission retained an engineering firm and a testing laboratory to aid in the investigation. The commission interviewed a number of eyewitnesses but suspended the interviews after about a dozen because there was no significant disagreement among the accounts (Granger et al. 1971).

The commission made a number of important findings:

- There were a number of irregularities in the issuance of the building permit. Key drawings were missing. Not a single drawing found in the file carried an architect's or engineer's registration stamp. The structural engineer refused to provide the calculations supporting his design to the commission. No principal or employee of the general contractor held a Boston builder's license. At the time, partial drawings could be used to obtain a building permit, with the understanding that final stamped drawings would have to be supplied before construction could begin.
- Ownership of the project changed a number of times, with changes in architects and engineers. This situation added to the overall confusion and contributed to the irregularities. Some of the key changes are discussed in King and Delatte (2004).
- The general contractor only had a single employee on site, and most subcontracts were issued directly by the owner to the subcontractors and bypassed the general contractor. At least seven subcontractors were involved.
- The structural concrete subcontract did not require any inspection or cold weather protection of the work, although the designer had specified these measures. There was no evidence of any inspection of the work by an architect or engineer, although the project specifications required this.
- The concrete mix designs were not prequalified. Such prequalification was a Boston Building Code requirement, which stipulated that the performance of the proposed concrete be verified by laboratory testing. Some concrete deliveries did not contain the required air entrainment. Calcium chloride was used as an accelerator for some of the concrete, although it was specifically prohibited by the designer's specifications. The designer's specifications included a water-reducing admixture, which was used

in only a small percentage of the concrete supplied. The Boston Building Code requirements for inspection and testing were not met on 65% of the days when concrete was delivered to the project. Chemical analyses also suggested that some samples had low cement content.
- The triggering mechanism of the collapse was punching shear at the roof slab around column E5, probably preceded by flexural yielding of the roof slab adjacent to the east face of the elevator core.

The commission examined the failure from three aspects:

- whether failure would have been expected if the construction had conformed to the design documents,
- whether the construction procedures and materials conformed to the design documents, and
- whether the design documents met the building code requirements.

The commission concluded that the failure would not have occurred if the construction had conformed to design documents and that the construction procedures and materials were deficient. The most significant deficiencies were a lack of shoring under the roof slab and low-strength concrete. The design documents specified a 28-day strength of 20 MPa (3,000 lb/in.^{2}). At the time of the failure, 47 days after casting, the concrete had yet to achieve the required 28-day strength.

There was some confusion as to whether the concrete at the point of the collapse had been cast on December 3 or 9, 1970. Both concrete placements had deficient strength, with 11 and 13 MPa (1,600 and 1,900 lb/in.^{2}) compressive strength at 47 and 53 days, respectively. The commission believed that low concrete strength and lack of shoring were the principal causes of the collapse.

However, the commission also found that the design did not meet code requirements for the slab thickness at column E5. The minimum thickness requirement was governed by deflection and not by strength, but a thicker slab would have provided a greater safety margin against punching shear (Granger et al. 1971).

Although the structural plans limited construction loading to 1.44 kPa (30 lb/ft^2), actual loads were estimated to approach 6.22 kPa (130 lb/ft^2). Some boilers and construction equipment were stored on the roof where the failure began. The locations of shores were specified on the structural plans, but these requirements were ignored (Granger et al. 1971). Witnesses reported that there were few shores.

The commission also noted a number of deficiencies and deviations with reinforcement placement. These problems included the following (Granger et al. 1971):

- Column ties were omitted in the bottom 1 m (39 in.) or so of several columns.
- Concrete cover for vertical column bars varied between 13 and 200 mm (½ to 8 in.), instead of the specified 48 mm (1⅞ in.).
- One collapsed column had only six instead of eight 19-mm (¾-in., U.S. #6) longitudinal reinforcing bars, with similar discrepancies in other columns.
- East–west top slab steel was specified to be four 19-mm (¾-in., U.S. #6) bars and five 13-mm (½-in., U.S. #4) bars. None of the larger bars were found, and the smaller bars were used consistently in place of the larger bars.
- The vertical position of the slab reinforcement was erratic. In some cases, the top steel was at or below the middle depth of the slab.
- The collapse occurred near a construction joint, and the structural plans specified additional steel dowels across such joints. The witness testimony indicated that these dowels were not placed.

The Punching Shear Mechanism

Punching shear is usually the critical failure mechanism for flat-slab reinforced concrete structures. This mechanism is illustrated in Fig. 5-4. With this type of failure, the column and part of the slab punch through the slab as it moves downward.

The force acting on the slab around a column overcomes the resistance, and the slab falls down around the column. A portion of the slab is left around the column, but the remainder of the slab falls down to the next floor. If the lower slab is unable to hold up both floors, then progressive collapse continues.

Also, punching shear redistributes forces acting on the failed slab to other columns. If the other columns cannot carry the added weight, then the slab starts punching through the surrounding columns as well. Punching shear at one column can initiate a complete failure of a building.

The punching shear strength of a flat slab (without shear reinforcement) depends on five factors (Ghosh et al. 1995):

1. concrete strength,
2. the relationship of the size of the loaded area to the slab thickness,
3. the shape of the loaded area,

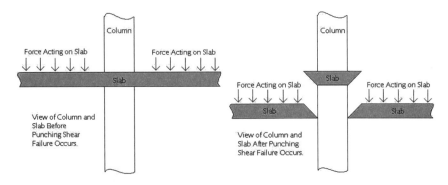

Figure 5-4. Punching shear.
Source: King and Delatte (2004).

4. the shape of the perimeter area, and
5. the ratio of shear force to moment at slab–column connection.

The punching shear strength, V_c, of a flat slab, for a simplified case of an interior column, may be expressed in U.S. customary units as (Ghosh et al. 1995, ACI 2005):

$$V_c = 4\sqrt{f'_c}b_0 d \qquad (5\text{-}1)$$

where f'_c = the 28-day cylinder compressive strength of the concrete, d = depth of the slab (measured from the bottom of the slab to the reinforcing steel location), and b_0 is the perimeter of the failure surface around the column measured at distance d from the face of the column. For SI units, the constant 4 changes, but the fundamental relationships among the variables remain the same.

Therefore, the punching shear strength of a flat slab depends on concrete strength and slab depth. Strength varies as the square root of the concrete strength. The effect of slab thickness is more than linear because increasing d also increases b_0 slightly. The lower concrete test strengths cited above would lead to a punching shear capacity reduction of 20–27%. The low placement of top steel bars in the slabs would lead to an even greater reduction of punching shear capacity.

Review of Causes of Failure

A week after the collapse, *Engineering News Record* reported that there were three possible causes of structural failure under investigation:

formwork for the penthouse floor slab collapsing onto the roof, a heavy piece of equipment falling from a crane and starting the progressive collapse, or failure of weak concrete placed during previous cold days (*ENR* 1971). After an extensive investigation, the mayor's commission concluded that many flaws contributed to the collapse.

The committee determined that punching shear failure at column E5 triggered the initial collapse. The major areas in which the construction deviated significantly from the design were shoring and concrete strength. Inadequate shoring under the roof slab on the east side of the building made it impossible for the roof to hold the freshly placed concrete for the mechanical room floor slab, construction equipment, and two boilers that were stored on that side of the building. Also, the concrete strength of the roof slab was well below the 20 MPa (3,000 lb/in.^{2}) specified in the design (Granger et al. 1971).

Deficient concrete strength could be attributed to poor-quality concrete, improper curing, or both. Tests indicated that the amount of cement in the concrete was sufficient, but records suggested that the maximum permissible slumps were consistently exceeded. The slump specification was only met in 37 of 240 tests. This result probably indicates too much water, leading to lower strength concrete.

Also, testimony indicated that the specifications for concrete protection in cold weather were not followed. Average curing temperature from the day of placement until the day of collapse was −4 °C (25 °F). The concrete was not protected against the effects of cold weather. The commission believed that poor curing seriously retarded concrete strength gain (Granger et al. 1971).

The commission also found that the reinforcing steel details did not provide for sufficient steel crossing columns or for sufficient development length. One important detailing error was that the bottom slab bars were not long enough to tie into the core walls. Furthermore, significant differences were found between the structural drawings and the location and amount of steel in the parts of the building that were recovered. In some locations, as little as one-half of the specified top slab steel was actually placed. There were no ties in column splice regions. Although these errors did not contribute to the initiation of the collapse, they probably influenced the speed and extent of the propagation of the failure (Granger et al. 1971).

Design and Detailing Concerns

Some of the reinforcing bars were not long enough to provide adequate development into the columns and walls as required by code, and placement of bars in some of the slabs was not sufficient to meet the ACI code at the time. ACI required that at least 25% of the negative slab reinforcement in each column strip pass over the column within a distance of the slab depth (d) on either side of the column face (Granger et al. 1971). This requirement was not fulfilled.

The commission found that the slab should have been 222 mm (8¾ in.) thick to satisfy the ACI 318-63 (ACI 1963) code requirements. However, this limit is based on deflection and not on strength (Granger et al. 1971). It should be emphasized that there were no indications of inadequate design. Rather, the builders failed to adhere to the plans and specifications, and the owner failed to provide for proper inspection of the work.

Procedural Concerns

There were many procedural concerns in the construction of 2000 Commonwealth Avenue. For all practical purposes, there was no supervision of the construction. Almost every step of construction was flawed (Kaminetzky 1991). Some of the major concerns included lack of proper building permit and field inspection, premature removal of formwork, and lack of construction control.

The investigating committee determined that if the construction had had a proper building permit and had followed codes, then the failure could have been avoided. Because numerous problems all played a part in the collapse, deciding whom to hold responsible for the collapse became difficult. Ownership changed hands many times, and most jobs were subcontracted.

Construction did not follow the structural engineer's specifications for shoring or formwork. Before removal of shores and forms, the concrete would have had to first reach 70% of its designated 28-day strength to meet that specification. It was the commission's opinion that the average strength of the concrete in the roof slab was only 13 MPa (1,900 lb/in.^{2}) after at least 47 days, not the required 14.5 MPa (2,100 lb/in.^{2}) for removal or the specified 20 MPa (3,000 lb/in.^{2}) required after 28 days. The reason for disregarding the tests was the difference between the curing conditions in the laboratory and at the project—the concrete on site would gain strength more slowly in cold weather, whereas laboratory specimens are cured at a specified temperature and humidity. No inspection or cylinder testing were performed for the east side of the building, so removal of formwork was based on values obtained from the west side of the building. Furthermore, there was no shoring under the roof slab below the freshly placed mechanical room (Granger et al. 1971).

Finally, there was little construction control on the site. There was no architectural or engineering inspection of the project, and the inspection done by the city of Boston was inadequate. The design plans specifically stated that certain aspects of the project needed to be approved by an architect, yet no architect or engineer was consulted. Instead, construction was based on arrangements made by the subcontractors. As mentioned before, there was only one representative from the general contractor, and this man was not a licensed builder. He did not direct, supervise, or inspect any of the work done by the subcontractors (Granger et al. 1971).

This case also illustrates the need for proper shoring of concrete construction. Many failures over the years have occurred because of insufficient shoring or premature removal of shoring and formwork.

Some of the causes of the failure and contributing factors are summarized below:

- The owner did not provide competent involvement of design professionals with knowledge of design and construction requirements.
- The contractor did not reshore the slab.
- Concrete, probably of poor quality, was not adequately protected against cold weather.
- Low top bars in the slab led to inadequate slab depth.
- Construction loads on the roof slab were too high.
- There was no inspection by an architect or engineer, only poor inspections by Boston's building inspector, and no inspection by the general contractor's representative (who was not a licensed builder). The owner did not provide quality control of the structural work, and the contractor did not comply with structural specifications.

Conclusions

Many lessons can be learned from the collapse of 2000 Commonwealth Avenue. The mayor's investigating commission made recommendations for improving the city of Boston's building codes. However, the commission also reported that if the construction of 2000 Commonwealth Avenue had fully complied with existing codes, then the collapse would not have occurred. The commission was dismayed that the project could have progressed through so many phases without the errors and omissions being found and corrected.

The commission made recommendations to prevent similar collapses in the future. These recommendations included changes in assigning responsibility and ensuring competence of design, construction, and inspection of major buildings, as well as additions to organization and staff competence of the Boston Building Department. At the time of the failure, the Building

Department had 130 employees but only two registered professional engineers, and no registered architect.

In addition, changes in building codes to prevent propagation of a local failure into a general collapse were recommended. This case and the cases of Bailey's Crossroads and Harbour Cay Condominium illustrate the importance and progressive nature of punching shear failure. This failure is a critical failure mechanism for concrete structures of this type. Structural safety depends on adequate slab thickness, proper placement of reinforcement, and adequate concrete strength.

Essential Reading

The most useful document to review on this case would be the report of the mayor's investigating commission by Granger et al. (1971). The report states, in its conclusions, that the findings should be broadly disseminated. Ironically, the report is difficult to find. A student's father copied the report at the Boston Public Library, which does not allow the report to circulate.

Skyline Plaza in Bailey's Crossroads

The progressive collapse of 2000 Commonwealth Avenue was similar to the later structural failures of buildings at Skyline Plaza in Bailey's Crossroads, Virginia, and Harbour Cay Condominium in Cocoa Beach, Florida. On March 2, 1973, the Skyline Plaza apartment building collapsed while under construction. The collapse extended vertically through the building from the 24th floor, leaving an appearance of the structure as two separate high-rise buildings with a gap between them. The collapse tore a 18.3-m (60-ft) wide gap through the building all the way to the ground. At the time of the collapse, two practically identical reinforced concrete towers had been built (Kaminetzky 1991, p. 64). Figure 5-5 shows an aerial photograph of the structure shortly before the collapse. The collapsed structure is shown in Fig. 5-6.

As in the case of 2000 Commonwealth Avenue, premature removal of shoring and insufficient concrete strength were suggested as the causes of failure. The National Bureau of Standards (NBS, now the National Institute of Standards and Technology, NIST) carried out an investigation on behalf of the Occupational Safety and Health Administration (OSHA). The NBS investigation team determined that punching shear failure at the 23rd floor caused a partial collapse that propagated to the ground. Fourteen workers were killed and 34 were injured (Carino et al. 1983).

Figure 5-5. Skyline Plaza before the collapse.
Courtesy National Bureau of Standards/National Institute of Standards and Technology.

Description of Structure and Construction

The building was a reinforced concrete flat-plate design, with a planned final height of 26 stories and a story height of 2.7 m (9 ft). The floor slabs were 203 mm (8 in.) thick. The slabs used sand and lightweight concrete with a specified compressive strength of 20.7 MPa (3,000 lb/in.^{2}). Normal-weight concrete was used for the columns, with the strength requirement varying with height in the structure. The rate of construction varied, but it was intended to be one floor per week. The fresh concrete was supported on wooden formwork.

At the time of collapse, the concrete on the 24th floor had been freshly placed. The only construction activity in the area was finishing work, and one finisher reported a large deflection near the subsequent collapse (Carino et al. 1983, pp. 36–37).

Formwork

The investigation quickly established the idea that determining the location and state of the formwork would be important for finding the cause

Figure 5-6a, b. Collapse of Skyline Plaza, Bailey's Crossroads.
Courtesy National Bureau of Standards/National Institute of Standards and Technology.

of the collapse. Both the architect and the engineer required that two floors remain shored at all times. Neither specified the required strength for form removal. Although there was conflicting testimony from the survivors, NBS concluded that stripping was in progress on the 22nd floor, which would not leave sufficient support under the fresh concrete on the 24th floor. Stripping the 22nd-floor shores would leave only one floor of shores under the fresh concrete. Photos taken just before the collapse also showed few or no shores under the 22nd floor. There were witness statements to the effect that the shores between the 22nd and 23rd floors came loose and toppled over, which would be consistent with the 22nd floor dropping after its support was removed (Carino et al. 1983, pp. 37–38).

Estimated Concrete Strength

Cold weather delays strength gain of concrete. Therefore, in cold weather, basing construction rates on a time, such as one floor per week, may be unsafe. Field-cured cylinders, which are stored next to the fresh concrete at roughly the same temperature, provide a better estimate of the concrete strength in the structure than laboratory-cured cylinders. Neither field-cured cylinders nor other methods of estimating in-place concrete strength were used on this project. Laboratory cylinders are typically cured at approximately 23 °C (73.5 °F), so if the field concrete is at colder temperatures, the laboratory test results would overpredict the strength of the concrete in the structure.

The NBS investigation team removed cores from the structure and tested them about two weeks after the collapse. Temperature records from nearby Washington National Airport were used to estimate the temperature at the construction site. The average temperatures for the 22nd and 23rd floor slabs during the days after placement were 5.6 and 7.2 °C (42 and 45 °F), respectively. An accelerating admixture, calcium chloride, was used to speed up strength gain.

On the basis of the length of time since casting, adjusted for the low temperatures, the estimated concrete compressive strength was only 6.6–9.9 MPa (960–1,440 lb/in.^{2}) at the time of the collapse. Materials tests on the reinforcing steel indicated that it met specifications (Carino et al. 1983, pp. 38–39).

Summary of NBS Findings

The NBS investigation team performed a structural analysis. The column strength was satisfactory. The flexural strength of the slabs was exceeded by about 10%. However, this situation should not have been enough to cause a collapse because moments would be redistributed to nearby understressed sections. Flexure is a ductile failure mechanism, implying a more gradual collapse and an opportunity for workers to escape.

As at 2000 Commonwealth Avenue, punching shear was the most likely culprit. As before, eq. 5-1 applies, but an adjustment factor reduces punching shear capacity by 15% for sand and lightweight concrete. Because the required shear capacity, V_u, the perimeter of the failure surface b_0, and the depth of the slab d are known, eq. 5-1 can be solved for the required concrete compressive strength f_c'. Four columns were analyzed. Around the columns, the required capacity was in the range of 11.9–13.2 MPa (1,724–1,920 lb/in.^{2}), considerably more than the estimated actual capacity when shores were removed (Carino et al. 1983, pp. 40–41). Unlike 2000 Commonwealth Avenue, there was no indication of incorrect detailing or placement of the reinforcing steel. "The NBS investigation concluded that the probable cause of the collapse was a punching shear failure of the 23rd floor.... The ... falling debris from the collapsing 23rd and 24th floors overloaded the 22nd floor slab and induced the progressive collapse of successive floors down to the ground" (Carino et al. 1983, p. 41).

Legal Repercussions

Kaminetzky (1991) notes that the

> collapse raised the issue of the extent of the engineer of record's responsibility for the success and safety of formwork design and inspection. While the engineer's contract specifically stated that he had no responsibility for field inspection, nevertheless a jury found him negligent. This was so because the pertinent code required that "a competent architect or engineer" must provide supervision "where requested by the building official." (p. 67)

Often, owners are not willing to pay engineers for the site visits.

Feld and Carper (1997, p. 243) observed that the federal jury found both the architect and engineer negligent and awarded large damages. The architect's and engineer's construction recommendations were not followed. The general contractor and building owner were not named, despite the fact that the failure obviously happened because of mismanaged construction. They were shielded because the plaintiffs had already received awards under the Virginia Workers' Compensation Act.

Lessons Learned

Kaminetzky (1991, p. 67) cites six lessons from this case:

1. The contractor should be responsible for preparing formwork drawings, including shores and reshores.
2. The contractor should prepare a detailed concrete testing plan for stripping forms, including cylinder tests.
3. Inspectors and other quality control agencies should verify that the contractor performs the above two items.
4. The EOR should make sure that he or she provides the contractor with all necessary design load data and other unique project information.
5. "Uncontrolled acceleration of formwork removal" may cause a total or partial collapse.
6. Continuous top and bottom slab reinforcement is necessary around the columns. Continuous reinforcement provides overall ductility.

If the contractor uses cylinder tests to determine when to strip forms during cold weather, the cylinders should be stored at the same ambient temperature as the structure. This procedure will prevent overestimation of the in-place concrete strength.

Essential Reading

The NBS report (Leyendecker and Fattal 1977) can be obtained through interlibrary loan or other sources. The findings of that report are summarized by Carino et al. (1983). This report focuses on the technical issues. Other discussions of this case are provided by Ross (1984), Kaminetzky (1991, pp. 64–67), and Feld and Carper (1997, pp. 242–245).

Harbour Cay Condominium

The collapse of the flat-plate Harbour Cay Condominium building in Cocoa Beach, Florida, on March 27, 1981, was caused by a punching shear failure that triggered a progressive collapse, much like those at 2000 Commonwealth Avenue in Boston and Skyline Plaza in Virginia. Eleven workers were killed and 23 were injured. Figure 5-7 shows the building before collapse, and Fig. 5-8 shows the structure afterward. As at Skyline Plaza, OSHA

150   BEYOND FAILURE

Figure 5-7. Harbour Cay Condominium before the collapse.
Courtesy National Bureau of Standards/National Institute of Standards and Technology.

asked NBS to investigate on its behalf. The NBS investigation team included two individuals who had investigated the Skyline Plaza collapse.

The collapse happened while the roof slab was being placed. Photos taken shortly after the collapse, while the debris was still intact, provided important evidence. Many of the columns were still standing, up to three or four stories high. The slabs were stacked vertically on top of each other. This evidence strongly suggested a punching shear failure mechanism, with no side sway during the collapse. The NBS team was not able to make a detailed examination of the debris in situ because that would have interfered with the ongoing rescue operations. Some samples were obtained for subsequent coring and materials testing (Lew et al. 1982, p. 64).

The NBS team determined that slab thickness of 203 mm (8 in.) did not meet the ACI code minimum of 280 mm (11 in.). Also, the top reinforcing steel was placed too low, reducing the slab effective depth d from 160 mm (6.3 in.) to 135 mm (5.3 in.). As a result, the calculated punching shear stresses exceeded capacity (Lew et al. 1982).

Kaminetzky (1991) also investigated. He reported some of the workers' eyewitness observations. As the workers were finishing the concrete, they

REINFORCED CONCRETE STRUCTURES   151

Figure 5-8a, b. Harbour Cay Condominium collapse.
Courtesy National Bureau of Standards/National Institute of Standards and Technology.

heard a loud crack that sounded like wood splitting. Accounts vary as to whether the center of the fifth or the fourth floor started to come down first.

> Another fact that was very disturbing was the indication that the design engineer was made aware of severe cracking of the floors and big deflections of the slabs several days prior to the collapse. The cracks were described as "spider-web-type cracks". . . . There were reports of deflections as large as 1¾ in. (44 mm) several days prior to the collapse. The structural engineer was requested to recheck his design, which he did, reporting back that it was "OK." (Kaminetzky 1991, p. 74)

Description of Structure and Construction

The structure was a five-story flat-plate residential building, with stairwells at north and south ends and a structurally detached elevator tower at the east end. The slabs spanned up to 6.75 and 8.43 m (22 ft 2 in. and 27 ft 8 in.) in the two directions. Interior columns were 254 × 457 mm (10 × 18 in.), and exterior columns were 254 × 305 mm (10 × 12 in.). The specified compressive strength of all above-grade concrete was 27.6 MPa (4,000 lb/in.^{2}).

The construction rate was approximately one floor per week, with each floor cast in two halves, two days apart. The roof slab was to be cast in one continuous placement and was roughly 80% complete when the collapse occurred.

Worker statements indicated that at the time of collapse, shores were in place on the fifth floor, and reshores were in place on the second, third, and fourth floors. Some of the reshores may have been removed by other trades, and it is possible that not all were replaced. There were no reshores under the second floor (Lew et al. 1982, pp. 65–66).

Summary of NBS Findings

Two principal NBS findings were the following:

1. Punching shear requirements of the applicable Code (ACI 318 1977), which would have controlled the thickness of the slabs, were not considered in the design of the building.
2. The use of chairs of insufficient height to support the top bars in the slab reduced the effective depth and corresponding punching shear capacity from what would have existed had the structural plans been followed (Lew et al. 1982, p. 65).

NBS test results showed that the compressive strength of the concrete slabs at the time of the collapse was at least 25.5 MPa (3,700 lb/in.²), which was close to the specified strength. The design called for 19 mm (¾ in.) of cover over the top reinforcing steel. With the 203-mm (8-in.) slab, this design provided an effective depth of 160 mm (6.3 in.). Structural analysis showed punching shear forces as high as 454 kN (102 kip) at some interior columns. Using eq. 5-1, with these forces and this strength of concrete, a 280-mm (11-in.) thick slab would be required to provide sufficient capacity.

The depth was further reduced because of a construction error. The top steel was placed on chairs only 108 mm (4¼ in.) high, as specified by the shop drawings. This situation increased the top cover to 41 mm (1⅝ in.). Therefore, the top reinforcing steel was placed too low, reducing d to 135 mm (5.3 in.) (Lew et al. 1982, pp. 67–72).

Design Versus Construction Error

NBS considered whether the inadequate thickness of the slab, a design error, or the inadequate effective depth of the reinforcing steel, a construction error, caused the collapse. If the slab had been designed and built correctly, the demand-to-capacity ratio would have been 0.72. With the design error of the 203-mm (8-in.) slab, the demand-to-capacity ratio became 0.90. With an increase of top cover to 41 mm (1⅝ in.) but adequate slab thickness (280 mm or 11 in.), the demand-to-capacity ratio was 0.84. The combination of the two increased the demand-to-capacity ratio to 1.13. Therefore, NBS determined that both factors contributed approximately equally. If only one of the errors had been made, the structure would probably not have collapsed (Lew et al. 1982, p. 72).

Professional Aspects

Incredibly, no punching shear calculation had been made for the concrete floor slabs. The chairs used to support the slab steel were 108 mm (4¼ in.) high, which, coupled with the thin slabs, led to a small effective depth (Kaminetzky 1991).

According to Feld and Carper (1997, p. 274), the structural engineer was a retired NASA engineer who hired another retired NASA engineer to do the calculations. This example illustrates the important point that structural engineering isn't rocket science. Evidently, it is considerably more difficult. Routine concrete design calculations were simply never performed. A discussion of licensure issues for professional engineers (P.E.s) in this and other cases is provided in Chapter 10.

The Bailey's Crossroads and Harbour Cay Condominium collapses resulted in 25 deaths and 58 injuries combined (Lew et al. 1982, Carino et al. 1983). These calamities could have been avoided if the engineers working in Virginia and Florida had learned the lessons of the 2000 Commonwealth Avenue collapse. Punching shear is hardly an unknown phenomenon. A punching shear failure in Indianapolis was reported as early as 1911 (Feld 1978).

Three factors were common to all three collapses: punching shear (which is critical for flat-slab buildings), improper formwork removal and reshoring procedures, and cold temperatures in winter (January through March) (Kaminetzky 1991, p. 76). These factors, however, cross different disciplines: design, construction management, and concrete technology and materials science. It is vitally important that the design and construction team members work together to eliminate these problems.

Lessons Learned

Kaminetzky (1991, pp. 77–78) cites six lessons from this case:

1. Punching shear is the most common failure mode for concrete flat-slab buildings and must be checked.
2. The minimum depth requirements of the ACI code must be met to provide adequate strength and limit deflections.
3. Top and bottom reinforcing bars should be placed directly within the column periphery to provide additional resistance to progressive collapse. This step should not add to construction costs.
4. Proper design of formwork by qualified professionals is an essential element of field construction control. Planning must address shoring and reshoring procedures and schedules, as well as methods for verifying that the concrete has achieved satisfactory strength.
5. Construction must stop immediately if there are warning signs of potential or impending failure. If necessary, the structure must be evacuated immediately.
6. During cold weather, special measures should be taken to estimate the in-place strength of the concrete. "It is also a fact that the level of construction carelessness increases in the winter months."

The most economical way to increase the punching shear capacity of the slabs would have been to increase the size of the columns, which would have increased b_0. This procedure would also have improved constructability by providing more space to cast concrete around the column reinforcement

(Kaminetzky 1991, p. 75). Increasing the column size would have required much less additional concrete than increasing the thickness of the slabs.

Given the magnitude of the errors, the question arose as to how the building was able to stand up for so long. Until the collapse, the gravity dead loads were completely carried by the shores and reshores and not by the structure itself (Kaminetzky 1991, pp. 75–76).

Feld and Carper (1997, pp. 273–274) note these problems, as well as some others:

- There were no checks for deflection or code minimum thickness provisions.
- There were no beam shear checks.
- There were no code checks for column reinforcement spacing.
- Calculations were based on grade 40 steel, although grade 60 was specified. This change to higher strength steel would actually increase safety.
- There was no actual calculation of effective depth of slab flexural reinforcement. Instead, a constant multiplier of computed moments was used. The 160 mm (6.3 in.) cited by NBS was determined by working backward from the design.
- The column steel was so congested that it prevented flow of concrete around the bars.

Essential Reading

The NBS report (Lew et al. 1981) can be obtained through interlibrary loan or other sources. The findings of that report are summarized by Lew et al. (1982). This report focuses on the technical issues. Other discussions of this case are provided by Ross (1984), Kaminetzky (1991, pp. 72–78), and Feld and Carper (1997, pp. 271–274). This case study is featured on the History Channel's Modern Marvels *Engineering Disasters 15* videotape and DVD.

Bombing of the Oklahoma City Murrah Federal Building

Extreme events, such as earthquakes, test structures near their ultimate capacity and sometimes lead to destruction. Explosions, whether intentional or accidental, do the same thing. On April 19, 1995, a truck bomb tore through the façade of the Murrah Federal Building in Oklahoma City. The blast destroyed a significant portion of the structure, killing 169 people.

156   BEYOND FAILURE

This event was a classic example of what is known as progressive collapse. *Disproportionate collapse* is now becoming the preferred term, because all collapses are inherently progressive. What disproportionate collapse means is that an event that should have been localized to one part of the structure instead causes most or all of the structure to collapse, out of proportion to the original damage. It implies a lack of structural redundancy. Ronan Point, covered in Chapter 4, is considered another case of disproportionate collapse.

The Murrah Federal Building

The Murrah Building was designed in 1974 and opened three years later. It conformed to all of the structural codes of the time (Wearne 2000, p. 117).

Some of the key structural features of the Murrah Building were reviewed by Hinman and Hammond (1997, pp. 3–5). The structure is shown in Fig. 5-9.

Figure 5-9. Oklahoma City Murrah Federal Building.
Source: Osteraas (2006).

- The building had a nine-story, reinforced concrete frame, 61 × 23 m (200 × 75 ft) in plan. It was 10 bays long and 2 bays wide. Typical floor-to-floor height was 4 m (13 ft).
- The rear of the building faced north and had curbside access for vehicles, with an indented loading zone.
- To improve natural lighting, the north façade had a full-height glass curtain wall. The glass was highly vulnerable to blast.
- Shear walls formed by the stairwell and elevator shafts provided the main lateral bracing system.
- Beams supporting the 150-mm (6-in.) thick floor system were 189 × 78 mm (48 × 20 in.) and dimensioned so that only nominal shear reinforcement was needed.
- The north elevation was supported by four large 142 × 78 mm (36 × 20 in.) columns extending up the first two floors. Large transfer girders, 12.2 m (40 ft) long, supported intermediate columns from the upper floors.

Attack and Recovery

At 9:02 A.M. on April 19, 2005, the Alfred P. Murrah federal office building in Oklahoma City, Oklahoma, was bombed. The weapon, consisting of a homemade bomb placed in the back of a rental truck, caused the partial collapse of the Murrah building and structural damage to numerous other buildings in the vicinity. The explosion caused the death of 167 persons, including 19 children, and injured 782 persons. (Hinman and Hammond 1997, p. xi)

The bomb had an estimated 2,177 kg (4,800 lb) of ammonium nitrate and fuel oil explosive, referred to as ANFO. The ammonium nitrate is commercially available as fertilizer and is relatively stable unless detonated by a high-grade explosive booster, such as dynamite and detonation cord. Despite alternate theories, it is now widely accepted that there was a single truck bomb. A similar vehicle bomb was used in the first World Trade Center bombing on February 26, 1993 (Hinman and Hammond 1997, pp. 1–3).

The explosion was roughly 3 m (10 ft) from the building, at column G20. The shock wave spread outward and upward across the face of the building. The explosion collapsed almost half of the building, essentially most of the north bay. Three of the columns (G16, G20, and G24) supporting the transfer girders collapsed. Column G20 failed because of *brisance*, or shattering of the concrete, which allowed the transfer girders to rotate inward. This rotation, in turn, led to the failure of the other two columns between the second and third floors, at the location of the transfer girders. This failure left the outer bay unsupported (Hinman and Hammond 1997, pp. 6–8). The G row of columns was along the north façade, and the F row was in the middle of the building between the north and south bays. The location of the explosion is shown in Fig. 5-10.

Floor slabs were blasted upward at the second through fourth floors.

> Floor slabs tend to be vulnerable to explosive effects due to their large surface area compared with other elements, such as columns. . . . Floor slabs are first subject to an upward and then a downward loading from the explosive forces. The upward loading causes cracking and weakening of the members in shear. This is followed by a downward pressure, causing collapse. As the slab deflects downward, it pulls the outer frame inward and causes this already weakened system to fail by buckling at the lower floors. Shear failure rather than flexural or bending failure occurs due to the extremely high intensity and short duration of the loading close-in to the explosion. In short, the slab is sheared off before it has a chance to bend. (Hinman and Hammond 1997, p. 8)

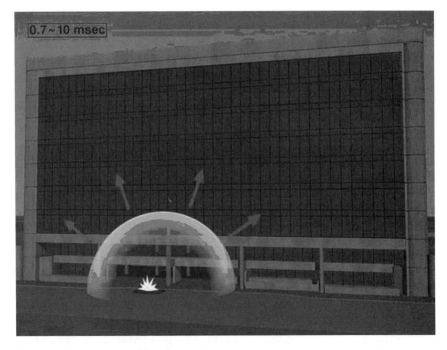

Figure 5-10. Location of bomb.
Source: Osteraas (2006).

Because the floor slabs brace columns, their failure may lead to column buckling and overall disproportionate collapse. The floor slab failure and its effect on the structural frame are shown in Fig. 5-11.

One of the inner columns, F24, failed, in addition to the three outer frame columns. Columns F20 and F22, closer to the blast, remained standing. It is possible that F24 was hit by debris, triggering buckling, or that the other two columns that did not fail were braced by the elevator core and stairways. The surviving columns near the blast area were examined to determine why they did not fail (Hinman and Hammond 1997, pp. 8–9).

The length of the surviving floor slabs generally corresponded to the length of the negative moment reinforcement provided, showing that the reinforcement was effective in arresting the slab collapse. On the other hand, the positive moment reinforcement broke off at the beam support (Hinman and Hammond 1997, p. 9).

Overall,

> The observed failure modes for the Murrah building were brittle and occurred mostly at the connections. This was because of the excessive forces of the explosion, progressive collapse, and the nonductile construction used in the design of this building. Other typical failure modes included the microcracking of members and the ripping out of reinforcing bars from the members. Only in regions in which the pressures were relatively low was there evidence of a ductile flexural response. (Hinman and Hammond 1997, p. 10)

The structure after the attack is shown in Fig. 5-12.

Structural engineers turned out to be important for the Urban Search and Rescue (US & R) effort led by the Federal Emergency Management Agency (FEMA) and other state, local, and federal authorities. An important task was to stabilize and support the wreckage during the US & R operations. A number of collapse and falling debris hazards were identified, and on occasion the US & R had to be halted temporarily. Pipe braces and confinement angles were used to brace columns and other elements. Some columns had lost the bracing from the floor slabs at the second and third floors and were braced by the debris piles. Bracing had to be installed before the debris piles could be removed (Hinman and Hammond 1997, pp. 16–21).

The single most dangerous slab was referred to as the "mother slab" or the "slab from hell." This 157-kN (35,000-lb) portion of the roof slab

160 BEYOND FAILURE

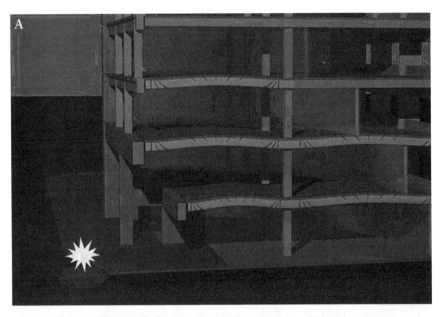

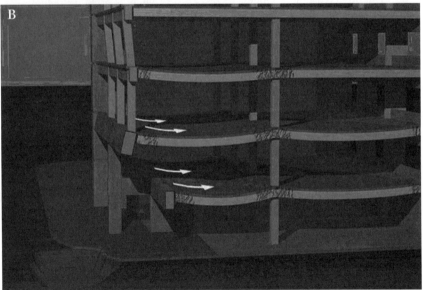

Figure 5-11a, b. Failure of the floor slabs.
Source: Osteraas (2006).

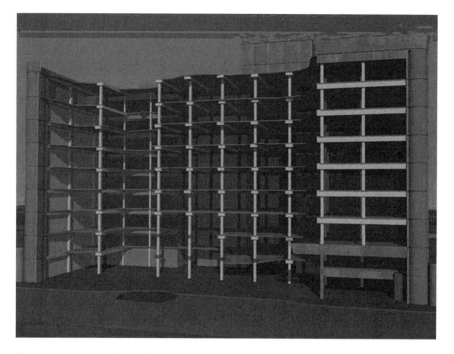

Figure 5-12. Murrah Building following the attack and partial collapse.
Source: Osteraas (2006).

hung from two 25-mm (1-in.) steel bars. It was over the portion of the building where much of the rescue and recovery work had to be carried out. One bar was cut, and then it was decided that it would be unsafe to cut the other. Instead, the slab was tied to the structure with steel cable loops and monitored for movement (Hinman and Hammond 1997, pp. 22–23).

Technical Lessons Learned

This case led to a shift in philosophy in structural design. Before this attack, it was generally thought that special detailing of reinforced concrete construction was necessary only in areas with significant seismic hazard.

Hinman and Hammond (1997, p. 34) recommended that reinforced concrete structures subject to blast and similar threats be designed and detailed with alternate load paths and with continuous top slab reinforcement. Columns, beams, and girders should have ties to confine the concrete and provide ductility. Reinforcement bars should be properly anchored.

Essentially, seismic design and detailing should be used to provide structural ductility and redundancy.

Essential Reading

One key reference is Osteraas (2006). The comprehensive work on the engineering aspects of the Oklahoma City bombing attack is *Lessons from the Oklahoma City Bombing: Defensive Design Techniques*, written by E. E. Hinman and D. J. Hammond (1997). This document has become an important reference for developing building designs that are more resistant to this type of attack. The case is also discussed in Chapter 6 of Wearne (2000, pp. 117–135).

The Pentagon Attack

The Pentagon terrorist attack of September 11, 2001, provides another case study of reinforced concrete building performance under extreme loading. The Pentagon performed quite well during the event, and as a result the damage was localized to the area of immediate impact.

Building Design and Construction

The Pentagon is a reinforced concrete structure built between September 1941 and January 1943. It has five sides and is arranged in five concentric rings, from the A ring in the interior to the E ring on the outside. It is one of the largest office buildings in the world, with a floor space of about 610,000 m^2 (6.6 million ft^2) (Mlakar et al. 2003, p. 3).

Most of the structure used concrete with a specified compressive strength of 17 MPa (2,500 lb/in.^{2}) and reinforcing steel with a specified yield strength of 276 MPa (40,000 lb/in.^{2}). Column sizes varied from 533 mm (21 in.) square in the first story to 356 mm (14 in.) square at the top story, with all columns that supported more than one level spirally reinforced. Some of the other columns were tied. Floor spans were short by modern standards, with 140-mm (5.5-in.) thick slabs spanning between beams 3 m (10 ft) on center. Beam spans were 3–6 m (10–20 ft) (Mlakar et al. 2003, pp. 5–6).

An important structural feature was that beam bottom reinforcement was continuous through the columns. This feature gave the floor slab and beam system considerable ductility. It was thought that after World War II the Pentagon would no longer be needed as an office building, so it was designed for record storage. As a result, the design floor live load was a relatively high 7.2 kPa (150 lb/ft^2) (Mlakar et al. 2003, pp. 8–9). In the late 1980s, renovations were planned for the building. Overall, the structure was found to be in sound condition (Mlakar et al. 2003, pp. 9–11).

By September 11, 2001, one-fifth of the minor renovation had been completed on the mechanical aspects of the building.

Attack and Building Performance

At 9:38 A.M. on September 11, 2001, a hijacked Boeing 757-200 airliner was deliberately crashed into the building. The impact hit the building about 43 m (140 ft) to the south of a boundary between the renovated section of the building and the next section scheduled for renovation. The aircraft continued into the section not yet renovated. The impact and fire killed the 64 people aboard the aircraft and 125 people in the building (Mlakar et al. 2003, p. 4).

Part of the building in the renovated section to the south of the expansion joint subsequently collapsed about 20 min after the aircraft struck the building. Photos taken between the aircraft impact and the collapse show that this portion of the building was sagging by about 0.45–0.6 m (1½–2 ft). All five floors of the building collapsed. Some of the blast-resistant windows stayed in place even after the collapse. The collapse was confined to the outer E ring (Mlakar et al. 2003, pp. 17–26). Figure 5-13 shows the building after the collapse of the portion adjacent to the expansion joint.

Investigation and Analysis

A number of investigations were carried out. The Pentagon Building Performance Study (BPS) team was granted access to the site for four hours on October 4, 2001. By this point, much of the structural debris had been removed. The BPS team classified the damage to the various structural elements remaining in the building (Mlakar et al. 2003, p. 24).

The upgraded windows were generally in place in their reinforced frames. In contrast, the windows near the impact that had not been upgraded were mostly broken. Almost all the interior damage that had been inflicted was on the first story. The path of structural damage was 23–24 m (75–80 ft) wide and about 70 m (230 ft) long, tapering as it went further into the building. Severe damage from either the impact or the subsequent fire included heavy cracking and spalling. Several columns were bent at midheight as much as three times the diameter of the spiral reinforcement, yet they remained attached at the ends (Mlakar et al. 2003, pp. 28–29). Figure 5-14 shows a damaged column.

164 BEYOND FAILURE

Figure 5-13. The Pentagon following the aircraft impact.
Source: Mlakar et al. (2003).

In the worst cases, the columns were severed from the structure or were missing completely. The columns outside the line of direct impact had little structural damage. Damage above the second floor seemed to be related to fire (Mlakar et al. 2003, p. 34). Detailed photographs and plots of damage to beams, columns, and floor slabs are provided in the BPS report (Mlakar et al. 2003, pp. 24–44, B1–B14).

An important element of the investigation was the performance of the structure in the fire. The damage was typical of severe fires in office buildings. Maximum temperature in some locations may have reached 950 °C (1,740 °F). The fire damage to structural elements was limited to cracking and spalling near the aircraft debris (Mlakar et al. 2003, pp. 41–43).

The investigation suggested three structural performance issues that required further analysis:

- response of the concrete columns to impact,
- load capacity of the floor system, and
- thermal response of columns and girders (Mlakar et al. 2003, p. 45).

Figure 5-14. Damage to a spiral column.
Source: Mlakar et al. (2003).

None of the columns appeared to have failed because of shear or reinforcing bar pull-out. Overall, the columns remained ductile unless they were torn from their supports. Where the supports failed, the bars had necked considerably before fracture. The spiral ties provided considerable shear resistance, ductility, and reserve capacity (Mlakar et al. 2003, pp. 45–48).

It is remarkable that the floor system was able to stay up for 20 min in the collapsed area despite the loss of most of the supporting columns. The eventual collapse may be explained by heat from the fire and possibly high-strain creep, along with the weight of the water pumped into the building to extinguish the fires (Mlakar et al. 2003, pp. 48–50).

Technical Lessons Learned

The report noted that,

> Despite the extensive column damage on the first floor, the collapse of the floors above was extremely limited. Frame and yield-line analyses attribute this life-saving response to the following factors:
>
> - Redundant and alternative load paths of the beam and girder framing systems;
> - Short spans between columns;
> - Substantial continuity of beam and girder bottom reinforcement through the supports;
> - Design for [7.2 kPa] 150 psf [lb/ft^2] warehouse live load in excess of service load;
> - Significant residual capacity of damaged spirally reinforced columns;
> - Ability of the exterior walls to act as transfer girders (Mlakar et al. 2003, p. 58).

The partial collapse 20 min after the impact was attributed to the effects of fire on the structural frame, particularly with protective materials and concrete cover removed by the impact.

The report's recommendations found,

> The Pentagon's structural performance during and immediately following the September 11 crash has validated measures to reduce collapse from severely abnormal loads. These include the following features in the structural system:
>
> - Continuity, as in the extension of the bottom beam reinforcement through the girders and bottom girder reinforcement through the columns;
> - Redundancy, as in the two-way beam and girder system;
> - Energy-absorbing capacity, as in the spirally reinforced columns;
> - Reserve strength, as provided by the original design for live load in excess of service.
>
> These practices are examples of details that should be considered in the design and construction of structures required to resist progressive collapse. (Mlakar et al. 2003, p. 59)

Recommendations were also made for further research and development into performance of reinforced concrete structures in extreme events.

Essential Reading

The comprehensive work on the engineering aspects of the Pentagon on September 11, 2001, is *The Pentagon Building Performance Report* by the Pentagon Building Performance Study Team (Mlakar et al. 2003). The case is also featured in five papers published in the proceedings of ASCE's Third Forensic Congress (Bosela et al. 2003).

Other Cases

Ronan Point

The collapse of Ronan Point is discussed in Chapter 4. It was a precast panel structure. The case illustrates the importance of workmanship in concrete construction, as well as the importance of tying building elements together.

L'Ambiance Plaza Collapse

One of the theories on the collapse of L'Ambiance Plaza is poor detailing of post-tensioned concrete floor slabs. The case is discussed in Chapter 4.

Willow Island Cooling Tower Collapse

The Willow Island Cooling Tower collapse represents yet another formwork failure. The case is discussed in Chapter 9.

Schoharie Creek Bridge

The Schoharie Creek Bridge collapsed on the morning of April 5, 1987, after three decades of service. The full case study is discussed in Chapter 8. The collapse of Pier 3 caused two spans to fall into the flooded creek. Five vehicles fell into the river, and 10 occupants died.

The main cause of the failure was scour, which undermined the foundation of Pier 3. The pier dropped into the scour hole and cracked, which caused the two steel spans resting on the pier to fall.

The detailing of the reinforced concrete pier may have contributed to the severity of the accident. The pier frames were constructed of two slightly tapered columns and tie beams. The columns were fixed within a lightly

168 BEYOND FAILURE

reinforced plinth, which was positioned on a shallow reinforced spread footing.

Shortly after construction was completed, in the spring and summer of 1955, the Schoharie Creek Bridge pier plinths began to form vertical cracks. The cracks occurred because of the high tensile stresses in the concrete plinth. The plinth could not resist the bending stresses between the two columns. The original designs called for reinforcement to be placed in the bottom portion of the plinth only because designers had confidence that the concrete in tension could resist the bending stresses without reinforcement.

In 1957, plinth reinforcement was added to each of the four piers to correct the problem of vertical cracking. The plinth may be seen as an upside-down, uniformly loaded beam, with the soil-bearing pressure providing the uniform loading and the two columns acting as supports. It becomes obvious that the top of the plinth represents the tension face of the beam and requires reinforcement. The plinth reinforcement is shown in Fig. 5-15.

However, to be properly anchored, the tension reinforcement must be extended past the supports—in this case, into the columns—for the full development length required by the ACI Code. Obviously, this extension was not done, and it would have been difficult to extend the reinforcement through the columns without replacing the columns. Ironically, because the added plinth reinforcement was not adequately anchored, it may have contributed to the brittle and sudden nature of the subsequent collapse by supporting the plinth until most of it had been undermined.

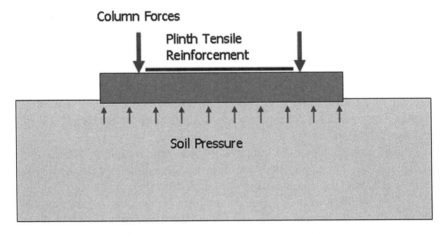

Figure 5-15. Schoharie Creek Bridge plinth reinforcement.

Sampoong Superstore

The collapse of the Sampoong Superstore in Seoul, South Korea, represents an example of a structural collapse of a reinforced concrete building attributed in large part to corruption. The full case study is discussed in Chapter 10. The store opened in December 1989. The first contractor, Woosung Construction, built the foundation and basement but balked at making some of the changes to the structure that the owner requested on the grounds that the proposed changes were not safe.

Sampoong completed the construction with its own engineers and construction workforce. In the process, they converted the use from an office block to a department store and changed the roller skating rink on the fifth floor to a traditional Korean restaurant. This change added significant dead load because at such a restaurant the diners sit on a heated floor, which was about 0.9 m (3 ft) thick and made of concrete.

At 6 P.M. on June 29, 1995, the store was full of shoppers. The center of the building collapsed, with a final death toll of 498 people.

The subsequent investigation revealed a number of important design and construction deficiencies:

- When the building had been converted from an office block to a department store, it had been necessary to cut large holes in the floor slabs for escalators and to remove some of the supporting columns.
- Because the change of use also required fire shutters, large chunks of the remaining concrete columns were cut away to allow shutter installation.
- The floor plan of the fifth-floor restaurant was not compatible with the lower floors. Therefore, the columns were not continuous from one floor to the next. The loads, which should have been transferred by continuous columns to the foundation, were instead transferred through the slab between the fourth and fifth floors.
- The roof system had about a quarter of the capacity required to support the water-cooling system.
- On the lower four floors, columns specified to be 890 mm (35 in.) wide were only 610 mm (24 in.) wide and had only 8 reinforcing bars rather than the 16 specified.
- Slab dead loads had been miscalculated, based on 100-mm (4-in.) thick slabs when some slabs were three or four times thicker.
- Some reinforcement had not been installed, and connections between slabs and walls were poor.

- To maximize sales space, the spans between the columns had been increased to almost 11 m (36 ft), which was much too large.
- The concrete strength used was only 18 MPa (2,600 lb/in.^{2}) rather than the specified 21 MPa (3,000 lb/in.^{2}), and the actual concrete strength from samples taken after the collapse ranged from 18.4 to 19.3 MPa (2,700 to 2,800 lb/in.^{2}). The originally specified value is rather low for structural concrete, and the further reduction in strength would reduce the slab punching shear capacity.
- The effective slab depth for negative moment areas had been reduced from the specified 410 mm to 360 mm (16 to 14 in.) because the reinforcement was improperly placed.

Some of the officers of Sampoong were convicted and sent to jail, and the building officials who had approved the changes were prosecuted.

Autoroute 19 de la Concorde Overpass

The de la Concorde Overpass over Autoroute 19 in Laval, Quebec, collapsed on September 30, 2006, killing five people and injuring six. The government of the province of Quebec convened a commission to investigate.

The bridge was an unusual side-by-side prestressed box girder configuration designed and built between 1968 and 1971. The ends of the girders rested on cast-in-place concrete cantilevers extending out from abutments. The bridge was an unusual structure, difficult to inspect, and no more bridges of this type were built in Canada after 1972.

The cantilever supports also doubled as expansion joints and had complex load-transfer mechanisms. The joints were difficult to seal and maintain and allowed water and de-icing chemicals to collect at the cantilever supports. Substantial repair work undertaken in 1992 may have caused some damage that later led to the bridge collapse.

The cantilever supports relied on a complex reinforcement detail to transfer bearing forces into the top bars through stirrups. However, the tops' stirrups were placed slightly below the top reinforcement, leaving a horizontal plane of weakness.

The investigating commission cited as primary physical causes the improper detailing of reinforcement (the top bars were not anchored), improper installation of reinforcement, and low-quality concrete. The concrete specification was confusing and appeared to allow the use of weaker and less durable concrete. At the time of the collapse, the concrete strength was a bit higher than the specified 27.6 MPa (4,000 lb/in.^{2}), but a higher strength would have been expected after 36 years. The air content and deicer scaling resistance of the concrete were also poor. The commission also cited the contributing physical causes of lack of shear reinforcement in the thick cantilever slabs, lack of waterproofing of the cantilever concrete, and possible damage caused during the 1992 repair work.

Just before the collapse, puddles of water and chunks of falling concrete were observed. Some drivers also noticed bumps at the expansion joints when they crossed the overpass. The concrete cantilever peeled off just below the top layer of reinforcement, and the prestressed box girders fell onto Autoroute 19 and two passing cars. Three cars and a motorcycle fell with the overpass.

The final report of the investigating commission was published (Commission 2007) and is available on-line in English at http://www.cevc.gouv.qc.ca/UserFiles/File/Rapport/report_eng.pdf.

6

Steel Structures

IN THE UNITED STATES, DESIGN OF STEEL STRUCTURES typically follows the American Institute of Steel Construction (AISC) Specification (2005). Topics such as composite columns, built-up sections, plate girders, plate stability, torsion, lateral-torsional buckling, and overall structural behavior (braced and unbraced frames) may be included in an introductory course or may be covered in a more advanced course.

Steel differs from reinforced concrete and other building materials in that local and global stability is almost always a concern because of the thin and slender sections, plates, and elements used. Therefore, column buckling, beam lateral-torsional buckling, local plate buckling, and bracing considerations are important.

In the United States, structural steel may be designed using two methods. Over the past few decades, steel design began to move from allowable stress design (ASD) to load and resistance factor design (LRFD). Both design methods are covered by the current AISC (2005) specifications. LRFD provides a more uniform level of safety for structures and is compatible with the strength design methods that reinforced concrete has used for almost half a century. For failure investigation, ASD methods may be more appropriate than LRFD methods because the actual rather than the factored loads are used. If LRFD is used to compare actual loads to available structural capacity, all load and resistance factors should be set equal to 1.

It is worth noting that structural steel courses typically cover hot-rolled and built-up steel sections. Cold-formed steel design is not covered. These sections use much thinner steel than hot-rolled and built-up sections and are even more prone to local and overall buckling. A different standard, published by the American Iron and Steel Institute (AISI 2007), applies to cold-formed steel, and the AISC specifications should not be used. Cold-formed steel analysis and design is rarely covered by university courses, even at the graduate level.

Hartford Civic Center Stadium Collapse

No one was killed or injured when the huge space truss roof of the empty Hartford Civic Center coliseum collapsed under a heavy snowfall at 4:19 A.M. on January 18, 1978. Had the failure occurred just a few hours before, the death toll might have been hundreds or thousands. The dramatic roof, designed with the aid of computers, had shown evidence of distress during construction, but the warnings had not been heeded. The building had been in service for five years when it collapsed (Levy and Salvadori 1992).

Design and Construction

The design engineers of the Hartford Civic Center in 1971 designed the roof truss system using a complicated, expensive, and innovative computer program. They were able to convince the city council to accept the results. The estimated savings from the refined state-of-the-art analysis compared to a more conventional design were half a million dollars (Wearne 2000, p. 21).

To save money, the engineers proposed an innovative design for the 91.4 × 110 m (300 × 360 ft) space frame roof 25.3 m (83 ft) over the arena. The proposed roof consisted of two main layers arranged in 9.14 × 9.14 m (30 × 30 ft) grids composed of horizontal steel bars 6.4 m (21 ft) apart. Diagonal members 9.14 m (30 ft) long connected the nodes of the upper and lower layers and, in turn, were braced by an intermediate layer of horizontal members. The 9.14-m (30-ft) members in the top layer were also braced at their midpoint by intermediate diagonal members (Fig. 6-1).

This design departed from standard space frame roof design procedures in five ways.

STEEL STRUCTURES 175

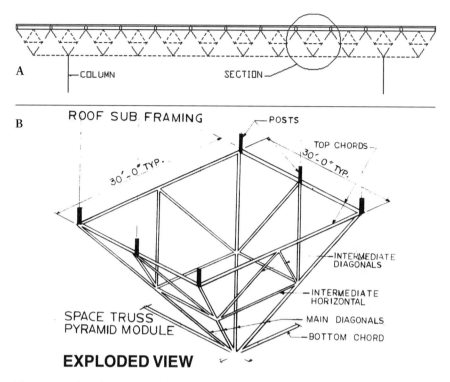

Figure 6-1a, b. Elevation and detail of Hartford Civic Center roof (1 ft = 0.305 m). Courtesy Lev Zetlin Associates.

1. The cross-sectional configuration of the four steel angles making up each truss member did not provide good resistance to buckling. The cross-shaped built-up section has a much smaller radius of gyration than either an I-section or a tube section configuration of the same structural members (Fig. 6-2). As a result, the buckling load for the cross-shaped section is much lower than that of the other configurations.
2. The top horizontal members intersected at a different point than the diagonal members rather than at the same point, making the roof especially susceptible to buckling because the diagonal members did not brace the top members against buckling.
3. The top layer of this roof did not support the roofing panels; short posts on the nodes of the top layer did. Not only were these posts meant to eliminate bending stresses on the top layer bars, but their varied heights also allowed water to be carried away to drains.

176 BEYOND FAILURE

Figure 6-2. Possible truss member configurations.

4. Four pylon legs positioned 13.7 m (45 ft) inside of the edges of the roof supported it instead of boundary columns or walls (Levy and Salvadori 1992).
5. The space frame was not cambered. Computer analysis predicted a downward deflection of 330 mm (13 in.) at the midpoint of the roof and an upward deflection of 150 mm (6 in.) at the corners (*ENR* 1978a).

Because of these money-saving innovations, the engineers used state-of-the-art computer analysis to verify the safety of the building. However, given the primitive state of computer structural analysis in 1971, there were understandable concerns about the accuracy of the solution.

A year later, construction began. To save time and money, the roof frame was completely assembled on the ground. While it was still on the ground, the inspection agency notified the engineers that it had found excessive deflections at some of the truss joints. Nothing was done.

After the frame was completed, hydraulic jacks located on top of the four pylons slowly lifted it into position. Once the frame was in its final position, but before the roof deck was installed, its deflections were measured and found to be twice those predicted by computer analysis. The engineers were notified. They, however, expressed no concern and responded that such discrepancies between the actual and the theoretical should be expected (Levy and Salvadori 1992).

When the subcontractor began fitting the steel frame supports for fascia panels on the outside of the truss, he ran into great difficulties because of the excessive deflections of the frame. On notification of this problem, the project manager "directed the subcontractor to deal with the problem or be responsible for delays." As a result, the subcontractor coped (or cut away material from) some of the supports and refabricated others to make the panels fit, and construction continued (*ENR* 1978a).

The roof was completed on January 16, 1973 (Feld and Carper 1997). The next year, a citizen expressed concern to the engineers regarding the large downward deflection he noticed in the arena roof, which he believed to be unsafe. The engineers and the contractor once again assured the city that everything was fine (Levy and Salvadori 1992).

During construction, the architect had twice tried to convince the owners to hire a qualified structural engineer for continuous on-site inspections of the truss construction and erection.

> The idea was rejected by the construction manager as a waste of money. He took on all the inspection responsibilities himself. Thus, at a stroke, the construction manager eliminated the system of checks and balances.... As the citizens' panel noted rather euphemistically: "inspection of his own work by the construction manager is an awkward arrangement." (Wearne 2000, p. 24)

Collapse

On January 18, 1978, the Hartford Arena experienced the largest snowstorm of its five-year life. At 4:19 A.M., the center of the arena's roof plummeted the 25.3 m (83 ft) to the floor of the arena, throwing the corners into the air. Just hours earlier, the arena had been packed. Luckily, it was empty by the time of the collapse (Ross 1984).

Causes of Failure

Hartford appointed a three-member panel to manage the investigation of the collapse. This panel in turn hired Lev Zetlin Associates, Inc. (LZA) to ascertain the cause of the collapse and to propose a demolition procedure (Ross 1984). LZA issued its report later that year (LZA 1978). LZA discovered that the roof began failing as soon as it was completed because of design deficiencies. A photograph taken during construction showed obvious bowing in two of the members in the top layer.

Three major design errors, coupled with the underestimation of the dead load by 20%, allowed the weight of the accumulated snow to collapse the roof. The estimated frame unit weight was 0.862 Pa (18 lb/ft^2), but the actual frame weight was 1.10 Pa (23 lb/ft^2) (*ENR* 1978a). The load on the day of collapse was 3.16–3.50 Pa (66–73 lb/ft^2), although the arena should have had a design capacity of at least 6.70 Pa (140 lb/ft^2) (*ENR* 1978b). The three design errors responsible for the collapse are listed below:

- The top layer's exterior compression members on the east and the west faces were overloaded by 852%.

- The top layer's exterior compression members on the north and the south faces were overloaded by 213%.
- The top layer's interior compression members in the east–west direction were overloaded by 72%.

In addition to these errors in the original design, LZA discovered that no midpoint braces were provided for the members in the top layer. The exterior members were only braced every 9.14 m (30 ft), rather than the 4.57-m (15-ft) intervals specified, and the interior members were only partially and insufficiently braced at their midpoints. The two members attached to the midpoint of the top chord were both in the same plane as the long axis of the chord, so they only provided bracing in one direction. The perpendicular direction was effectively unbraced for the full 9.14-m (30-ft) length. This situation significantly reduced the load that the roof could safely carry. In addition, certain perimeter top chord members with a post landing at midpoint were subjected to bending stress from the roof load applied through the post. Because the members were not designed for bending, this design led to considerable overstress (LZA 1978).

Table 6-1 compares some of the original details to actual designs used in the building, demonstrating the reduction in strength that these changes caused. Connection A was typically used on the east–west edges of the roof, whereas connection B was used on the north–south edges. Most of the interior members used connection C, and a few used connection D. The key difference between the original and the as-built details may be seen by comparing the top and bottom rows of the table. The diagonal members were attached some distance below the horizontal members. Thus, the flexibility of the connection reduced the effectiveness of the bracing by introducing a "spring brace" instead of the hard brace that had been assumed. The effect of a spring brace instead of a hard brace is illustrated in Fig. 6-3.

The most overstressed members in the top layer buckled under the added weight of the snow, causing the other members to buckle. This buckling changed the forces acting on the lower layer from tension to compression, causing them to buckle also in a progressive failure. Two major folds formed, initiating the collapse (*ENR* 1978a). These faults were not the only errors that LZA discovered. Listed below are the other factors that contributed to but probably were not solely responsible for the collapse.

- The slenderness ratio of the built-up members violated the American Institute of Steel Construction (AISC 1970) code provisions. The spacer plates separating the individual angles were placed too

TABLE 6-1. *Comparison of Original Design and Actual As Built Connections*

	Connection A	Connection B	Connection C	Connection D
Original Design	Assumed Brace Allowable force: 710 MN (160,000 lb) Allowable moment: 0	Assumed Brace Allowable force: 821 MN (185,000 lb)	Allowable force: 2.78 GN (625,000 lb)	Allowable force: 2.51 GN (565,000 lb)
As-Built	Allowable force: 68.6 MN (15,440 lb) Allowable moment: 12,800 N·m (9,490 lb-ft)	Allowable force: 262 MN (59,000 lb)	Allowable force: 1.61 GN (363,000 lb)	Allowable force: 2.51 GN (565,000 lb)

Source: Martin and Delatte (2001).

far apart in some of the four-angle members, allowing individual angles to buckle.
- The members with boltholes exceeding 85% of the total area violated the AISC code requirements for section reduction of tension members (*ENR* 1978b).
- Some diagonal members were misplaced (Feld and Carper 1997).

Loomis and Loomis, Inc., also investigated the Hartford collapse. They agreed with LZA that gross design errors were responsible for the progressive collapse of the roof, beginning the day that it was completed. They, however, believed that the torsional buckling of the compression members, rather than the lateral buckling of the top chords, initiated the collapse.

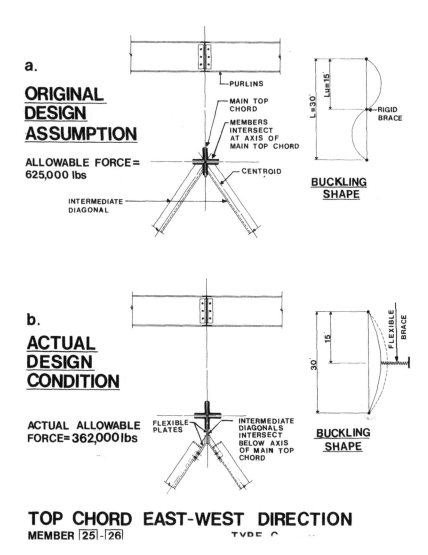

Figure 6-3. Effect of assumed (hard) versus actual (spring) bracing condition. Courtesy Lev Zetlin Associates.

Using computer analysis, Loomis and Loomis found that the top truss members and the compression diagonal members near the four support pylons were approaching their torsional buckling capacity the day before the collapse. An estimated 0.575–0.718 Pa (12–15 lb/ft^2) of live load would cause the roof to fail. The snow from the night before the collapse constituted a live load of 0.670–0.910 Pa (14–19 lb/ft^2). Because torsional buckling is uncommon, it is often an overlooked mode of failure (*ENR* 1979a). The cruciform shape is particularly weak against torsional buckling.

Hannskarl Bandel, a structural consultant, completed an independent investigation of the collapse for the architect's insurance company. He blamed the collapse on a faulty weld connecting the scoreboard to the roof. This opinion conflicts with the opinions of all the other investigators (*ENR* 1979b). The LZA report's findings were also disputed by the engineer (*ENR* 1978c).

Technical Aspects

The engineers for the Hartford Civic Center depended on computer analysis to assess the safety of their design. Computer programs, however, are only as good as their programmers and may tend to offer engineers a false sense of security (Shepherd and Frost 1995). The LZA report noted

> ... the computer model used by the structural engineer only included the top and bottom chords and the main diagonals. Roof loads were only applied at top chord main panel points. If the computer model had represented the intermediate diagonals and horizontals and had included the roof loads at the mid-point, sub-panel points at the top chord, the instabilities and primary bending moments would have been detected by the designer. (LZA 1978)

Instead of the cruciform shape of the rods, a tube or I-bar configuration would have been more stable and less susceptible to bending and twisting. The cruciform shape has the advantage of making the members much easier to connect. Also, if the horizontal and diagonal members intersected at the same place, the bracing would have increased the buckling capacity in these members. The LZA report noted that "apparently, the choice of the typical member as a cruciform, a section that is weak in bending and torsion, was based on the design assumption that such bending and torsion would be negligible in the space truss" (LZA 1978).

The LZA report further noted that "space trusses and/or space frames are valid and safe structural systems" and pointed out that some of the construction work was not done correctly (LZA 1978).

Professional and Procedural Aspects

The Hartford Civic Center contract was divided into five subcontracts coordinated by a construction manager. Not only did this fragmentation allow mistakes to slip through the cracks, but it also left confusion over who was responsible for the project as a whole. Even though the architect recommended that a qualified structural engineer be hired to oversee the construction, the construction manager refused, saying that it was a waste of money and that he would inspect the project himself. After the failure, he disclaimed all responsibility on the grounds that a design error had caused the collapse. He asserted that he was only responsible for ensuring that the design was constructed correctly and not for the performance of the project (Feld and Carper 1997, p. 202).

It is important that the responsibility for the integrity of the entire project rest with one person. Feld and Carper (1997, pp. 202–204) offer an excellent discussion of the role that procedural deficiencies played in this collapse.

As a result of the construction manager's refusal to hire a structural engineer for the purpose of inspection, no one realized the structural implications of the bowing members. This collapse illustrates the importance of having a structural engineer, especially the designer, perform the field inspection. The designer understands the structure that is being built and would best be able to recognize the warning signs of poor structural performance and rectify them before they grow to catastrophic proportions. The LZA report noted that the quality control procedures had been inadequate.

Finally, the Hartford department of licenses and inspection did not require the project peer review that it usually required for projects of this magnitude. If a second opinion had been obtained, the design deficiencies responsible for the arena's collapse probably would have been discovered (LZA 1978). Peer reviews are an essential safety measure for high-occupancy buildings and structures experimenting with new design techniques (Feld and Carper 1997). Today, Connecticut is one of the few states that require peer review of certain buildings.

Ethical Aspects

The excessive deflections apparent during construction were brought to the design engineer's attention several times. The engineer, confident in his design and the computer analysis that confirmed it, ignored these warnings and did not take the time to recheck his work. The engineer should pay close attention to unexpected deformations and should investigate their causes. They often indicate structural deficiencies and should be investigated and corrected immediately. Unexpected deformations provide a clear signal that the structural behavior is different from that anticipated by the designer.

Kaminetzky (1991) quotes at length from a story in *The Philadelphia Inquirer* from May 28, 1978, about this incident, headlined "Why The Roof

Came Tumbling Down." The story suggests that the ironworkers knew from observing the deformations during construction that the building was a death trap and had vowed never to enter it once it was completed. It also questions why the worker's warnings were not listened to.

Educational Aspects

Petroski discusses this case in terms of the need for engineers to be able to reason out whether computer results make sense, through hand calculations and knowledge of structural behavior and performance.

> Because the computer can make so many calculations so quickly, there is a tendency now to use it to design structures in which every part is of minimum weight and strength, thereby producing the most economical structure. This degree of optimization was not practical to expect when hand calculations were the norm, and designers generally settled for an admittedly overdesigned and thus a somewhat extravagant, if probably extra-safe, structure. However, by making every part as light and as highly stressed as possible, within applicable building code and factor of safety requirements, there is little room for error—in the computer's calculations, in the part manufacturers' products, or in the construction workers' execution of the design. Thus computer-optimized structures may be marginally or least-safe designs, as the Hartford Civic Center roof proved to be. (Petroski 1985, p. 199)

In the years since Petroski wrote these words, despite tremendous advances in computing power and software, there is no sign that computer programs will soon be able to envision failure modes that the designer has not foreseen or to check their work.

According to technology journalist Mitch Ratcliffe, "Computers have enabled people to make more mistakes faster than almost any invention in history, with the possible exception of tequila and hand guns." Engineers should not uncritically assume that computer results are correct.

As a class example or homework problem, students may compare the moment of inertia for the cruciform, I, and tube configurations of four angles, as shown in Fig. 6-2. Angle legs ranged from 89 to 203 mm (3½ to 8 in.) long and were 8–22 mm ($5/16$–$7/8$ in.) thick, depending on loads, and the angles were separated by spacers 19–22 mm ($3/4$–$7/8$ in.) thick (LZA 1978). For numerical examples, $127 \times 127 \times 8$ mm ($5 \times 5 \times 5/16$ in.) angles may be used. The torsional stiffness of these configurations may also be calculated and compared.

Conclusions

A useful lesson from this case is that computer software is only an analytical tool and that computed results must be checked by the designer with a careful eye. Users must understand the theoretical foundations of the programs and the associated limitations. This case serves as a lesson for engineering students and practicing engineers concerning the difficult technical, professional, procedural, and ethical issues that may arise during the design and construction of a complex, high-occupancy structure. There is no substitute for a thorough knowledge of structural behavior, coupled with a healthy skepticism toward the completeness and accuracy of computer software solutions to unusual problems.

It is interesting to compare this case to the collapse of the Quebec Bridge in 1907, discussed in Chapter 3. The errors in concept, design, and construction management are strikingly similar. Some of the parallels include:

- high confidence on the part of the designers,
- pushing design well beyond the limits of the established state of the practice without taking sufficient care against errors,
- lack of effective on-site field inspection,
- ignoring excessive deformations (which signaled impending buckling failure),
- serious underestimation of dead loads, and
- failure by buckling of steel truss compressive struts.

Essential Reading

The case is discussed by Levy and Salvadori (1992, pp. 68–75) as well as by Wearne (2000, pp. 17–25). This case study is also featured in the History Channel's Modern Marvels *More Engineering Disasters* and *Engineering Disasters 9* videotapes and DVDs.

Mianus River Bridge Collapse

In the late 1950s, a simple span bridge was built over the Mianus River in Cos Cob, Connecticut, as part of Interstate 95 and the Connecticut Turnpike. This bridge design included pin and hanger style connections commonly used at the time. The Mianus River Bridge was a multispan structure carrying more than 100,000 vehicles per day (Levy and Salvadori 1992, p. 138). On the morning of June 28, 1983, at around 1:30 A.M., an eastbound section of

STEEL STRUCTURES   185

the bridge fell into the chilly waters of the river, leaving three people dead and others injured.

Each bridge span was 30 m (100 ft) long, three travel lanes wide, with a 53° skew. Each section had a 2.7-m (9-ft) deep plate girder on each side, on top of a 190-mm (7.5-in.) concrete deck topped with 50 mm (2 in.) of asphalt. The mass of the bridge section was approximately 500 tonnes (500 tons). The drop to the tidal flats was 20 m (70 ft) (Feld and Carper 1997, p. 145).

Investigation

An investigation began immediately after the collapse. At that point, the only evidence of a possible reason for the bridge failure was a missing pin that was found in the river. The pin had been sheared completely in half and separated from the rest of the bridge. Once this was discovered, a more in-depth study was required to determine if there were any other factors involved. The state asked John W. Fisher, an engineer and a Lehigh University professor, to investigate the collapse (Levy and Salvadori 1992, p. 139).

Besides Fisher, three outside engineering firms and the National Transportation Safety Board were brought in to study the collapse and determine the reason. Investigations began by looking at the pin and hanger assembly to see if there was any way to identify the cause of the failure of the bridge.

Figure 6-4 shows the pin and hanger assembly. Washers were used to separate hangers from the steel girders. Locking caps were used to keep the hanger from slipping off of the pin. According to the experts, the caps were thin and flimsy. "Calculations proved the caps' thickness was less than one half that required by the design regulations in force at the time of construction (1957)" (Levy and Salvadori 1992, p. 139). The thin pin caps could be easily bent.

Another factor was the excessive amount of rust found on the hanger assembly. The rust and the dishing of the pin cap could have been seen with the naked eye, but the rust behind the pin cap could not be inspected. At the time, bridge inspection did not require the removal of the pin cap. The rust on the steel increased the stress on the pin cap because of the reduced cross section of the hanger plate.

Where did all the rust come from?

> It was discovered that ten years before the accident the storm drains had been paved over, allowing water, salt, and dirt to flow through the joint between the cantilevered and suspended spans, dropping directly onto the hanger assembly and causing accelerated rusting of its steel. (Levy and Salvadori 1992, p. 140)

186   BEYOND FAILURE

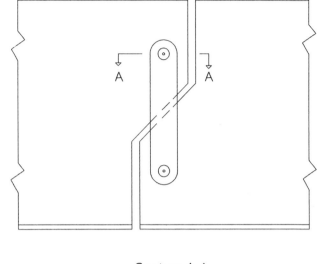

Section A-A

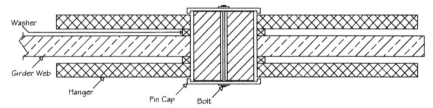

Pin and Hanger Elevation and Cross Section of Assembly

Figure 6-4. Pin and hanger assembly.

There was another important factor, suggested by investigator Lev Zetlin. Zetlin argued that sideways forces generated by the skew of the bridge caused the collapse. Zetlin also pointed out that there was no excuse for the deplorable conditions of the hangers and other portions of the bridge (ASCE 1985).

To support Zetlin's theory, Fisher noted that the collapse was caused by the pin being overloaded due to the sideways loading caused by the skew of the bridge. "The collapse could have been predicted if the skew forces, coefficients of friction, slippage, creep, corrosion, cohesion, and indentations had been known in advance" (ASCE 1985). Perhaps if the pins and plates had been heavier, then the collapse could have been prevented. Lateral displacement of the assembly was what ultimately brought the section of the Mianus River Bridge down (Feld and Carper 1997, p. 147).

Changes to Standards and Procedures

After the collapse of the Mianus River Bridge, the Connecticut Department of Transportation focused inspections on other similar bridges and began to retrofit them with steel slings. The steel slings were used to add redundancy to the remaining spans of the bridge. A temporary road section was also put into place so that the section of I-95 could be reopened for traffic.

This retrofit detail helped to save Boston from a similar situation when inspectors found two broken hangers on the Harvard Bridge over the Charles River (Schlager 1994, p. 230).

> Even before the final jury decision in 1986, ConnDot had embarked on a ten-year infrastructure renewal program. Expected to cost $5.5 billion over the decade, the program includes inspection and maintenance of all bridges and highways. (Schlager 1994, p. 234)

On July 19, 1984, the National Transportation Safety Board published the highway accident report for the Mianus River Bridge collapse. The NTSB proposed regulations and other recommendations to make bridges safer and to give better guidelines as to how they should be inspected. A recommendation was made:

> Conduct detailed inspections of the Mianus River Bridge and other representative bridges having a skewed and nonskewed suspended span design with pin and hanger assemblies to determine whether there is a significant difference between the two designs in terms of the movement of hangers on pins due to either dead or live loading and whether such movement is acceptable (Class III, Longer Term Action) (H-84-49). (NTSB 1983)

The Connecticut Department of Transportation reopened the paved-over drains so that water could run off. Regular maintenance of these pipes was also required. The Federal Highway Administration developed an integrated bridge inspection procedure to ensure that all bridges in the United States were inspected properly and along common guidelines. Other organizations, such as the American Institute of Steel Construction, the U.S. Department of Transportation, and the American Association of State Highway and Transportation Officials, also instituted guidelines and changes (NTSB 1983).

Lessons Learned

Like the Point Pleasant Bridge, the Mianus River Bridge on the Connecticut Turnpike (I-95) had collapsed after decades of service. The bridge's skewed simple spans were suspended on pins and hangers. Ten years before the accident, the bridge's storm drains had been paved over. This paving allowed water, silt, and dirt to drop directly onto the hanger assembly and accelerate corrosion. The end caps that were intended to keep the hangers from slipping off the pins were only 7.5 mm (0.3 in.) thick (Feld and Carper 1997, pp. 145–148).

The NTSB report concluded that the collapse of the bridge probably occurred because of "deficiencies in the State of Connecticut's bridge safety inspection and bridge maintenance program" (NTSB 1983). Many similar bridges were retrofitted to prevent similar collapses (Levy and Salvadori 1992, p. 148).

Essential Reading

The key reference for this case is the NTSB (1983) highway accident report *Collapse of Suspended Span of Route 95 Highway Bridge over the Mianus River, Greenwich, Connecticut, June 28, 1983*. The case is also discussed in Chapter 9 of Levy and Salvadori (1992). This case study is featured in the History Channel's Modern Marvels *Engineering Disasters 8* videotape and DVD.

Cold-Formed Steel Beam Construction Failure

During a concrete placement on the second story of a building under construction, the cold-formed steel beams supporting the decking and concrete collapsed. Four workers were injured, one fracturing his hip. Approximately two-thirds of the deck had been placed. The project structural engineer had been at the site earlier but had left before the collapse.

The collapse occurred while concrete was being placed onto steel decking on the second floor of the structure. The steel decking was supported by 203-mm (8-in.) deep cold-formed 1.21-mm (0.0478-in., 18 gauge) steel beams without shoring. Some of the workers raised concerns about the safety of the structure with the project structural engineer. He assured the contractor and workers that shoring was unnecessary and that the beams were rated for more than enough capacity to support the concrete.

The testimony of the workers and the photographs available indicated that good construction practices were followed with respect to placing and finishing the concrete. The project structural engineer contended that the failure occurred because workers allowed the concrete to form a pile on the formwork, thus increasing the loading.

Description of the Structure

The structure had three parallel masonry walls, each nominally 305 mm (1 ft) wide, with cold-formed steel beams crossing the walls. The second level was approximately 7.62 × 28.3 m (25 × 93 ft) in plan. The spans across the masonry walls were 3.4 and 4 m (11 and 13 ft) center to center, and the beams were continuous across the two spans. The cold-formed steel beams were spaced 610 mm (2 ft) on center, and they supported metal decking on which the concrete was to be placed. Welded wire mesh reinforcing fabric was also placed on the decking before the concrete.

The steel grade and section properties of the 203-mm (8-in. deep) cold-formed 1.21-mm (0.0478-in., 18 gauge) steel beams were not provided in the documents reviewed. Therefore, it was necessary to assume section properties based on the available information from the design drawings, as well as tables from the American Iron and Steel Institute (AISI) Manual (AISI 1996).

In the *Cold-Formed Steel Design Manual* (AISI 1996), the CS designation refers to C-sections with lips. Therefore, the section was assumed to be 8CS1.625×045, which is the only 1.21-mm (0.0478-in., 18 gauge) thick, 203 mm (8-in.) deep section listed. A yield stress of 228 MPa (33 kip/in.^{2}) was used in the computer calculations for the roof truss, which also used cold-formed steel. Therefore, it was most likely that the beams were 228 MPa (33 kip/in.^{2}). However, because this section may be made of steel with a yield point of either 228 or 379 MPa (33 or 55 kip/in.^{2}), both grades of steel were investigated.

One of the workers noted that some of the beams had been bent earlier as a heavy point load of decking was placed on them. The top flanges of the beams were damaged. They were repaired, straightened, and re-used, rather than being replaced. A short length of intact beam was screwed to each straightened beam. These beams may have had local buckling or residual stress from the bending, resulting in reduced load-carrying capacity. Damage to the beam compression flange occurring in the midspan at the point of maximum positive moment may significantly reduce the bending strength of a beam. The damaged beams may have triggered the progressive collapse.

Collapse

Immediately before the collapse, concrete was being placed from a pump onto the decking. An experienced worker was using the pump nozzle to spread the concrete. The workers started at one end, moving toward the other end of the second floor. One worker claimed that the deck was vibrating during the concrete placement.

When approximately two-thirds of the concrete had been placed, the decking on the longer 4-m (13-ft) span gave way suddenly, and five of the workers fell. Two workers were able to grab wire mesh and avoid falling the entire distance. The others fell onto the first floor. One fell onto a plumbing fixture pipe and broke his pelvis. Photographs taken immediately after the collapse showed that the beams were bent at the interior wall support and at about the midpoint of the longer 4-m (13-ft) span. The damaged beams hung downward from the interior wall.

Investigation

A number of documents and records were obtained and reviewed. Most of the project records and reports were available. The attorneys for the plaintiff and for the defendants obtained depositions from a number of individuals.

Because the building owner was listed as a defendant, it was not possible to arrange a site visit during the preliminary analysis. Instead, the investigator was asked to prepare an analysis based on the available documents and records. Those documents included the depositions, photos taken before and after the collapse, and the building plans. After the initial analysis, a site visit was to be arranged. However, the case was settled before trial.

Loads During Concrete Placement

The loads to be considered for concrete placement may be found in a number of sources. These sources include the Standard Building Code (SBC 1997), as well as the American Concrete Institute (ACI) publication *Formwork for Concrete* (Hurd 1995).

Using these sources and the 100-mm (4-in.) thick deck slab shown on the plans, a combined dead and live load of 31.4 kN per linear meter (200 lb per linear foot) of concrete and workers was used for the calculations. All live loads should be distributed so as to cause "maximum effect," as specified by Section 1604.4 of the Standard Building Code (1997, p. 215). To cause maximum effect, they may be placed on one span or on both.

Structural Analysis

To determine the bending moments in the two-span continuous beam, slope-deflection equations were used. Three load cases were considered:

- Case I: Concrete and live load spread uniformly across both spans. This uniform load distribution would be impossible to achieve at all times during construction because concrete cannot be simultane-

ously placed across the entire 7.32-m (24-ft) length of a beam from a single pump nozzle.
- Case II: Concrete spread uniformly across both spans, with the live load on the longer (4-m, 13-ft) span only. This load combination is commonly used for design, with dead load assumed along the full length of the beam and live load only where it causes the greatest effect, in accordance with the Standard Building Code (1997). However, it is not possible to place concrete in this way.
- Case III: Concrete and live load spread uniformly across the longer 4-m (13-ft) span only. This would be a prudent load combination for the designer to consider because the concrete must be placed on one area of the beam, and the workers need to be where the concrete is to work with it. This is also the only load combination that considers the "unbalanced loads" referred to in Chapter 5 of the ACI publication (Hurd 1995).

Bending moments for these three load cases are shown in Table 6-2 and Fig. 6-5. The greatest absolute value of the moment is always the positive moment near the midspan of the longer beam, for the load cases considered.

Therefore, the bending failure in the beams would occur near the center of the 4-m (13-ft) span and at the support between the 4- and 3.4-m (13- and 11-ft) spans. This result agrees with the damage shown in the photographs that accompanied the various depositions.

Possible Failure Modes

According to Chapter 4 of Yu (1991), C-sections used as beams can fail through bending, shear, combined bending and shear, lateral-torsional buckling, and web crippling. The centroid and the shear center of such a singly symmetric section do not coincide, leading to torsion stresses.

The notation for the engineer's computer calculations for the roof truss indicated the use of the American Institute of Steel Construction (AISC) *Steel Construction Manual* (AISC 1998). This specification is only applicable to

TABLE 6-2. *Maximum Moments by Loading Case*

Case	w_1 kN/m	plf	w_2 kN/m	plf	M_N kN-m	ft-kips	Distance to M_{max} m	ft	M_{max} kN-m	ft-kips
I	2.92	200	2.92	200	−3.31	−2.45	1.69	5.56	4.20	3.09
II	1.46	100	2.92	200	−2.68	−1.99	1.75	5.73	4.44	3.29
III	0	0	2.92	200	−2.06	−1.53	1.80	5.91	4.72	3.50

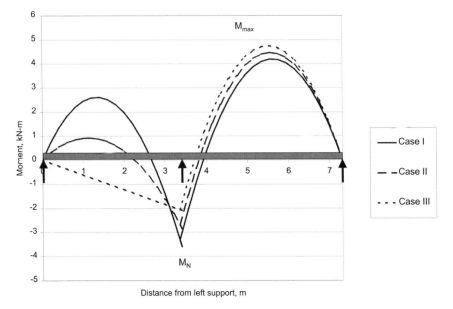

Figure 6-5. Bending moment diagrams.
Source: Delatte (2005).

hot-rolled steel sections, not to cold-formed steel. Therefore, this was not the correct specification to use for this design. The AISI *Cold-Formed Steel Design Manual* (AISI 1996) and the AISI *Specification for the Design of Cold-Formed Steel Structural Members* (AISI 2007) should have been used instead.

For bending, the nominal moment capacity in the 8CS1.625×045 section is 4.27 kN·m (3.16 ft-kip) for 228 MPa (33 kip/in.^{2}) steel, and 5.49 kN·m (4.07 ft-kip) for 379 MPa (55 kip/in.^{2}) steel from Table II-1, p. II-3, of AISI (1996). These capacities are shown in Table 6-3. These capacities are based on the effective section modulus, adjusted for local buckling of the beam compression flange.

For allowable stress design, these nominal capacities must be reduced by the appropriate factor of safety. For bending, the required factor of safety (Ω_b) is 1.67 (AISI 1996, Section C3.1.1, p. V-45). Therefore, the allowable moment is 2.56 kN·m (1.89 ft-kip) for 228 MPa (33 kip/in.^{2}) steel and 3.29 kN·m (2.44 ft-kip) for 379 MPa (55 kip/in.^{2}) steel.

Under all loading conditions considered here, the moment in positive bending exceeded the allowable moment regardless of the grade of steel used. In fact, for 228 MPa (33 kip/in.^{2}) steel, the moment exceeded the nominal moment for beams 1.21 mm thick (0.0478 in. thick, 18 gauge). Actual factors of safety are shown in the last column of Table 6-3. Factors

TABLE 6-3. *Beam Cross-section Moment Capacities*

Metal thickness	Flange width, mm (inches)	Steel yield strength, MPa (ksi)	Nominal moment capacity, kN-m (foot-kips)	Factor of safety (moment capacity/M_{max})
1.21 mm (0.0478 inch, 18 gauge)	41.3 (1.625)	228 (33)	4.27 (3.16*)	0.90
		379 (55)	5.49 (4.07*)	1.16
	63.5 (2.5)	228 (33)	4.19 (3.10)	0.89
		379 (55)	6.98 (5.17)	1.48
1.52 mm (0.0598 inch, 16 gauge)	41.3 (1.625)	228 (33)	5.32 (3.94*)	1.13
		379 (55)	8.48 (6.28*)	1.80
	63.5 (2.5)	228 (33)	5.82 (4.31)	1.23
		379 (55)	9.71 (7.19)	2.06

of safety less than 1 indicate that the design loads exceed design capacity, with a risk of failure under service load conditions.

Because cold-formed steel sections are not universally standardized (Yu 1991, p. 24), it is possible that the section used had a larger flange than the 41.3 mm (1.625 in.) assumed. Therefore, section properties for a channel with a 63.5-mm (2.5-in. flange) were calculated and are shown in Table 6-3. Thus, even with a larger flange, the 203-mm deep, 1.21-mm thick (8-in. deep, 0.0478-in. thick, 18 gauge) beam would be overstressed with 228 MPa (33 kip/in.^{2}) steel for all load cases, and for 379 MPa (55 kip/in.^{2}) steel and unbalanced concrete and live load.

For other construction materials, such as hot-rolled steel and reinforced concrete, multiple span beams have reserve capacity because of the formation of plastic hinges. However, cold-formed steel sections, such as those investigated in this chapter, are not compact and may have local buckling; they cannot be relied on to form plastic hinges (Yu 1991). Failure of the system occurs when any part of the beam is overstressed.

All of these stresses occur because of bending only. Torsional stresses, which were not calculated, would add to the bending stresses. The other potential failure modes, e.g., shear and web crippling, were not analyzed because the photographs of the collapsed structure strongly suggested bending failure.

Reconstruction

After cleanup, the slab decking was rebuilt using 1.52-mm (0.0598-in., 16 gauge) steel beams to replace the damaged thinner beams. Beams that had not been damaged were not replaced. This time, the beams were shored. The concrete placement occurred without mishap. The building was completed and put into service.

Discussion

The structural integrity of the beams and decking was questioned, but the structural engineer provided assurances that they were adequate. No supporting documents were available.

The structural engineer contended that the collapse occurred because of poor construction practices, leading to concrete piling up and causing unbalanced loading. However, the testimony of the workers indicated that the concrete placement was carried out in accordance with good practice. There was no testimony from the workers or observers present that the workers allowed the concrete to pile up at any point on the decking. In fact, this would have made screeding and finishing the concrete much more difficult.

It is well known that the structural integrity of formwork for concrete is important. Hanna writes, "Partial or total failure of concrete formwork is a major contributor to deaths, injuries, and property damages within the construction industry" (Hanna 1999, p. 6).

The investigation was hampered because it was not possible to access the site and because the failed structure had been removed before the investigation started. Unfortunately, material samples had not been retained. Therefore, it was necessary to analyze the failure solely from the available documents and records.

Summary and Conclusions

There was only one engineer qualified by training, experience, and professional licensure on the project. The structural engineer should have consulted the proper references and performed the necessary structural calculations to ensure that the structure would be safe against collapse, under the load combinations prescribed by building codes. He should have analyzed the beam under an unbalanced load of concrete and live load and compared the calculated moments to the section capacities provided in the AISI manual (AISI 1996).

There are important differences between design procedures for hot-rolled structural steel, which are taught in most civil engineering undergraduate programs, and those for cold-formed steel. Designing with cold-formed steel requires a knowledge of failure modes, such as local buckling, that can often be safely ignored with hot-rolled steel. Engineers designing with this material should take care to obtain the proper codes and design documents. This case study was previously published (Delatte 2005).

The World Trade Center Attacks

The response of reinforced concrete buildings to terrorist attacks is discussed in Chapter 5, with the cases of the Oklahoma City Murrah Federal Building and of the Pentagon attack on September 11, 2001. The collapse of the twin World Trade Center (WTC) towers on September 11, 2001, provides some insight into the vulnerabilities of steel structures against such attacks. After the event, the Federal Emergency Management Agency (FEMA) conducted an investigation (2002).

Design and Construction

The two World Trade Center towers were the tallest of six buildings on the WTC Plaza Complex in New York City. Construction started on August 5, 1966, and steel erection began in August 1968. The north tower (WTC-1) was occupied starting in December 1970, and the south tower (WTC-2) in January 1972. Each tower was 110 stories tall (FEMA 2002, p. 1-2).

The towers were of similar height, 417 m (1,368 ft) for WTC-1 and 415 m (1,362 ft) for WTC-2 at the roof. WTC-1 also supported a 110-m (360-ft) television and radio transmission tower. The buildings were square, a little more than 63 × 63 m (207 × 207 ft) on a side, providing almost 0.4 hectare (1 acre) of floor space on each level. Each building also had a rectangular service core at the center, 26.5 × 42 m (87 × 137 ft), housing three exit stairways, elevators, and escalators. The service core in WTC-1 was oriented east to west, and that in WTC-2 north to south (FEMA 2002, p. 2-1). Figure 6-6 shows a typical floor plan.

The basic structural form for the building was a tube of closely spaced box columns, with about 59 per face of the building. At each floor, they were connected by 1.3-m (52-in.) deep spandrel plates. The outer steel frame was made of overlapping three-story-tall segments. Splices between segments were staggered, so that no more than one-third of the splices were on any one story. Plate thicknesses and grades of steel were varied to accommodate different loads. Under wind loading, the tube acted similarly to a box beam in flexure, with windward and leeward walls acting as compression and tension flanges connected by a Vierendeel truss web (FEMA 2002, pp. 2-2–2-3). Figure 6-7 shows the outer steel frame, and Fig. 6-8 shows the structural behavior under lateral loading.

196 BEYOND FAILURE

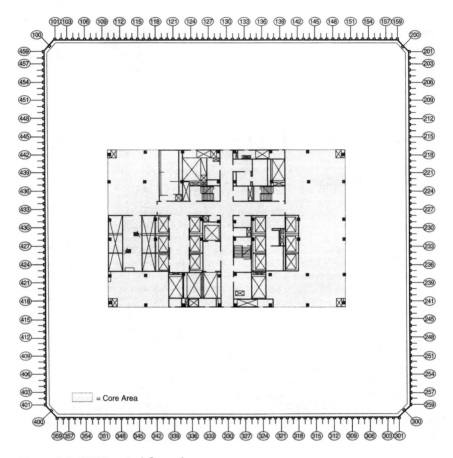

Figure 6-6. WTC typical floor plan.
Source: FEMA (2002).

The floor slab and framing system connected the outer tube to the inner core. The floor was 100-mm (4-in.) thick lightweight concrete on 38-mm (1½-in.) noncomposite steel deck. The floor rested on a series of composite floor trusses. The trusses were similar to open web joists but had more redundancy and better bracing. They were placed in pairs, approximately 2 m (6 ft 8 in.) apart. Transverse trusses ran between these main trusses. Truss spans were approximately 18 m (60 ft) to the sides and 10.7 m (35 ft) to the ends of the central cores. The truss top chords were supported at the outer wall in bearing by seats attached to alternate columns, and on seats at the central core (FEMA 2002, pp. 2-3–2-4). Figure 6-9 shows the trusses and end connection details.

STEEL STRUCTURES 197

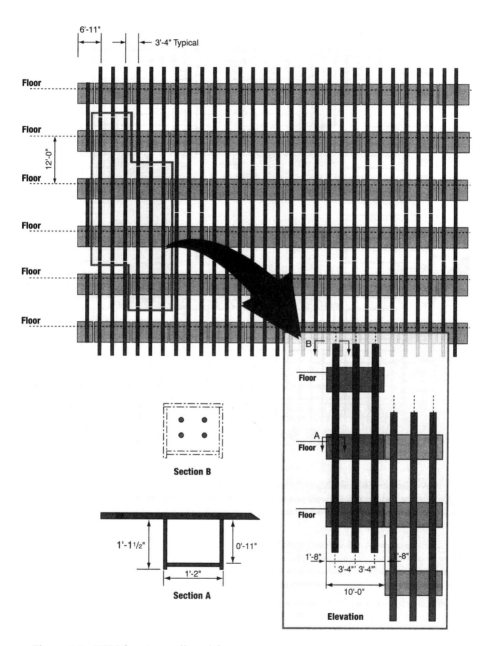

Figure 6-7. WTC bearing wall steel frame.
Source: FEMA (2002).

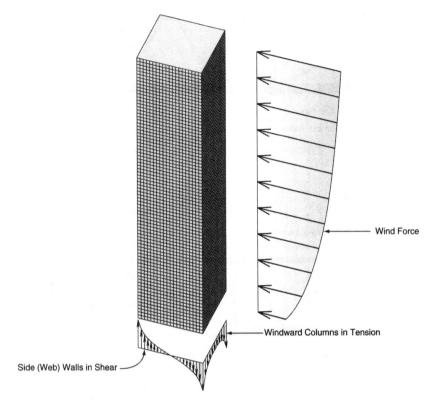

Figure 6-8. WTC structural behavior under lateral loading.
Source: FEMA (2002).

The building was stiffened further against wind loads by a diagonal brace truss system between the 106th and 110th floors. This system coupled the outer tube and the core. On WTC-1, the truss also supported the transmission tower (FEMA 2002, pp. 2-5–2-10).

For most tall buildings, wind forces are the controlling design issue. Ordinary structures are designed using code-proscribed wind loads. However, for important tall buildings, wind tunnel studies are often performed to predict more accurately the wind loads. WTC-1 and WTC-2 were among the first structures designed using wind tunnel studies (FEMA 2002, p. 1-15).

Model design codes do not consider loads that occur because of acts of war or terrorism (FEMA 2002, p. 1-15). In the wake of Oklahoma City and the September 11 incidents, these types of attacks are likely to be considered in the future for some government buildings, particularly overseas, and other potential highly important targets.

STEEL STRUCTURES 199

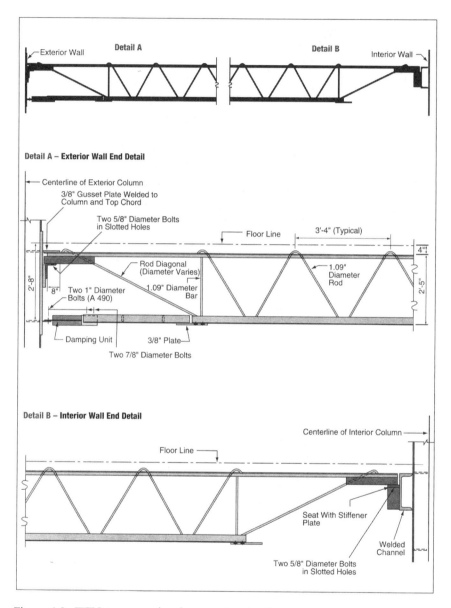

Figure 6-9. WTC trusses and end connection details.
Source: FEMA (2002).

Fire protection is an important element of buildings codes. The first line of defense is automatic sprinkler systems, which are effective for small fires but less so for larger fires, such as those that engulfed WTC-1 and WTC-2 after the aircraft impacts. The second line of defense is firefighters. The effectiveness of firefighting efforts is reduced if fires are high up in the building or if elevators are damaged. The final defense is the inherent fire resistance of the building materials themselves. In particular, the fire resistance of the structural steel is important for preventing collapse during fire. All of these defenses were overwhelmed by the scale of the attack on WTC-1 and WTC-2 (FEMA 2002, p. 1-16).

In WTC-1, a spray-applied fireproofing product containing asbestos had been used up to the 39th floor. The asbestos was later replaced or encapsulated. On the other floors of WTC-1 and throughout WTC-2, an asbestos-free mineral fiber spray product was used. The initial thickness was 19 mm (¾ in.), which was scheduled for upgrade starting in the mid-1990s to 38 mm (1½ in.) as individual floors became vacant. However, at the time of the incident, only 31 floors had been upgraded. Spandrels and girders were specified to have a three-hour fire protection rating, and stair and elevator shafts and stairwells a two-hour rating. The towers had originally been built without sprinkler systems, which were installed starting in 1990. Tanks on the 41st, 75th, and 110th floor provided water into the standpipe system (FEMA 2002, pp. 2-12–2-13).

The towers had actually been designed for an aircraft impact.

> The WTC towers were the first structures outside of the military and the nuclear industries whose design considered the impact of a jet airliner, the Boeing 707. It was assumed in the 1960s design analysis for the WTC towers that an aircraft, lost in the fog and seeking to land at a nearby airport, like the B-25 Mitchell bomber that struck the Empire State Building on July 28, 1945, might strike a WTC tower while low on fuel and at landing speeds. However, in the September 11 events, the Boeing 767-200ER aircraft that hit both towers were considerably larger with significantly higher weight, or mass, and traveling at substantially higher speeds. The Boeing 707 that was considered in the design of the towers was estimated to have a gross weight of [1.16 MN/119 Mg] 263,000 pounds and a flight speed of [290 km/h] 180 mph [mi/h] as it approached an airport; the Boeing 767-200ER aircraft that were used to attack the towers had an estimated gross weight of [1.22 MN/124 Mg] 274,000 pounds and flight speeds of [756 to 950 km/h] 470 to 590 mph on impact. (FEMA 2002, p. 1-17)

The kinetic energy that is transferred to the structure on impact is ½ mv^2. With the small increase in mass and considerable increase in velocity, the kinetic energy of the Boeing 767-200ER aircraft was 7 to 11 times that assumed in the design for a slow-moving Boeing 707. Calculations of impact forces are discussed in Chapter 2.

A deep foundation extended under WTC-1 and WTC-2 and the rest of the WTC plaza. The western part, under the towers, was 21 m (70 ft) deep with six underground levels. It was surrounded by a slurry wall, which formed a bathtub to keep water from the Hudson River out. The slurry wall was stabilized by tieback anchors. The subterranean floor slabs provided lateral support to the bathtub structure (FEMA 2002, pp. 2-10–2-11).

The Attack

The FEMA report describes the attack:

On the morning of September 11, 2001, two hijacked commercial jetliners were deliberately flown into the WTC towers. The first plane, American Airlines Flight 11 . . . crashed into the north face of the north tower (WTC 1) at 8:46 A.M. The second plane, United Airlines Flight 175 . . . crashed into the south face of the south tower (WTC 2) at 9:03 A.M. (FEMA 2002, p. 1-4)

The north tower was hit by a jetliner traveling at approximately 756 km/h (470 mi/h) between floors 94 and 98. The impact caused a huge fireball, spreading jet fuel and igniting fires over several floors. The north tower burned until it collapsed at 10:29 A.M., or 1 hour and 43 minutes after the impact (FEMA 2002, p. 1-4).

The south tower was hit by a jetliner traveling at approximately 950 km/h (590 mi/h) between floors 78 and 84. Thus, the impact on WTC-2 was faster and lower in the structure than that on WTC-1. The south tower also caught fire and collapsed first, at 9:59 A.M., or 56 min after impact. It was estimated that the complex held about 58,000 people at the time of the collapse, and almost everyone below the impact areas was able to escape. The total loss of life was 2,830, including 2,270 building occupants, 157 in the aircraft, and 403 emergency responders (FEMA 2002, p. 1-4).

The collapse of the two structures also severely damaged other nearby buildings, as well as underground services and utilities. One nearby building, the 47-story WTC-7, caught fire and collapsed after burning for seven hours (FEMA 2002, p. 1-8).

The FEMA Investigation and Results

The FEMA Building Performance Study (BPS) team investigation examined the evidence and sequence of events and thoroughly reviewed the structural performance of WTC-1 and WTC-2 during the event, as well as that of other nearby buildings.

Damage and Response of WTC-1

Each tower was subjected to three loading events: the initial aircraft impact, the simultaneous ignition and growth of fires over several floors of the building, and, finally, a progressive sequence of failures leading to total collapse. The impact to WTC-1 broke loose at least five of the three-column assemblies. An estimated 31–36 columns were destroyed over four stories of the building. It also appears clear that the building core experienced significant but undetermined damage. Some aircraft debris passed completely through the structure (FEMA 2002, pp. 2-15–2-16).

Because of the structure's high degree of redundancy, the area of immediate collapse was limited to the general area of the impact. The loads previously carried by the destroyed columns were transferred to alternate load paths, through the Vierendeel truss. The most heavily loaded columns were probably near, but not over, their ultimate capacities. The inherent robustness of the structural system allowed it to remain standing for 1 hour and 43 minutes after the impact (FEMA 2002, pp. 2-16–2-21).

The fires, however, would prove fatal. Each of the aircraft contained approximately 38,000 L (10,000 gal) of jet fuel at the time of impact. Some of the fuel was consumed in a huge fireball on impact, and some remained within the building to fuel the fires. Damage from the impact created openings that provided oxygen for the fires. The impact probably also damaged and disrupted sprinkler and fire standpipe systems. At any rate, so many sprinklers were opened by the fires that the system would have quickly depressurized and become ineffective (FEMA 2002, pp. 2-21–2-23).

The fires imposed structural effects on the damaged building. It is likely that the impact knocked off and damaged some of the insulation protecting the structural steel. As mentioned earlier, some columns were heavily loaded because of the redistribution of forces after the impact. Also, some of the floor framing beneath the partially collapsed area was probably carrying considerable additional weight from the debris (FEMA 2002, p. 2-24).

The specific chain of events leading to the structural collapse will probably never be known, but certain structural effects of fire are likely to have played a part:

- As the floor framing and slabs were heated, they expanded. This effect alone may have caused some structural members or connections to fail.
- With increasing temperature, the floor and slab assemblies became less stiff and sagged into catenary action. Loading from debris would have increased the forces caused by sagging. The sagging imposed tensile forces on horizontal framing and floor elements, possibly causing end connections to fail. Because the floor and slab assemblies braced the exterior columns, as these connections failed, the unbraced length of the columns would have increased and their buckling loads would have decreased substantially.
- As the temperature of steel increases, its yield strength and modulus of elasticity decrease. Therefore, the elastic and inelastic buckling strength is decreased (FEMA 2002, pp. 2-24–2-25).

The final structural collapse was rapid. Although much of the debris stayed within the building footprint, some was scattered as far as 120–150 m (400–500 ft) from the tower base and heavily damaged some adjacent structures (FEMA 2002, p. 2-27).

Damage and Response of WTC-2

WTC-2 was subjected to the same loading events as WTC-1, but it collapsed more quickly. At the location of impact, six three-column assemblies were broken loose. An estimated 27–32 columns were destroyed over five stories of the building on the south building face, with more perhaps at the southeast corner as well. As with WTC-1, some aircraft debris passed completely through the structure, and the building core may also have been badly damaged. The building stood for 56 minutes after the impact (FEMA 2002, pp. 2-27–2-31).

There were, however, important differences between the aircraft impacts on WTC-2 and WTC-1. The higher speed of the aircraft hitting WTC-2 imposed about 60% more energy to the structure, which would have resulted in more severe damage. Also, the area of impact was closer to the corner of the building, and thus damaged two adjacent faces. In addition, the impact on WTC-2 was about 20 stories lower than that on WTC-1, so the columns were carrying substantially higher gravity loads. As a result, the overall structural effect of the impact on WTC-2 was more severe (FEMA 2002, pp. 2-31–2-32).

Before the impact, the outer columns of WTC-2 were estimated to be loaded to 20% of capacity because of gravity loads, and the interior columns to 60%. Wind and deflection were the design considerations for the outer structural frame, not gravity loads. After impact, fires spread through WTC-2 in a similar manner to those in WTC-1 (FEMA 2002, pp. 2-33–2-34).

Damage to Substructure

With the collapse of the two buildings, almost 600,000 tonnes (600,000 tons) of debris fell. The impact punched through the plaza and several of the six levels of substructure, which was partially filled with debris. The damage degraded the support provided to the slurry wall bathtub by the floor slabs. A significant engineering effort proved necessary to tie back and stabilize the wall during debris removal (FEMA 2002, pp. 2-35–2-36).

Overall FEMA Study Findings

The FEMA report assessed overall performance:

> The structural damage sustained by each of the two buildings as a result of the terrorist attacks was massive. The fact that the structures were able to sustain this level of damage and remain standing for an extended period of time is remarkable, and is the reason that most building occupants were able to evacuate safely. Events of this type, resulting in such substantial damage, are generally not considered in building design, and the ability of these structures to successfully withstand such damage is noteworthy. (FEMA 2002, p. 2-36)

Although the buildings withstood the initial attack, they were not able to withstand the severe structural effects caused by the fire loading. The burning fuel was not enough by itself to cause the buildings to collapse, but this fire ignited the building contents. The burning building contents, over time, combined with the structural damage to cause the collapses (FEMA 2002, pp. 2-36–2-37).

Some features of the building design helped the buildings stand long enough and aided the evacuation of most of the inhabitants. These features were the overall robustness and redundancy of the steel frames; the provision of adequate, well lighted and marked egress stairways; and prior emergency exit training for the building occupants (FEMA 2002, p. 2-38).

The FEMA BPS team identified a number of potential design features that might have improved building performance if they had been used. These features were:

- the type of steel floor truss system used in the two buildings and the system's structural robustness compared to other floor systems,
- impact-resistant enclosures around egress stairways and paths,
- resistance of fireproofing to blast and impact, and
- the position of egress stairways in the central core, as opposed to dispersed over the building floor plan (FEMA 2002, p. 2-38).

The FEMA BPS team was not able to determine the effect of these design features on the collapse, and they recommended further study. They also did not recommend any building code changes because there remains insufficient data about specific threats to specific buildings (FEMA 2002, pp. 2-38–2-39).

Some additional studies were recommended:

- detailed modeling to evaluate the condition of the interior structure of the towers after impact and before collapse;
- more detailed modeling of the fires within the structures;
- modeling of the complexity and redundancy of the floor framing system, as well as the connections to the structural frame;
- fire performance of steel trusses with sprayed-on fireproofing;
- susceptibility of spray-applied fireproofing to damage from blast and impact; and
- whether, given the size and weight of the towers, there were feasible design and construction features that might have been able to arrest the damage (FEMA 2002, pp. 2-39–2-40).

Recommendations to Improve Robustness and Safety

The FEMA BPS team made some general observations, findings, and recommendations based on the study of WTC-1 and WTC-2, as well as the other nearby structures:

- Structural framing systems need redundancy and/or robustness, so that alternative paths or additional capacity are available for transmitting loads when building damage occurs.
- Fireproofing needs to adhere under impact and fire conditions that deform steel members, so that the coatings remain on the steel and provide the intended protection.

- Connection performance under impact loads and during fire loads needs to be analytically understood and quantified for improved design capabilities and performance as critical components in structural frames.
- Fire protection ratings that include the use of sprinklers in buildings require a reliable and redundant water supply. If the water supply is interrupted, the assumed fire protection is greatly reduced.
- Egress systems currently in use should be evaluated for redundancy and robustness in providing egress when building damage occurs, including the issues of transfer floors, stair spacing and locations, and stairwell enclosure impact resistance.
- Fire protection ratings and safety factors for structural transfer systems should be evaluated for their adequacy relative to the role of transfer systems in building stability. (FEMA 2002, pp. 8-1–8-2)

Essential Reading

The Federal Emergency Management Agency published *World Trade Center Building Performance Study: Data Collection, Preliminary Observations, and Recommendations*, FEMA report No. 403 (FEMA 2002). This report is available online in PDF format at http://www.fema.gov/rebuild/mat/wtcstudy.shtm. In addition to providing a thorough discussion of WTC-1 and WTC-2, the report also discusses the performance of the other nearby buildings, such as WTC-7. Free printed copies of FEMA reports may be obtained directly from FEMA at www.fema.gov.

Pittsburgh Convention Center Expansion Joint Failure

Joints and connections are often weak points in structures. They have to fulfill the competing needs of allowing movement and preventing stress buildup and, on the other hand, of transferring loads.

The David L. Lawrence Convention Center

The David L. Lawrence Convention Center in downtown Pittsburgh was built between 2000 and 2003 for the Sports & Exhibition Authority of Pittsburgh and Allegheny County, or SEA. It is a four-level structure, roughly 265 × 174 m (870 × 570 ft) in plan. An expansion joint along column line

X9, roughly 146 m (480 ft) from the west end, splits the center approximately in two. The connections of the expansion joint are exposed to ambient temperatures (WJE 2008).

Collapse

At about 1:30 P.M. on Monday, February 5, 2007, a tractor-trailer was parked on the second-floor loading dock of the convention center. The trailer had just hitched its bumper to the loading dock. Under the weight, a 6.1 × 18 m (20 × 60 ft) section of concrete slab and the steel beam supporting it collapsed. There were, fortunately, no injuries. The ambient temperature at the time was about −19 to −14 °C (−3 to −7 °F). Problems with 18 misaligned portions of the column foundations had halted construction work in November 2001, and the collapse had occurred in the vicinity of the shifted columns. Work was resumed once repairs had been made to some precast concrete beams. The $370 million building opened in 2003. The collapse led to the cancellation of the Pittsburgh International Auto Show (Ritchie and Houser 2007, WJE 2008).

At the time, a reporter for the *Pittsburgh Tribune-Review*, Mark Houser, called me and sent me a picture of the collapse and asked for comment. From the photo, it was difficult to attempt to figure out what was going on. He noted, however, that the failure had occurred at an expansion joint. Because it had been a cold day, it was possible that the expansion joint detail had played a role. In the newspaper article, I observed,

> If you have a bolt holding something on, and it contracts due to cold ... then you come along and load it up, the combination of the two may be enough to cause a failure where neither one by itself would. ... An expansion joint will be open the most when it's coldest. The more an expansion joint opens up, the harder it's going to be to hold up the gravity load. And at some point, if an expansion joint opens up too much, it may actually slip off (its support). (Ritchie and Houser 2007)

It was, of course, a somewhat wild but educated guess based on limited information. However, the investigation determined that the expansion joint was in fact the culprit.

Investigations

Several independent investigations were carried out. The owner hired the Wiss, Janney, Elstner Associates Cleveland office and Leslie E. Robertson

Associates, and the architect hired Thornton Tomasetti. A follow-up story was printed in the *Pittsburgh Tribune-Review* after the investigators briefed the public on their initial findings. It was disclosed that a beam had failed at a similar connection in 2005, causing the beam to drop 64 mm (2½ in.) before it was stopped by a column, but the earlier collapse had not been disclosed to city and county officials. Another failure had occurred in February 2002 during construction. In the 2002 collapse, an ironworker was killed when incorrect nuts were used to connect some of the steel structural elements (Houser and Ritchie 2007). The failed expansion joint detail is shown in Fig. 6-10.

The expansion joint essentially divided the building into two large sections. Twenty-five slotted expansion joint connections were provided along the expansion joint.

Greg Luth, who wrote an award-winning paper on the Hyatt Regency collapse (Luth 2000), discussed the connection with the *Pittsburgh Tribune-Review* and called it "structurally unreliable."

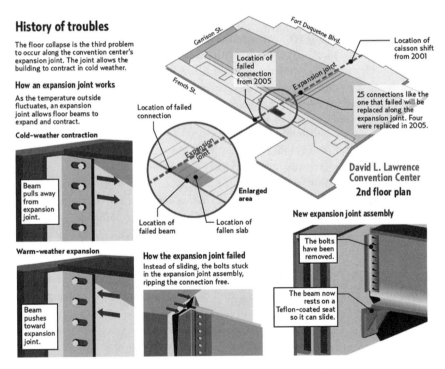

Figure 6-10. Pittsburgh Convention Center expansion joint detail.
Courtesy *Pittsburgh Tribune-Review*.

"I've never talked to a structural engineer that thought this was a reliable way to do an expansion joint," said Gregory Luth, whose Santa Clara, Calif., firm designed a casino and racetrack outside Harrisburg and a Disneyland hotel. Luth was not involved in the inspection at the convention center, but reviewed diagrams of the connection provided by engineers. "In my opinion, this would not be first-class engineering," he said. . . . "I've only seen the slotted hole connection used one other place in an expansion joint in 30 years of doing engineering. And it fell in that place, as well," Luth said. That was at a Hyatt Regency hotel in Kansas City, in 1979. (Houser and Ritchie 2007)

This was the atrium partial collapse that occurred just before opening and almost two years before the walkway collapse.

The WJE investigation addressed issues with design, materials, fabrication, and construction. The main design issue was that the slotted hole expansion joint was almost guaranteed to fail because of significant friction and insufficient room for thermal contraction. Also, the design drawings did not prohibit bolt threads on the bearing surface, which increased friction further. Other design errors did not contribute significantly to the collapse; these errors included inadequate length of the slot and no limitation on bolt torque (WJE 2008).

Materials and fabrication issues included steel with too high a strength—ASTM A 992, not A 36. This high strength kept the angles from bending and caused them instead to tear away at the weld. Other problems were bolt threads bearing on the slot surface and slots fabricated with a bump in the middle. There was little evidence of the bolts actually sliding within the slot; instead, the threads seem to have worn away at the same spot. Washer plates were added, although they were not shown on the drawings (WJE 2008).

The construction issues included the use of the wrong type of steel angles, with high-strength steel. A drift pin was lodged tightly into one of the boltholes, helping to lock the connection. Other construction errors that did not contribute significantly to the collapse were a missing bolt and the torque of the bolts (WJE 2008).

Expansion Joint Detailing and Forces

The amount of displacement caused by temperature change δ_T is

$$\delta_T = \alpha(\Delta T)L \qquad (6\text{-}1)$$

where α = the coefficient of thermal expansion, ΔT = the change in temperature, and L is the length of the element. The amount of displacement δ caused by load is

$$\delta = \frac{PL}{AE} \qquad (6\text{-}2)$$

where P = force, A = cross-sectional area, and E = modulus of elasticity.

Therefore, if thermal deformation is restrained, the force built up in the element is

$$P = \alpha(\Delta T)AE \qquad (6\text{-}3)$$

WJE estimated the amount of free movement required at the expansion joint as 41 mm (1.6 in.), based on a thermal coefficient for steel of 0.00001 mm/mm/°C (0.000006 in./in./°F), a temperature change of 28 °C (50 °F), and L equal to half the length of the building: 133 m (435 ft or 5,220 in.). The WJE finite element analysis estimated that the distortion of the ASTM A 992 high-strength angles imposed a force of 630 kN (140,000 lb) of tension at the connection welds, with approximately 8 mm (0.3 in.) of displacement. With lower strength A 36 steel, the force on the welds would have been reduced by 40%.

There are two obvious problems with a slotted hole expansion joint of this type. The first is that the slots must be long enough, and the bolt must be centered, so that the bolt can move freely back and forth in the slot without bearing against either edge. If the beam contracts and the bolt bears against the edge of the slot, the connection is locked and will behave as a fixed connection and pull apart when any additional contraction occurs. The second problem is that the bolts must be loose enough to keep from locking the joint. If the bolts are tightened, which can easily happen during construction, the joint won't work. Corrosion, paint, and debris can also lock the joint. For this particular type of joint to work, it must be built and maintained perfectly, and that may not happen in the field.

During cold weather contraction, as shown in Fig. 6-10, the beam pulls away from the connection. The two slotted angles were welded only at the outer edges, which made them weak in tension. When they pulled free, the connection failed.

A more reliable detail for this type of connection is a low-friction supporting bracket. To retrofit the connections at the convention center, the bolts were removed and Teflon-coated supporting seats were added. This detail was designed by Thornton Tomasetti and is shown in Fig. 6-10.

The failed expansion joint did not conform to the American Institute for Steel Construction (AISC) *Manual of Steel Construction* (1998). Charles Carter, an AISC engineer, observed that the manual recommends two ways to build an expansion joint. One way is to provide a double line of structural columns, one on each side of the joint, and in essence create two buildings that can move independently. The other is a low-friction sliding connection, such as the shelf support connection developed for the retrofit. Carter noted that the original detail, with the sliding steel bolts, would create a lot of friction and would probably not be an effective joint. The manual's recommendations are not mandatory but are considered good practice. The recommended retrofit of the 25 expansion joint connections was estimated to cost $350,000 (Houser 2007).

Reading and Follow-Up

This partial collapse was, on the whole, a relatively minor and local case. However, it illustrates the principle that many useful failure and near-failure case studies can be found in the newspaper every year. This expansion joint failure occurred near the start of a course on strength of materials that I was teaching, and the newspaper article handouts provided useful material for classroom discussions. A recent connection failure is of more direct interest to students than the case of the Hyatt Regency walkways, which is now approaching three decades old.

The Wiss, Janney, Elstner Associates, Inc., final report can be found online at http://www.pgh-sea.com/images/DLLCCCollapseFinalReportFeb%2008.pdf.

Minneapolis I-35W Bridge Collapse

On August 1, 2007, an eight-lane, three-span section of the I-35W Bridge across the Mississippi River in downtown Minneapolis collapsed at approximately 6:00 P.M. It was crowded with rush hour traffic. The final death toll was 13. The bridge was the main north–south route through the city. Figure 6-11 shows the bridge before the collapse. Eyewitness reports suggested that a span at the south end collapsed first and that the failure propagated across the bridge to the other two spans in turn.

Much of the wreckage remained above the water, but about 50 vehicles with their occupants went into the river. At the time, repairs were being made to the bridge deck, guardrails, and lights. Figure 6-12 shows the wreckage of the collapsed bridge.

A view of the west side of the deck truss portion of the bridge, looking northeast.

Figure 6-11. Minneapolis I-35W Bridge before the collapse.
Source: National Transportation Safety Board.

The Bridge

The total length of the three spans was approximately 300 m (1,000 ft), and the bridge deck was about 18 m (60 ft) above the river. The bridge had been built in 1967 and was about 40 years old. There are approximately 466 bridges of similar design in the United States.

The total length of all 14 spans of the bridge was 581 m (1,900 ft). The 11 north and south approach spans were a combination of steel multibeam and concrete slab simple spans. The three main continuous spans over the Mississippi River consisted of a pair of steel deck trusses, each with two 80.8-m (265-ft) approach spans and a 139-m (456-ft) central span. The two main trusses also had 11.6-m (38-ft) cantilever spans at the north and south end, with 27 perpendicular floor trusses spaced at 11.6 m (38 ft). These floor trusses cantilevered out past the main trusses approximately 5 m (16 ft). Truss member connections included both rivets and bolts and had numerous poor welding details (classified as Category D or E). Some corrosion of the connections had been noted on inspection reports. The entire deck truss was supported on four piers, three with rollers and one with a

STEEL STRUCTURES 213

Figure 6-12. Wreckage of the collapsed Minneapolis I-35W Bridge.
Source: National Transportation Safety Board.

fixed connection. The bridge deck was 32.9 m (108 ft) wide from gutter to gutter (O'Connell et al. 2001, pp. 25–26).

This bridge was the subject of a research project on fatigue evaluation completed in 2001. The report's executive summary noted,

> The approach spans have exhibited several fatigue problems, primarily due to unanticipated out-of-plane distortion of the girders. Although fatigue cracking has not occurred in the deck truss, it has many poor fatigue details on the main truss and floor truss systems. Concern about fatigue cracking in the deck truss is heightened by a lack of redundancy in the main truss system. The detailed fatigue assessment in this report shows that fatigue cracking of the deck truss is not likely. Therefore, replacement of this bridge . . . may be deferred. (O'Connell et al. 2001, p. xi)

One of the authors of the report, the late Robert Dexter of the University of Minnesota, was a well-known expert in the field of fatigue of steel structures.

The report also noted that cracks had been discovered before the collapse (O'Connell et al. 2001, pp. 28–29).

Other cracks were found and repaired with similar retrofits. These cracks considerably reduced stress ranges, which should have reduced the probability of further fatigue cracking.

The bridge was reportedly rated "structurally deficient" in 2005, with a rating of 50 out of 100 for structural stability. This is not particularly uncommon. About one in eight U.S. bridges are considered structurally deficient. As a whole, the bridge was structurally nonredundant.

The NTSB Investigation

The National Transportation Safety Board (NTSB) immediately began an investigation, as with most bridge collapses. The investigation team arrived the day after the collapse.

At first glance, this collapse seemed to share some parallels with the Point Pleasant Bridge collapse of 1967 (discussed in Chapter 3) and the Mianus River Bridge failure of 1983 (reviewed earlier in this chapter). As with the Point Pleasant Bridge, the Minneapolis bridge was 40 years old and collapsed during rush hour with many vehicles on the span. In fact, the Minneapolis bridge was completed just a few months before the Point Pleasant Bridge failure. Both the Point Pleasant and the Minneapolis I-35W Bridge had three spans that all fell together.

As of summer 2008, the cause of the I-35W Bridge collapse remains under investigation, and the final reports have yet to be written. However, considerable work has been done, and a critical design flaw that may well have been the cause of the collapse has been identified. The U10 and L11 gusset plates, shown in Fig. 6-13, were only about 13 mm (½ in.) thick when they should have been 25 mm (1 in.) thick.

At the time of the collapse, the contractor working on the bridge had stockpiled aggregates and heavy construction equipment on the site. Careful reconstruction of the traffic and other loading by the Federal Highway Administration (FHWA) and others has found that the U10 and L11 gusset plates were considerably overstressed at the time of the collapse (Holt and Hartmann 2008).

Discussion

All states owning similar bridges have been encouraged to review them for similar flaws. Two bridges of this type are along Interstate 90 in northeast Ohio, and I have driven over them many times. At this point, the gusset plate design error seems to have been isolated to the I-35W Bridge.

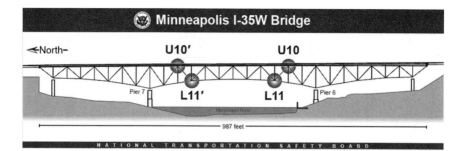

Figure 6-13. Minneapolis I-35W Bridge U10 and L11 gusset plates.
Source: National Transportation Safety Board.

This sort of design defect would not be noted during routine bridge inspections because the inspectors would merely note the condition of the gusset plate. They would not have the documents or the time to check its adequacy. In this respect, then, one final similarity to the Point Pleasant Bridge collapse remains: a hidden flaw, not detectable to normal inspection procedures, that can lie dormant for 40 years and then bring a structure crashing down.

Further Reading

The NTSB has posted the Holt and Hartmann (2008) report and other reports, illustrations, and documents on a website at http://www.ntsb.gov/dockets/Highway/HWY07MH024/default.htm.

Other Cases

Dee Bridge

Beams can fail through lateral-torsional buckling, as well as by bending. An early failure of this type was the cast iron Dee Bridge in 1847, designed by Robert Stephenson, one of the most famous early British engineers. This collapse is described by Petroski (1994, pp. 83–93).

Because cast iron is weaker in tension than in compression, the tension flanges of girders were larger than compression flanges by a ratio of 16:3, following the ratio of material strengths. The Dee Bridge had a relatively long simple span for the day of roughly 29 m (95 ft). The girder cross section is shown in Fig. 6-14. The bridge was also designed with a relatively low factor of safety of 1.5.

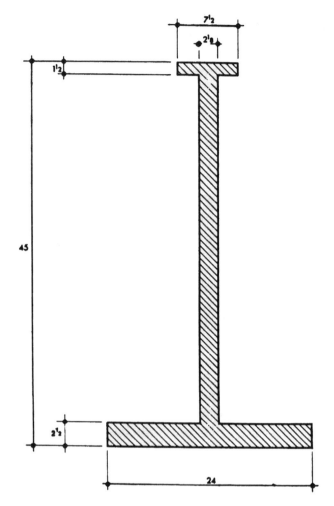

Figure 6-14. The Dee Bridge girders (1 in. = 25.4 mm).
Source: Petroski (1994), courtesy Cambridge University Press.

To strengthen the bridge, an inverted arch of wrought iron tension bars was used to truss the girder. This arch added compressive stress to the span. Two girders supported wooden decking, ballast, and rails from their bottom flanges. The top flanges were not braced.

The bridge was completed in September 1846. In May 1847, 125 mm (5 in.) of ballast was added on top of the wooden decking, in part to reduce the risk of fire from the sparks of the locomotives. When the first train crossed the bridge after the installation of the ballast, the bridge sank beneath it. The locomotive reached safety, but five cars fell into the river. Five people were killed, and 18 were injured.

"The most likely cause of failure was a torsional buckling instability to which the bridge girders were predisposed by the compressive loads introduced by the eccentric diagonal tie rods on the girder" (Petroski 1994, p. 90). In effect, the top flange of the girder became a column under the compressive force and buckled out of plane, causing the collapse.

The lengthening of cast iron spans caused the governing mode of failure to change from bending to lateral-torsional buckling.

> The Dee Bridge failed because torsional instability was a failure mode safely ignored in shorter, stubbier girders whose tie rods exerted insufficient tension to induce the instability that would become predominant in longer, slenderer girders prestressed to a considerable compression by the trussing action of the tie rods. (Petroski 1994, p. 95)

Subsequently, Stephenson pointed out the need to study failures to learn from them.

Lateral-torsional buckling remains an important failure mode for steel beams, unless the top flanges are braced. The problem is generally less severe than it was with iron because steel is equally strong in tension and in compression, and thus top and bottom flanges are of equal size. This type of buckling, however, remains an important limit state that must be checked.

Quebec Bridge

The 1907 collapse of the Quebec Bridge remains a landmark in the development of steel design and construction. The full case study is presented in Chapter 3. To some extent, the lessons of the Quebec Bridge collapse are similar to those of the Hartford Civic Center failure, which is discussed earlier in this chapter.

A number of elements of the Quebec Bridge collapse are relevant to a course in the design of steel structures. These include the following:

- Determination of loads: The dead loads turned out to be considerably higher than the design assumptions.
- Structural ductility and redundancy: Overall, structures should be designed to provide warning of impending failure and to provide for load redistribution in the event of a member failure.
- Buckling of columns and bars: The buckling strength of compression elements depends on the slenderness ratio. Cooper's allowable

buckling stress, as a function of slenderness ratio, was higher than that allowed by the contemporary AISC (2005) specification, with newer and better steel.
- Analysis and design of built-up members: This point follows from the previous discussion of buckling of columns and bars. Many existing steel bridges use built-up members, and engineers involved in assessing and rehabilitating such structures need to know how to evaluate member capacity and likely failure modes.
- Importance of structural deformation: If deformations exceed predictions, the cause for the excessive deformations should be determined. This is a clear warning that something is wrong.

Point Pleasant Bridge Collapse

The Point Pleasant Bridge collapse of 1967 illustrated the importance of structural redundancy and ductility, as well as the design of steel structures for fatigue loading. The full case study is presented in Chapter 3. One fact relevant to steel design that was not discussed in detail is the lower ductility and fatigue resistance of high-strength steel. High-strength steel can be used successfully for structures, but there is a need to take special precautions against fatigue failure. Connections and details that are susceptible to fatigue failure should be avoided for bridges whenever possible.

Hyatt Regency Walkway

The immediate cause of the 1981 Kansas City Hyatt Regency walkway collapse was the failure of a hanger rod to box beam connection. The full case study is discussed in Chapter 2. The actual failure mechanism was a failure of the material between the nut attached to the hanger rod and the box beam. The poor quality of the box beam welded connection, without stiffeners, substantially reduced the bearing capacity of the beam. It would be difficult to provide a proper weld to connect the two channels for this type of box beam.

The connection had an unacceptably low factor of safety. One of the possible design alternatives, suggested after the fact, was the use of a bearing plate to distribute the bearing stress against the box beam. This is, of course, a gross oversimplification of the case study, and the reader is encouraged to review the complete narrative. This critical connection was never designed, so the bearing stress was never considered. The ethical implications of this case are also important.

L'Ambiance Plaza Collapse

L'Ambiance Plaza, discussed in detail in Chapter 4, was a lift-slab project that collapsed during construction, killing 28 workers. Lift-slab construction uses long steel columns, supported on footings, which experience their heaviest loads during the lift-slab jacking operations. Individual columns may buckle (in nonsway mode), or the structure as a whole may buckle in sway mode. Overall sway buckling is one theory offered for the collapse.

Analysis of the overall structure is complicated by several factors. The columns approximate a fixed-free condition, with an effective length factor (K) of 2, but the footings allow some rotation. Therefore, the K factor may exceed 2. A procedure for determining the K factor is provided by Zallen and Peraza (2004, pp. 39–51).

The slab shear heads do not provide a moment connection, unless, as at the Cleveland Pigeonhole Parking garage, the building sways 2.1 m (7 ft) out of plumb and the shear heads jam against the columns (which may be too late to be helpful). Therefore, the building must be stabilized during construction by shear walls, cross bracing, k bracing, or a combination of bracing systems. Unfortunately, the bracing systems cannot be installed until the slab packages have been lifted past their attachment points, and the heavy slab packages high on the columns contribute to instability.

Agricultural Product Warehouse Failures

Structural damage to two agricultural product warehouses is examined in a case study discussed in Chapter 4. Some of the structural members of the warehouses were hot-rolled steel, and some were cold-formed steel. The base plates at the bottom of the hot-rolled steel frames and columns transmitted a moment to the unreinforced concrete foundation, causing severe cracking. The load of the stored cottonseed, which provided lateral pressures against the warehouse walls, was not properly addressed in the design.

Citicorp Tower

The Citicorp Tower, which averted collapse, is discussed in Chapter 10. The diagonal brace member connections were designed for net compression forces only, not to develop the full strength of the members. This deficiency turned out to be a potentially fatal design flaw, and it required an expensive retrofit of the connections.

7

Soil Mechanics, Geotechnical Engineering, and Foundations

GEOTECHNICAL ENGINEERING IS A SUBDISCIPLINE OF CIVIL engineering that has important implications for structures, pavements, dams, and other facilities, pretty much any facility that sits on the earth. Much of an introductory geotechnical engineering course relies on concepts from strength of materials, such as stress resultants and Mohr's circle. Interactions between soil and water are also important.

Current geotechnical engineering relies in large part on failure analysis. By definition, subsurface conditions have a high level of uncertainty. Morley (1996) notes that the study of cases such as the Leaning Tower of Pisa and the Transcona Grain Elevator helped change the model of geotechnical and foundation failures as "acts of God" and to instead put the discipline on a sound theoretical footing.

Morley further states that for the Transcona Grain Elevator, as well as for the Leaning Tower of Pisa,

> A large amount of data was known about the structure and the soil, making failure analysis a realistic endeavor. Failures such as these pose a number of questions for theorists and practitioners. How did the failure happen? What does the failure reveal about materials' performance, construction process, and especially in these cases, site conditions? Could existing theory have predicted the failure?

How useful is this theory for analyzing and understanding the failure? How does the failure reveal the imperfection of existing factual knowledge and the weaknesses of theoretical assumptions? Does the failure present an opportunity to verify or check new theoretical statements? How can the failure be corrected or future failures avoided? ... A subsequent failure must ... be analyzed in terms of the theoretical assumptions, tests conducted, and the initial design concept. The results of this analysis may, in turn, call for modifications in assumptions, in the choice of tests or interpretation of results, or in the design concept. (1996, pp. 27–28)

These thought processes, of course, are useful in all fields of engineering. All fields of engineering are subject to similar unknowns and uncertainties.

Dam case studies cross the disciplines of geotechnical engineering, hydraulic and water resources engineering, and structural and materials engineering. Therefore, some of the case studies in Chapter 8 will also be of interest for geotechnical engineering students and courses.

Dams are often designed and built to hold water resources and produce electrical power. However, dam failures may be triggered by deficiencies in soil used to build earthen dams, or soil and rock supporting concrete dams. The geotechnical conditions under and around the dam and reservoir should be thoroughly investigated. Even a difficult to identify thin clay seam may lead to a dam failure, as at the Vaiont and Malpasset dams in Europe.

Throughout history, dams have failed and lives have been lost. Levy and Salvadori (1992) point out that a dam failure in Grenoble, France, was recorded as early as 1219. For dams built in the United States before 1959, on the average 1 in 50 failed.

Levy and Salvadori (1992) describe the failures of the South Fork Dam, which led to the Johnstown Flood, and the Malpasset Dam in detail. The failure of the South Fork Dam on May 31, 1889, released a wall of water 12 m (40 ft) high traveling at 32 km/h (20 mi/h) that killed almost 3,000 people in Johnstown, Pennsylvania, and other towns. More recently, the Malpasset concrete arch dam in France failed on December 2, 1959, when the abutment shifted because of a weak seam in the rock. Almost 400 people lost their lives (Levy and Salvadori 1992). Dam failures stand as a warning against overconfidence and hubris. "As every dam engineer knows, water also has one job, and that is to get past anything in its way" (Macauley 2000, p. 93).

Popescu and Popescu discussed using two library research assignments in an undergraduate geotechnical engineering course. In the first assignment, students discuss a prominent personality in geotechnical engineering.

The second assignment is to report on a failure case history. "The list of proposed case histories includes:

- Pisa Tower;
- Lower San Fernando Dam failure;
- Carsington Dam failure;
- Kettleman Hills Waste Landfill slope failure;
- Desert View Drive Embankment failure;
- La Conchita slide;
- Landslide Dam on the Saddle River (Alberta);
- Transcona Grain Elevator failure;
- Buffalo Creek Disaster, West Virginia; and
- Vaiont Dam Landslide, Italy" (Popescu and Popescu 2003, p. 43).

Teton Dam

Some of the factors influencing dam safety and performance can be reviewed through the case of the Teton Dam. The failure of the Teton Dam in eastern Idaho during initial filling of the reservoir on June 5, 1976, killed 14 people and caused hundreds of millions of dollars in property damage downstream. A thorough investigation identified the causes of the failure and suggested improvements for the design and construction of earthen dams.

Design and Construction

The Teton Dam was situated on the Teton River, 5 km (3 mi) northeast of Newdale, Idaho. It was designed to provide recreation, flood control, power generation, and irrigation for more than 40,000 hectares (100,000 acres) of farmland. The Office of Design and Construction of the U.S. Bureau of Reclamation (USBR), at the Denver Federal Center, designed the dam; the construction contract was awarded to the team of Morrison–Knudsen–Kiewit in December 1971.

The preparations for this dam project had been underway for many years. The first active site investigation in the area occurred in 1932. Between 1946 and 1961, eight alternate sites within about 16 km (10 mi) of the selected site were investigated. Between 1961 and 1970, approximately 100 borings were taken at the site (Independent Panel 1976).

The design of the foundation consisted of four basic elements: 21-m (70-ft) deep, steep-sided key trenches on the abutments above the elevation of 1,550 m (5,100 ft); a cutoff trench to rock below the elevation of

1,550 m (5,100 ft); a continuous grout curtain along the entire foundation; and the excavation of rock under the abutments (Independent Panel 1976). These elements for the foundation were important because the types of rock located in this area, basalt and rhyolite, are not generally considered acceptable for structural foundations.

The embankment itself consisted of five main zones. Zone 1 was the impervious center core, which formed the water barrier of the dam. Zone 2 overlaid Zone 1 and extended downstream to provide a layer to control seepage through the foundation. Zone 3 was downstream, and its main function was to provide structural stability. Zone 4 consisted of the storage areas downstream from the control structure and the temporary enclosures built to permit the work to be done. Finally, Zone 5 was the rockfill in the outer parts of the embankment (Independent Panel 1976). Some of these features are shown in Fig. 7-1.

Construction of the dam began in February 1972. When complete, the embankment would have a maximum height of 93 m (305 ft) above the riverbed and would form a reservoir of 356 million m^3 (288,000 acre-ft) when filled to the top. The dam was closed and began storing water on October 3, 1975, but the river outlet works tunnel and the auxiliary outlet works tunnel were not opened (Arthur 1977).

Because these sections were incomplete, the water was rising at a rate of about 1 m (3 ft) per day, which was higher than the predetermined target rate set by the U.S. Bureau of Reclamation of 0.3–0.6 m (1–2 ft) per day for the first year. However, the increased rate was considered acceptable by the Bureau of Reclamation as long as seepage and the water table downstream of the dam were measured more frequently (Independent Panel 1976).

The Failure

On June 3, 1976, several small seepages were noticed in the north abutment wall. Pictures were taken, and these leaks were reported to the Bureau of Reclamation. This finding led to more frequent inspections of the dam. It was now to be inspected daily, and readings were to be taken twice instead of once a week. On June 4, 1976, wetness was observed in the right abutment along with small springs (Independent Panel 1976).

On June 5, 1976, the first major leak was noticed between 7:30 and 8:00 A.M. The leak was flowing at about 500–800 L/s (20–30 ft^3/s) from rock in the right abutment. By 9:00 A.M., the flow had increased to 1,100–1,400 L/s (40–50 ft^3/s), and seepage had been observed about 40 m (130 ft) below the crest of the dam (Arthur 1977).

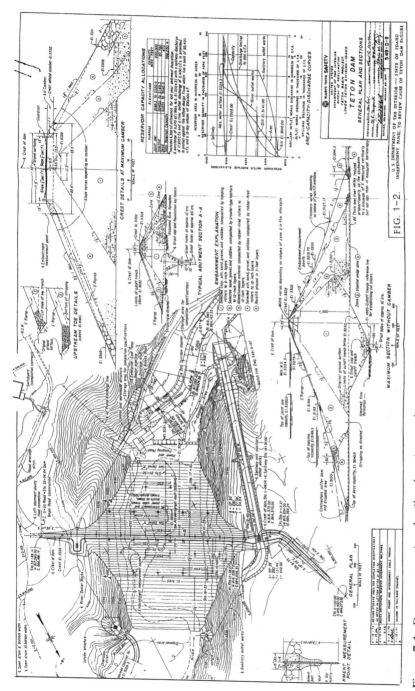

Figure 7-1. Dam construction details, plan view, and cross section.
Source: Independent Panel (1976).

At 11:00 A.M., a whirlpool was observed in the reservoir directly upstream from the dam, and four bulldozers were sent to try to push riprap into the sinkhole near the dam crest (Independent Panel 1976). Two of the bulldozers were swallowed up by the rapidly expanding hole, and the operators were pulled to safety by ropes tied around their waists. Figure 7-2 shows a cross section of the dam with the approximate locations of the seepage, the sinkhole, and a whirlpool in the reservoir noted.

Figure 7-3a shows the location of the initial leak as it expanded up the dam. Between 11:15 and 11:30 A.M., a 6 × 6 m (20 × 20 ft) chunk of dam fell into the whirlpool, and within minutes the entire dam collapsed (Independent Panel 1976). The failure sequence is illustrated in Figs. 7-3a through 7-3d.

At 10:30 A.M., dispatchers at the Fremont and Madison county sheriffs' offices were notified that the dam was failing. An estimated 300 million m^3 of water (80 billion gal) headed down the Upper Snake River Valley. The towns in its path included Wilford, Sugar City, Rexburg, and Roberts. More than 200 families were left homeless. The final toll was 14 killed directly or indirectly and an estimated $400 million to $1 billion in property damage.

Investigating Panel

After the failure, the governor of Idaho and the U.S. Secretary of the Interior selected an independent panel to review the cause of the failure.

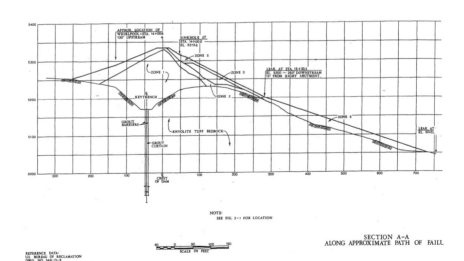

Figure 7-2. Initial failure indications.
Source: Independent Panel (1976).

SOIL MECHANICS, GEOTECHNICAL ENGINEERING, AND FOUNDATIONS   227

Figure 7-3a. Failure sequence, initial erosion.
Source: Independent Panel (1976).

Figure 7-3b. Failure sequence, seepage widens.
Source: Independent Panel (1976).

228 BEYOND FAILURE

Figure 7-3c. Failure sequence, beginning of final failure.
Source: Independent Panel (1976).

Figure 7-3d. Failure sequence, failure of dam.
Source: Independent Panel (1976).

This independent panel was made up of prominent civil and geotechnical engineers, including Wallace L. Chadwick, a former president of the ASCE, and the eminent geotechnical engineers Ralph B. Peck, H. Bolton Seed, and Arthur Casagrande. The investigators quickly developed a detailed plan, including field excavation of the failed dam down to the grout curtain and extensive laboratory testing. The panel began work almost immediately and issued its report in December 1976 (Independent Panel 1976).

Panel Investigation and Results

The panel considered all possible causes of failure and tried to establish the sequence of events leading to the failure. During the investigation, conditions favoring erosion and piping were evaluated. Levy and Salvadori (1992, p. 162) define piping as "the development of tubular leak-causing cavities." In effect, a pipe forms within the earth fill and gradually enlarges through erosion.

One of the first possible mechanisms considered was excessive settlement of the structure under the weight of the structure and the water, which would have led to cracking. It was determined that this did not contribute to the failure because the tunnel below the spillway would also have been cracked by the settlement. Furthermore, earthen dams are relatively flexible and tolerant of differential settlements. The failure hypotheses eliminated included seismic activity, reservoir leakage, and seepage around the end of the grout curtain, as well as differential settlement.

Conditions favorable for erosion and piping existed in Zone 1 (see Fig. 7-1), where the primary materials were highly erodible silts. Wherever this material was in contact with flowing water, it could be attacked and washed away. This contact could have occurred in three possible ways. First, seepage through the material could have caused backward erosion. This was determined not to play a major role in the failure because this process occurs slowly. Second, erosion by direct contact could have occurred where water was in contact with open joints, and third, where there was direct contact through cracks in the fill itself. It was determined that these last two were possible and were probably occurring simultaneously (Independent Panel 1976).

The key trench contained a grout cap, overlying a grout curtain that was intended to stop the flow. The investigation found openings and windows in the grout curtain near the failure section. The review panel also found that the construction of the grout curtain differed from the original design. The intended grouting procedure was to first grout the row of holes downstream, then grout the row of holes upstream, and then grout the center row of holes. This procedure was not followed during construction, and the closure between the two outer rows, the center row of grout, was left out. Also, the spacing between the holes was not as specified, and gaps were likely to be present (Independent Panel 1976). There was no way to determine if the spacing had an effect on the erosion. Another effect on the erosion was that the topography near the key trench showed that the foundation was probably poorly compacted, which meant more rapid erosion could occur (Arthur 1977).

An additional possible cause of failure that was investigated was hydraulic fracturing near the leaks in the dam. Hydraulic fracturing causes cracking when the sum of the normal and tensile stresses exceeds the pore water pressure. It was determined that because of the cracks that had already existed, the pressure beneath the key trench was less than full reservoir pressure, and it was not possible to build up enough pressure for hydraulic fracturing. In other words, because of the fact that the grout curtain was not fully effective, the ultimate failure was probably not caused by hydraulic fracturing. Hydraulic fracturing, though, may have been a factor in the initial breaching of the key trench fill (Independent Panel 1976).

The failure is best explained overall in the report to the U.S. Department of the Interior as, "... caused not because some unforeseeable fatal combination existed, but because (1) the many combinations of unfavorable circumstances inherent in the situation were not visualized, and because (2) adequate defenses against these circumstances were not included in the design" (Independent Panel 1976, pp. 12–18).

The panel summarized its conclusions (Independent Panel 1976, pp. vii–ix):

1. The predesign site and geological studies were "appropriate and extensive."
2. The design followed well-established USBR practices but without sufficient attention to the varied and unusual geological conditions of the site.
3. The volcanic rocks of the site are "highly permeable and moderately to intensely jointed."
4. The fill soils used, "wind-deposited nonplastic to slightly plastic clayey silts," are highly erodible. The soil classification was ML, low plasticity silt.
5. The construction was carried out properly and conformed to the design, except for scheduling.
6. The rapid rate of filling of the dam did not contribute to the failure. If the dam had been filled more slowly, "a similar failure would have occurred at some later date."

7. Considerable effort was used to construct a grout curtain of high quality, but the rock under the grout cap was not adequately sealed. The curtain was nevertheless subject to piping; "too much was expected of the grout curtain, and . . . the design should have provided measures to render the inevitable leakage harmless."
8. The dam's geometry caused arching that reduced stresses in some areas and increased them in others and "favored the development of cracks that would open channels through the erodible fill."
9. Finite element calculations suggested that hydraulic fracturing was possible.
10. There was no evidence of differential foundation settlement contributing to the failure.
11. Seismicity was not a factor.
12. There were not enough instruments in the dam to provide adequate information about changing conditions of the embankment and abutments.
13. The panel had quickly identified piping as the most probable cause of the failure, then focused its efforts on determining how the piping started. Two mechanisms were possible. The first was the flow of water under highly erodible and unprotected fill through joints in unsealed rock beneath the grout cap and thus development of an erosion tunnel. The second was "cracking caused by differential strains or hydraulic fracturing of the core material." The panel was unable to determine whether one or the other mechanism occurred, or a combination of the two.
14. "The fundamental cause of failure may be regarded as a combination of geological factors and design decisions that, taken together, permitted the failure to develop."

The panel further described the geological factors as numerous open joints in the abutment joints and scarcity of better fill material than the highly erodible soil. The design decisions resulted in "complete dependence for seepage control on a combination of deep key trenches filled with windblown soils and a grout curtain," a geometry of the key trench that "encouraged arching, cracking, and hydraulic fracturing" in the backfill, using compaction of the fill as the only protection against piping and erosion, and failure to provide for the inevitable seepage (Independent Panel 1976, pp. ix–x).

Another factor was the poor compaction of the aeolian (wind-transported) silt fill material. It was compacted at less than the optimum moisture content. The "material, as compacted in the dam, permitted continuous erosion channels (pipes) to be formed in the core without any evidence of their existence becoming visible" (Independent Panel 1976, pp. 7–14).

Technical Aspects

The failure of the Teton Dam could have been avoided. Early investigations into the geology of the site showed that the rocks in the area were almost completely of volcanic origin. These volcanic rocks consisted of basalt and rhyolite. In the footnotes to the geological survey of January 1971, the rhyolite is defined as "lightly to locally highly fractured and jointed, relatively light weight" (Independent Panel 1976, pp. 4–7). This state was also the condition for other possible sites located upstream of the site where the dam was constructed. These materials are usually avoided because of a history of erosion and deposition. The reason for the extensive foundation was the poor quality of the underlying material, including the grout curtain. The grout curtain failed to do its job of preventing these materials from being washed away.

The panel noted that the design did not provide for downstream defense against cracking or leakage and did not ensure sealing of the upper part of the rock under the grout cap. The grout curtain was not constructed in three rows, and the reliance on a single curtain was judged to be "unduly optimistic." The dam and foundation were not instrumented sufficiently to warn of changing conditions.

Professional and Procedural Aspects

At the first sign of a problem, the people at the dam site informed the Bureau of Reclamation. The bureau did not immediately inform the public because they feared panic and because there were initially no signs of imminent danger, but the public was warned about 45 min before the collapse (Arthur 1977). It was determined that the people involved acted responsibly. However, the failure of the Teton Dam brought about changes in dam construction and operation by the federal Bureau of Reclamation to ensure safety.

On the 25th anniversary of the disaster, Ken Pedde, the acting regional director of the Pacific Northwest Region of the USBR, reviewed the lessons that had been learned. Some of the changes in the process of dam design and construction include peer review of dams, special treatment for fractured rock foundations, and frequent site visits during construction by the design engineer. Also, redundant measures were implemented to control seepage and prevent piping (USBR 2001).

This failure also made federal agencies review their dam safety activities. Congress passed several acts that authorized a national dam safety program. These reviews and programs brought about annual dam inspections and the installation of instruments to monitor dams. Also, the Reclamation Safety of Dams Act of 1978 provided funds to analyze and modify existing structures that were determined to be potentially unsafe (USBR 2001).

Educational Aspects

This case demonstrates the importance of engineering geology and geotechnical engineering for civil engineering students. Engineering geology is important for evaluation of the suitability of foundation and borrow or fill materials. In this instance, both the rhyolite under the dam and the fine aeolian silt used as a fill material were deficient. Ironically, better borrow material was available in the valley downstream from the dam site, but the USBR decided that using it would be environmentally disruptive. It is also important to compact fill materials to maximum density at or near the optimum moisture content. In this case, the material was too dry and was not sufficiently compacted (Independent Panel 1976).

The static hydraulic pressure may also be determined using the height of the dam of 93 m (305 ft) and the height of the water of 86.1 m (282.4 ft). The pressure increased continually as the dam was filled. The static pressure at the bottom of the reservoir was 844 kPa (122 lb/in.^{2}). Also, this case can be used to review aspects of professional practice, such as the responsibility of reviewing designs and using redundant measures to ensure the safety of the public.

From the failure of Teton Dam, many lessons were learned. One of these lessons is that a dam site must have solid foundation material. A grout curtain may be an effective way of dealing with this unsuitable material, but only if there is a way to check if the grout curtain forms a solid barrier once it is in place. Also, this failure showed that dams must be designed so that pressure can be decreased if necessary. This option was especially important in this case because the dam was allowed to begin filling while other parts were still under construction.

Summary and Conclusions

The case may be best summarized in the words of the panel report,

> The Panel concludes (1) that the dam failed by internal erosion (piping) of the core of the dam deep in the right foundation key trench, with the eroded soil particles finding exits through the channels in and along the interface of the dam with the highly pervious abutment rock and talus, to point at the right groin of the dam, (2) that the exit avenues

were destroyed and removed by the outrush of reservoir water, (3) that openings existed through inadequately sealed rock joints, and may have developed through cracks in the core zone of the key trench, (4) that, once started, piping progressed rapidly through the main body of the dam and quickly led to complete failure, (5) that the design of the dam did not adequately take into account the foundation conditions and the characteristics of the soil used for filling the key trench, and (6) that construction activities conformed to the actual design in all significant aspects except scheduling. (Independent Panel 1976, pp. iii–iv)

In the design and construction of earthen dams, it is necessary to select proper materials that are sufficiently resistant to piping and to ensure that they are compacted to the proper density. If a grout curtain is used, it is necessary to ensure that it is continuous and forms a seal with the underlying rock. The design should incorporate adequate defense against cracking and leakage. Finally, dams must have sufficient instrumentation to provide early warning of piping and impending failure.

Essential Reading

The most valuable single reference on this case study is the independent panel's 1976 report, "Report to the U.S. Department of the Interior and State of Idaho on Failure of Teton Dam." The case study has also been published by Solava and Delatte (2003).

Vaiont Dam Reservoir Slope Stability Failure

A slope stability failure is more commonly known as a landslide, particularly among nonengineers. This type of failure occurs when the weight of a soil mass overcomes the soil's shear resistance along a failure plane. Water within soil increases its unit weight while reducing the shear strength. As a result, water and water pressures often play a role in triggering a slope stability failure.

The Vaiont Dam disaster of 1963 was a classic slope stability failure. This is also called the Vajont Dam in some references; in Italian they are pronounced the same. Ironically, the dam itself did not fail and still stands today. The dam is a thin concrete wedge in a narrow gorge. A vast soil mass falling into the reservoir triggered a massive wave that blew over the dam and destroyed villages downstream.

Design, Construction, and Operation

The Vaiont Dam was part of an extensive system of dams, reservoirs, and hydroelectric powerhouses located in the Piave River Valley, high in the Italian Alps. The elements of this system were linked by tunnels and pipelines (Ross 1984, p. 131). The Piave River Valley is roughly 100 km (60 mi) due north of Venice, near the Austrian border.

The thin arch dam was 262 m (858 ft) high, but the arc around the crest was only 190 m (625 ft), and the chord was 169 m (555 ft). The dam was a double curvature type, 22 m (73 ft) thick at the base but only 3.4 m (11 ft) thick at the crest. It was the highest arch dam in the world at the time of construction, exceeded only by the 284-m (932-ft) Grand Dixence concrete gravity dam in the Swiss Alps (Ross 1984, pp. 132–134). The full reservoir was intended to contain a volume of 169 million m^3 (6 billion ft^3 or about 138,000 acre-ft) of water (Genevois and Ghirotti 2005).

The designer of the dam, Carlo Semenza, had reservations about the dam site as early as June 1957, when the dam owner proposed increasing the dam height by 30% to triple storage and power generation capacity. The owner was the private power company SADE, for Società Adriatica di Elettricità (Adriatic Electric Society). A March 22, 1959, landslide at the nearby Pontesei Reservoir of 3 million m^3 (106 million ft^3) of rock had killed one person (Wearne 2000, pp. 206–207).

The Pontesei landslide started slowly, but then began moving very rapidly. The flow over the Pontesei Dam was only a few meters deep and caused no damage to the valley below. The Vaiont Dam was already at an advanced state of construction, so this landslide caused some concern. Leopold Müller was responsible for studying the stability of the Vaiont Reservoir, in light of the concerns (Semenza and Ghirotti 2000).

Müller asked Carlo's son Edoardo, a recently graduated geologist, to investigate the site. Carlo identified an "uncemented mylonitic zone" running about 1½ km (1 mi) along the gorge, as well as the site of an ancient landslide. He identified layers typical of river sedimentation down to about 30 m (100 ft), showing where the ancient landslide had buried a riverbed. Edoardo took borings as deep as 171 m (561 ft). Edoardo was young, however, and the utility was reluctant to take his opinions seriously. They sought another opinion from a Professor Calois, who claimed that there were only 10–20 m (33–66 ft) of loose slide material over firm in situ rock (Wearne 2000, pp. 208–209).

Clearly, the issue was whether the reservoir was the site of an ancient landslide. If it was, the landslide could move again under the right conditions. The size of the landslide and its speed of movement would determine the extent of the damage that the slide would cause. Edoardo Semenza in fact identified a number of ancient landslides, but he considered only one to be potentially dangerous. The ancient landslide had pushed uphill on the other side of the valley, and the river channel had subsequently cut off and isolated a hill from the original slide. Edoardo Semenza cited this hill as proof of the ancient slide (Semenza and Ghirotti 2000).

Edoardo Semenza identified the following features:

- the 1.5-km (1-mi) zone of uncemented cataclasites along the base of the left wall of the valley, along with solution cavities, sinkholes, and springs;
- ancient landslide masses that had filled the valley and then had been cut into two by the new Vaiont stream;
- the southern slope of Mt. Toc, which had a "chair like" structure of bedding planes, dipping steeply at the top and more shallowly near the base; and
- a fault separating the in situ rock mass from the ancient landslide (Genevois and Ghirotti 2005).

Genevois and Ghirotti state that the dam's designers concluded that a landslide was not likely to occur, "mainly because of both the asymmetric form of the syncline ... and the good quality of in situ rock masses" (2005, p. 41). In other words, because the form of the landslide was broken up and difficult to make out, it was thought that it would probably not move as a mass again.

In 1960, SADE began slowly filling the reservoir and monitoring earth movements. Elevations of 594 m (1,950 ft) and 650 m (2,133 ft) were reached. Throughout September, the movement rate increased from 5 to 10 mm/day (0.2 to 0.4 in./day), reaching 20–40 mm/day (¾–1½ in./day) in early October (Genevois and Ghirotti 2005).

On November 4, 1960, 750,000 m^3 (27 million ft^3) of rock fell into the Vaiont reservoir after a week of heavy rain. The landslide caused a 2-m (7-ft) wave in the reservoir, but no one was injured. Creep of the soil mass was observed over a large area. The recommendation was to lower the reservoir to slow the slide and to add drainage tunnels under the slide mass to reduce water pressures (Wearne 2000, pp. 209–210).

Müller had been asked to study the problem and propose remedial measures. Measures such as draining the mountainside, removing millions of cubic meters of soil, cementing the sliding mass along the failure surface, and buttressing the foot of the slide were considered but rejected as impractical. Müller believed that it would not be possible to completely stop the slide but that its movements could be monitored and controlled (Semenza and Ghirotti 2000).

SADE implemented the plan. A 1.6-km (1-mi) bypass tunnel 5 m (16 ft) wide was built. The purpose of this tunnel was to ensure water flow to the dam even if part of the reservoir were blocked by a small landslide. A grid of sensors was installed to monitor earth movement. Drill holes more than 90 m (300 ft) deep explored the mountainside, and two drainage tunnels were provided. The water level was dropped to 600 m (1,968 ft), and creep slowed. At about this time, Carlo Semenza died, and his voice of caution was lost. As the reservoir was gradually raised again, a pattern seemed to be established—higher reservoir level, more movement. Therefore, it seemed that any future movement could be handled safely by lowering the reservoir (Wearne 2000, p. 211).

Müller observed that when the movement exceeded 15 mm/day (0.6 in./day) on the second filling, the lake level was 100 m (330 ft) higher than it had been when that amount of movement had occurred on first filling. He therefore hypothesized that the initial saturation of the materials was causing the movement and that it was safe to raise the lake level (Semenza and Ghirotti 2000).

In March 1963, SADE was nationalized and absorbed into the national power grid. The new owners, the Italian electric monopoly ENEL (for Ente Nazionale per L'Energia Elettrica), emphasized increasing electrical power production. It was assumed that the reservoir was safe up to 700 m (2,268 ft) and dangerous at 715 m (2,346 ft). However, by July 1963, the water in the reservoir was muddy, and people reported noises from within the mountain. Movements of as much as 570–700 mm (22½–27½ in.) per day were measured. Rain increased the level of the reservoir to 720 m (2,361 ft). Obviously, this was of concern. Movements were approximately 200 mm (8 in.) per day. The new owner began lowering the reservoir approximately 1 m (3 ft) per day (Wearne 2000, pp. 212–213). It is possible that the transfer of ownership delayed the decision to lower the reservoir (Semenza and Ghirotti 2000).

Failure

On October 9, 1963, at 10:41 P.M., approximately 270 million m^3 (9,535 million ft^3) of rock fell into the reservoir, moving as fast as 25 m/s (82 ft/s). A tremendous wave of water blew over the dam, virtually the entire reservoir, sending a 70-m (230-ft) wall of water down Vaiont gorge. It destroyed the town of Longarone downstream, with a population of 4,600, and severely damaged or destroyed the hamlets and villages of Villanova, Codissago, Pirago, and Fraseyn. There were 2,043 people killed, including 58 of the utility's employees (Wearne 2000, pp. 213–214). The flood also knocked out many access routes, hampering rescue operations (Ross 1984, p. 132).

The slide moved a 250-m (820-ft) thick mass of rock about 300–400 m (980–1,300 ft) horizontally. It pushed the old slide mass up the far slope. Trees and soil along the Vaiont Valley were removed as high as 235 m (770 ft) above the reservoir level (Hendron and Patton 1985, p. 8)

The dam, however, stood. It had withstood a force of approximately 4 million metric tons (4 million tons) of water, roughly eight times the force for which it had been designed. The dam is still there, but there is no water behind it (Wearne 2000, pp. 217–219). There was a small gouge in the concrete about 1.5×9 m (5×30 ft) along the crest near the left abutment (Ross 1984, p. 132).

The volume of the slide was slightly larger than the working volume of the Vaiont Reservoir. It was more than twice the volume of the largest earthen dam ever built, the Fort Peck Dam on the Missouri River in Montana. Because of the great volume, it was not practical to remove the material from the reservoir to restore the dam's function (Ross 1984, p. 134).

Investigations and Repercussions

About a week after the disaster, *Engineering News Record* (ENR) reported that the new owner, ENEL, had anticipated a slide but only expected about 19 million m^3 (670 million ft^3). They claimed that it was impossible to foresee that the slide would be so large. The engineer in charge of the dam had reportedly phoned the electric company asking for permission to evacuate the entire area on account of the earth movements. *ENR* noted that the electric company was said to have told him "to stay calm and sleep with his eyes open" (Ross 1984, pp. 132–133).

Investigations started almost immediately. An Italian government committee of inquiry with four members was charged to determine:

- whether the hydrogeological examination of the dam area was given proper consideration in planning and construction, and whether the previous landslides in the area were taken seriously,
- whether the dam's testing was still continuing at the time,
- the level of the reservoir in the 10 days before the disaster, and whether safety recommendations for the level were followed,
- whether a previous landslide in the area a few days before the disaster should have warranted an evacuation order downstream, and
- whether officials acted properly (Ross 1984, p. 133).

The commission's report was released four months after the failure. It blamed "bureaucratic inefficiency, muddled withholding of alarming information, and buck-passing among top officials." The prime minister of Italy suspended a number of public officials, including the province chiefs and civil engineers of Belluno and Udine (Ross 1984, pp. 134–135).

More than four years after the disaster, the public prosecutor of the province of Belluno charged 11 men with crimes ranging from manslaughter to negligence. By that time, two had already died. A third committed suicide the day before the trial was to begin. The charges included ignoring consultants' cautions and failing to fully investigate the earlier earth movements in the slide area. The prosecutor asserted that each of the men charged could have controlled the situation and prevented the disaster. The charges for each individual were outlined in detail in the December 7, 1967, issue of *Engineering News Record* (Ross 1984, pp. 135–137).

Three of the 11 charged were found guilty, and two served short jail terms. The proceedings, however, focused more on assigning blame than on determining the technical cause of the failure. Technical papers were written, but the conclusions did not agree (Wearne 2000, pp. 217–218).

Engineering Analyses

Although a landslide had been feared, the size and velocity were surprising. The mass and velocity pounded considerable kinetic energy into the reservoir, which was the main cause of the height of the wave in the lake and the extent of the destruction downstream of the dam. Müller's hypothesis that the movements only took place when the material was first saturated was obviously disproved. The phenomena involved proved to be complex.

The following causes have been suggested for the slide's triggering mechanism:

- the creation of the lake basin, as well as the variations in the level of the reservoir;
- the clay seam along the failure surface;
- the ancient landslide;
- the geological structure;
- seismic action; and
- a confined aquifer behind and below the failure surface (Semenza and Ghirotti 2000).

Seismic action has generally been ruled out, but the other factors all seem to have been involved in the landslide.

Müller wrote several papers over the five years after the landslide, contending that once a certain limit velocity had been achieved, some type of thixotropy must have occurred. Perhaps water in the joint had made the mass of the slide buoyant. Müller continued to favor the hypothesis of a first-time slide and argued that the slide that took place had not been predictable (Genevois and Ghirotti 2005). *Thixotropy* is shear-thinning material behavior, a decrease in shear flow resistance once a certain strain rate is achieved. The phenomenon is seen when ketchup suddenly flows quickly from a bottle, drenching a plate.

The lack of agreement among the studies of the Vaiont slide was a cause of concern for the engineering geology community.

Despite the substantial literature on the topic, no study had previously taken account of:

- the three-dimensional shape of the slide surface,
- actual laboratory shear strengths of material from the site, and
- piezometric levels taking into account rainfall and reservoir levels (Hendron and Patton 1985, p. 2).

Frank Patton, a consulting engineering geologist, and his colleagues began investigating the slide in 1975 and visited the failure plane. They found a layer of plastic, low-strength clay (also known as fat clay) at the base of the slide, between 13 and 100 mm (½ and 4 in.) thick. The clay would have reduced friction along the failure plane and also acted as a membrane to hold back water (Wearne 2000, pp. 218–220).

Surprisingly, many of the earlier investigations had reported that there was no clay layer. The Hendron and Patton (1985) report contains dozens of photographs of the soft clay layer, along with a map providing the locations. The clay showed up everywhere they looked along the failure plane. They noted that where the clay was exposed, it was rapidly eroded by rainfall. Also, for the first few years the clay had been covered up by slide debris that was gone by the time Hendron and Patton surveyed the hillside. There was some confusion because of the different terms used in technical writings for the soft clay layer.

> The basic structures affecting the slide are: (a) the steep back of the slide which provided the driving forces, (b) the pronounced eastward dip of the seat of the slide, (c) the continuous layers of very weak clays within the bedded rocks, and (d) the faults along the eastern boundary of the slide. (Hendron and Patton 1985, p. 21)

It had been observed that movements increased as the reservoir level increased, but the reservoir level also increased when it rained. Therefore, two possible causes for the increased movement were higher water pressures caused by higher water levels within the reservoir, or increased pressures within the mountain from rainfall against the fat clay. To determine the cause, it would be necessary to know water pressures behind the clay before the failure (Wearne 2000, pp. 220–221).

Hendron and Patton (1985, pp. 51–58) plotted the correlations among reservoir level, precipitation, and rate of movement. They discussed Müller's contention that the movement was greatest on first wetting, pointing out, "The erroneous assumption which led to the conclusions . . . was that all other factors were remaining constant and the reservoir level was the main variable controlling the stability of the slide. In fact, rainfall was significant and was not remaining constant" (Hendron and Patton 1985, p. 54).

Periods of high rainfall preceded all of the major slide movements, but obviously the reservoir level often also rose at the same time. At those times when the reservoir was at the same level, however, the difference in movement correlated with the amount of rainfall. It proved possible to plot a failure envelope of combinations of reservoir elevation and precipitation, with points plotted above the envelope indicating instability of the soil mass.

Patton reviewed Edoardo Semenza's reports, finding that Edoardo had reported this clay layer as mylonite. Edoardo Semenza had made four boreholes, one of which penetrated to the base of the slide. This borehole found water pressures equivalent to 70–90 m (230–295 ft) above the reservoir. Unfortunately, the tunnels that had been installed to reduce water pressure were too high up on the mountain. The heavy rains the first week in October had infiltrated the mountain and increased pressures behind the clay layer. When the water level in the reservoir had been lowered just before the landslide, it had reduced the pressure holding back the slide and might in fact have triggered it. The nature of the clay might also have accelerated the disaster; its kinetic coefficient of friction was much smaller than its static coefficient of friction. Tests later showed that the shear strength decreased as much as 60% when slip exceeded 100 mm (4 in.) per minute. In fact, the fat clay acted as a lubricant (Wearne 2000, pp. 221–223). The test results that identified the loss of shear strength had been carried out by Tika and Hutchinson (Genevois and Ghirotti 2005).

Patton and co-author Hendron started from Edoardo Semenza's findings, with the following results:

- The existence of the old landslide was confirmed.

- The clay layer was up to 100 mm (4 in.) thick, with a residual friction angle between 8° and 10°.
- They found the probable existence of two aquifers, one above and one below the clay layer.

The lower, confined aquifer was fed mostly by precipitation on Mt. Toc (Semenza and Ghirotti 2000). Figure 7-4 shows a plan view of the mass that slid to the north. To observers on the opposite side of the valley, it appeared to be M-shaped.

Using aerial photographs available at the time, Hendron and Patton (1985, p. 31) identified a pattern of apparent kettles or sinkholes, along with shallow slide areas. These are shown in Fig. 7-5. These features could have been used to identify a possible landslide area for further ground-based investigation.

These sinkholes readily allowed rainwater or snowmelt to infiltrate into Mt. Toc, and to reach the base of the slide as shown in Fig. 7-6. With increasing precipitation, pressure would build behind the clay layer. During his surveys in 1959 and 1960, Edoardo Semenza had observed moist areas and springs below the eventual slide plane, which was consistent with the hydrogeology shown in Fig. 7-6 (Hendron and Patton 1985, pp. 34–35).

Soil samples were tested at laboratories in the United States, Canada, and Italy. The grain size distribution was 51% clay, 36% silt, 7% sand, and 6% gravel. Earlier studies had reported 52–70% clay. Atterberg limit tests placed the clay in two groups. The first was a CL/ML/MH (low plasticity clay/low plasticity silt/high plasticity silt) with liquid limits from 33 to 60 and plasticity indices from 9 to 27. The second was a CH (high plasticity clay) with liquid limits from 57 to 91 and plasticity indices from 30 to 61. Clay mineral analyses indicated that up to 80% of each sample was clay minerals. Overall, "such clay minerals have an expanding lattice, are associated with low shear angles, and exhibit swelling properties when stresses are reduced and water is present" (Hendron and Patton 1985, pp. 37–39).

Hendron and Patton (1985, pp. 59–65) included the reservoir level, precipitation findings, and the soil properties in a revised three-dimensional stability analysis. "The shear strength along the base of the slide was assumed to be related more to the residual shear strength of the multiple layers of clay found along the basal surface of sliding than to the higher shear strengths of the rock-to-rock contacts," which had been used in previous stability analyses (Hendron and Patton 1985, p. 59). Laboratory tests suggested a residual angle of shearing resistance ϕ_r of 5°–16°, with no cohesion. Such a low angle had puzzled previous investigators because it meant that the slide mass should have never been able to stay in place at all. The

SOIL MECHANICS, GEOTECHNICAL ENGINEERING, AND FOUNDATIONS 243

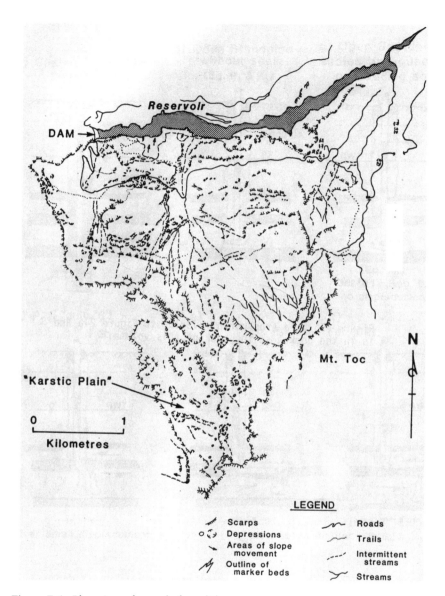

Figure 7-4. Plan view of mass before slide.
Source: Hendron and Patton (1985).

residual angle of shearing resistance ϕ_r was adjusted to a mean value of 10°–12°, based on the number of rock-to-rock contacts and the irregular surface. The angle of shearing resistance within the mass (β) was estimated at 30°–40°. This shearing resistance was mobilized when adjacent slices of

Figure 7-5. Aerial photograph of site before slide.
Source: Hendron and Patton (1985).

the mass moved relative to each other. Water pressures were also considered, with the reservoir pressure against the front of the slide and higher pressures, up to 90 m (295 ft), at the back.

Hendron and Patton (1985, pp. 66–79) noted that because the failure had occurred, it was necessary for stability analyses to demonstrate a factor of safety near 1.0 under the failure conditions. It was also necessary to demonstrate factors of safety near 1.0 at the times when significant movement was observed, as well as somewhat greater for the periods where

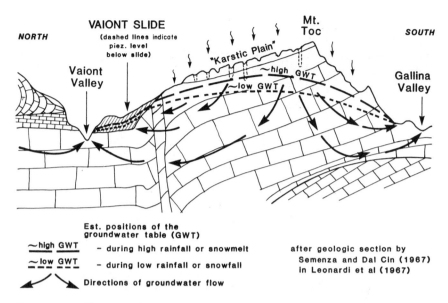

Figure 7-6. Infiltration of water through the mountain.

movements were insignificant. The periods of time when the factor of safety should be near 1.0 are the prehistoric landslide, the major movement of October 1960 when the cracks formed, and October 9, 1963. The important input parameters for the model are the soil shear strength, which is relatively constant under all conditions, and the pore pressure, which varies with different reservoir levels and rainfall conditions. The angle of shearing resistance within the mass was also important. Two- and three-dimensional slope stability analyses were performed for no reservoir and for reservoir levels of 650 and 710 m (2,130 and 2,330 ft), under low and high rainfall conditions. A two-dimensional analysis indicated that the slope would be unstable for much of its history, which could clearly not be correct. The three-dimensional analysis found factors of safety ranging from 1.21 for no reservoir with low rainfall to 1.00 for 710 m (2,330 ft) and high rainfall, the conditions under which the slide occurred. The second lowest factor of safety, 1.08, occurred at 650 m (2,130 ft) with high rainfall. The final soil parameters used were ϕ_r of 12° and β of 40° between slices and 36° along the eastern surfaces of the slide.

Once the slide began to move, of course, the problem became more complicated. Because the mass of the slide and the driving force remained the same, a considerable reduction in the resisting force was necessary to explain the velocity reached by the slide. This velocity could be due to reduced shear strength of the soil, increased pore pressure, or both. Shearing resistance could also be decreased as the toe of the slide began to ride into and over the water of the reservoir because water has no shear strength. This last factor was not found to be significant (Hendron and Patton 1985, pp. 80–82).

Hendron and Patton stated,

> The strength losses along the sliding surface which resulted in the unexpected high maximum velocity of the slide probably originated from three mechanisms: (a) a displacement induced reduction in the friction angle, β, between adjacent vertical surfaces of the sliding mass, especially at the back of the slide at the abrupt change from a steep to a flat failure plane; (b) a reduction of peak to residual shear strength along the eastern side of the slide where the sliding surface did not follow the bedding planes but sheared across the bedding; and (c) a reduction in shear strength along the basal sliding plane parallel to the bedding caused by heat-generated increases in the water pressure along this plane. (1985, p. 83)

In their analysis, they found that the combined action of all three phenomena could explain the velocity reached by the slide mass if the reduction in the clay's shear strength was about 50% (Hendron and Patton 1985, p. 90). The explanation for the high velocity of the slide was that the shear resistance of the clay layer decreased substantially once movement started. This change could have occurred due to either frictional heat or an inherent property of the material. Because it would have taken time for the heat to build up, the alternate explanation of the loss of frictional strength with shear strain rate is most likely (Semenza and Ghirotti 2000).

Lessons Learned

The Vaiont Dam case study shows the need for a thorough geotechnical investigation for a construction project, particularly one as important as a massive dam. It is necessary in particular to locate any thin clay seams that represent potential weak failure planes. Genevois and Ghirotti state,

> The catastrophic 1963 landslide failure has demonstrated to professionals and researchers in the fields of civil engineering and engineering geology the importance of performing detailed geologic investigations of the rim of narrow steep-walled valleys, which are planned as the reservoir for large dams. The failure mechanism of a large landslide mass

may be very complex and difficult to evaluate and even leading experts may fail to reach correct conclusions if they do not fully understand all factors affecting the mechanism and the evolution of the landslide. (2005, p. 43)

At the time, geotechnical investigations were much less thorough and relied on more primitive tools than those available today. The full implications of the confined aquifer were not recognized, and the speed of the slide was much more than expected. The Vaiont landslide has played an important role in the subsequent development of engineering geology (Semenza and Ghirotti 2000). In fact, at the time of the Vaiont project, reservoir slope stability analysis was usually not included in the design. The available geological studies, carried out decades before, did not identify the ancient landslide (Genevois and Ghirotti 2005).

Interestingly, the local population seemed to be aware of the unstable nature of the slope. Edoardo Semenza pointed out that the name Mt. Toc means "crazy" in the local dialect (Hendron and Patton 1985, pp. 64–65).

The low strength and lubricating properties of high plasticity fat clays are important. Another important property of clay material is impermeability: Because it is difficult for water to pass through, it is easy for high pressures to build up under or behind clay layers.

Slope stability and the factors that affect the weight of the soil mass, as well as the frictional shear resistance along the failure plane, are important. Overall, water tends to increase the soil mass and reduce the shear resistance of many soils. In hindsight, it might have been possible to reduce pore pressures, drain the slope, and reduce the risk of a landslide, but the two drains installed were too high up the slope and did not penetrate behind the clay layer.

The engineers responsible for the dam's operations also failed in their duty to inform the public about the danger of a dam failure. Although the landslide itself was sudden, there had been plenty of concern and warning beforehand. The fact that the reservoir was being lowered shows that there was an appreciation of the risk. Even if the anticipated slide were 19 million m^3 (670 million ft^3), that was more than enough to warrant an evacuation order.

Later that same year, the Baldwin Hills Reservoir near Los Angeles, California, failed. Fortunately, the recent memory of the Vaiont disaster led the engineers to order an evacuation, perhaps saving hundreds of lives. On the other hand, the Malpasset Dam failure less than four years before should have been adequate notice for the engineers operating the Vaiont Dam. The Malpasset case study is discussed in Chapter 8.

An October 24, 1963, editorial published in *Engineering News Record* made some comments on the Vaiont Dam landslide:

- The main lesson is obvious—get out of the way if a landslide threatens a reservoir. An evacuation should have been ordered at the first signs of trouble, which were roughly 10 days before the slope stability failure. The potential, and in the end actual, loss of life greatly outweighed any inconvenience from a possible false alarm.
- The engineers operating the dam were confident that the slide would only be about a tenth of what actually occurred, but they were wrong. They were also confident that the dam would hold, and it was known that arch dams had considerable reserve strength. Given the uncertainty on both elements, not evacuating the downstream population was a considerable gamble.
- This was the first time in history that a reservoir had been completely filled by a massive landslide (Ross 1984, pp. 138–139).

As lessons learned, Genevois and Ghirotti state, "By its nature, any specific landslide is essentially unpredictable, and the focus is on the recognition of landslide prone areas . . ." (2005, p. 50). Hendron and Patton (1985, pp. 96–97) cautioned that cursory studies of important events such as the Vaiont Slide could be misleading. "Previous studies of the Vaiont Slide vary from useful factual accounts to misleading fiction. . . . The most misleading accounts in the literature have generally been given by those who have not visited the site or who are not familiar with the geology."

Essential Reading

The key document on this case study is *The Vaiont Slide: A Geotechnical Analysis Based on New Geological Observations of the Failure Surface, Volume I, Main Text* (Hendron and Patton 1985). It may be difficult to find except on loan from some large university engineering libraries.

The Vaiont Dam case study is covered in pp. 206–225 of *Collapse: When Buildings Fall Down* by Wearne (2000) as well as pp. 127 and 130–139 of Ross (1984). Wearne's chapter contains accounts from engineers and survivors of the disaster. *Engineering News Record* reported on the case in the October 17, 1963, and December 7, 1967, issues and published an editorial in the October 24, 1963, issue.

A similar mechanism caused the 2005 Bluebird Canyon landslide near Los Angeles, which is featured in the History Channel's Modern Marvels *Engineering Disasters 17* videotape and DVD.

The Transcona and Fargo Grain Elevators

The Transcona Grain Elevator collapsed in 1913, and the Fargo Grain Elevator collapsed in 1955, with almost identical failure mechanisms. They were similar structures built on similar types of soil. The two cases provided important confirmations of soil-bearing capacity calculations.

The Transcona Grain Elevator

This collapse was investigated four decades after the event by Ralph B. Peck, a geotechnical engineer and professor of civil engineering at the University of Illinois, who was also a native of Transcona's neighbor Winnipeg (Morley 1996, p. 27). Construction of the elevator started in 1911. It consisted of a work house and a bin house. The work house was 21 × 29 m (70 × 96 ft) in plan, and 55 m (180 ft) high, with a raft foundation 3.7 m (12 ft) below the surface. The bin house had 13 bins, each approximately 28 m (92 ft) high and 4.3 m (14 ft) in diameter. The bins rested on a reinforced concrete raft foundation, 23.5 m (77 ft) wide and 59.5 m (195 ft) long, also at a depth of 3.7 m (12 ft) below the surrounding soil. The underlying soil was a layer of stiff blue clay, roughly 6–11 m (20–35 ft) thick (Peck and Bryant 1953).

Small plate load tests were performed before construction. These tests indicated that the soil should be able to bear a pressure of 383–479 kPa (4–5 ton/ft^2). The total pressure with the warehouse filled with grain would be no more than 316 kPa (3.3 ton/ft^2) (Morley 1996, p. 26).

The structure was finished in September 1913 and began to be filled with grain as uniformly as possible. On October 18, 1913, the bins were about 88% full, and settlement was observed, increasing within an hour to a uniform 300 mm (1 ft). Over the next 24 hours, the structure tilted to the west by almost 27°. Surprisingly, the bin house remained intact through the rotation. Wash borings were taken immediately after the failure, and they determined the thickness of the layer of clay as well as several underlying layers (Peck and Bryant 1953).

> An eyewitness account of the failure states that on October 18, 1913, the bin house began to move almost imperceptibly, but after an hour, 0.31 m [1 ft] of vertical settlement had occurred. It then began to tilt to the west, and when it stopped moving the next day, it was resting at an angle of 27° from the vertical. The east side had risen 1.52 m [5 ft] above the original ground level, and the western side was about 8.84 m

[29 ft] below. . . . Displaced soil all around the structure had risen up to at least 1.52 m [5 ft] above the ground. (Morley 1996, p. 27)

It proved possible to restore the grain elevator to service because it had subsided monolithically without cracking. The bin house was gradually righted using jacks. By October 17, 1914, the elevator was upright and back in service, although now 4.27 m (14 ft) below grade (Morley 1996, p. 27).

Two additional borings were obtained in 1951 to apply subsequently developed soil-bearing capacity theories to the failure. The soil was identified as a tan and gray slickenslided clay with an average water content of 45% and an unconfined compressive strength of 108 kPa (1.13 ton/ft²). The liquid limit and plastic limit values were about 105% and 35%, respectively, and the soil was classified as an inorganic high plasticity clay (CH). The underlying layer was a weaker gray silty clay, with a 62 kPa (0.65 ton/ft²) unconfined compressive strength. For both layers, the angle of internal friction was assumed to be 0 (Peck and Bryant 1953).

Peck and Bryant (1953, p. 205) calculated the weight of the grain and the dead load of the structure and determined a uniform load of 293 kPa (3.06 ton/ft²). Assuming a unit weight of the soil removed for the foundation at 1,900 kg/m³ (120 lb/ft³), the net unit load at the bottom of the foundation was 224 kPa (2.34 ton/ft²).

Peck and Bryant (1953, p. 206) next calculated the soil-bearing capacity. The bearing capacity q_n is a function of the soil unconfined compressive strength q_u or the shear strength s and the dimensions of the foundation:

$$q_n = \frac{1}{2} q_u N_e = s N_e \qquad (7\text{-}1)$$

The factor N_e depends on the length L, width B, and depth D of the foundation. The equation works equally well with SI and U.S. customary units because it uses ratios.

$$N_e = 5\left(1 + \frac{B}{5L}\right)\left(1 + \frac{D}{5B}\right) \qquad (7\text{-}2)$$

Using the dimensions of the Transcona Grain Elevator in meters and feet, respectively:

$$N_e = 5\left(1 + \frac{23.5}{5 \times 59.5}\right)\left(1 + \frac{3.6}{5 \times 23.5}\right) = 5\left(1 + \frac{77}{5 \times 195}\right)\left(1 + \frac{12}{5 \times 77}\right) = 5.56$$

Peck and Bryant (1953, p. 207) used a weighted average of 89 kPa (0.93 ton/ft^2) based on the properties of the two clay layers. This figure gives a bearing capacity of 246 kPa (2.57 ton/ft^2), which agrees quite well with the bearing stress at the time of failure of 224 kPa (2.34 ton/ft^2). This example indicates a factor of safety of 1.09, which is close enough to 1 to indicate failure, particularly given the uncertainty in the input parameters for the equations.

The development of soil mechanics after the Transcona failure eventually provided a basis for computing the ultimate bearing capacity of soils. It was subsequently realized, therefore, that the Transcona failure served as a "full-scale" check of the validity of such computations. (Shepherd and Frost 1995, p. 5)

Fargo Grain Elevator

The Fargo Grain Elevator collapsed 42 years after the Transcona Grain Elevator. Major filling began in April 1955, and the collapse occurred on June 12, 1955. The structure broke apart and was completely destroyed in the collapse. It collapsed in the northward direction, forming a mass of concrete rubble and grain with soil heaved up on the south side as much as 1.8 m (6 ft). Like the Transcona Grain Elevator, the Fargo structure was on part of the ancient Lake Aggassiz clay deposits that stretch across much of the northern Great Plains of North America (Nordlund and Deere 1970). Figure 7-7 shows the elevation and plan of the failed grain elevator.

The Fargo Grain Elevator was a reinforced concrete structure made up of 20 circular bins and 26 small interstitial bins. The circular bins, arranged in two rows of 10 bins each, were 5.8 m (19 ft) in diameter and 37 m (122 ft) high. The foundation was a reinforced concrete raft 16 m (52 ft) wide, 66 m (218 ft) long, and 0.7 m (2 ft 4 in.) thick. It was thus narrower but longer than the Transcona Grain Elevator, with about 75% of the capacity. The edge of the foundation was thickened, and the bottom of the raft was 1.8 m (6 ft) below grade, excluding the thickened edges. Steel sheet piling was also driven around the edges of the foundation (Nordlund and Deere 1970).

Settlement readings were taken on the raft foundation starting on May 10, 1955, a month before the collapse. Seven elevation benchmarks were established, and readings were taken once per week. There was no record of any possible settlement before the benchmarks were established. Settlement readings are shown in Table 7-1. Settlements increased dramatically over the week and a half before the collapse, reaching 222–308 mm (8.75–12.13 in.).

252 BEYOND FAILURE

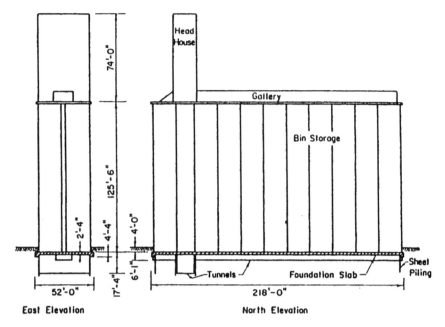

ELEVATION VIEWS OF ELEVATOR

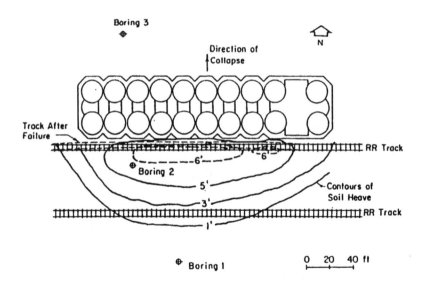

PLAN OF STRUCTURE AND SITE

Figure 7-7. Fargo Grain Elevator.
Source: Nordlund and Deere (1970).

TABLE 7-1. *Settlement Readings*

Date of observation	Settlement readings, mm (in.)						
	BM 1	BM 2	BM 3	BM 4	BM 5	BM 6	BM 7
May 10, 1955	0 (0.00)	0 (0.00)	0 (0.00)	0 (0.00)	0 (0.00)	0 (0.00)	0 (0.00)
May 18, 1955	15 (0.60)	3 (0.12)	18 (0.72)	24 (0.96)	30 (1.20)	40 (1.56)	37 (1.44)
May 25, 1955	49 (1.92)	27 (1.08)	52 (2.04)	58 (2.28)	73 (2.88)	89 (3.49)	76 (3.00)
June 1, 1955	125 (4.92)	119 (4.68)	137 (5.40)	140 (5.52)	150 (5.89)	152 (6.00)	150 (5.89)
June 8, 1955	241 (9.48)	222 (8.75)	265 (10.42)	277 (10.92)	293 (11.52)	308 (12.13)	287 (11.30)

Source: Nordlund and Deere (1971).

"It is clear that a day-to-day plot of these data would have given clear warning of the impending failure" (Nordlund and Deere 1970, p. 589).

For the investigation, Nordlund and Deere (1970, pp. 587–588) divided the elevator footprint into nine sections. Within each section, they calculated pressures ranging from 188 to 262 kPa (3.92 to 5.47 kip/ft^2) with an average of 227 kPa (4.75 kip/ft^2). There was significant eccentricity to the loading.

Three borings were made at the site and samples were removed for laboratory analysis. Penetration and torque measurements were also obtained. Four strata were labeled A, B, C, and D from the surface down.

- Stratum A was a 0.9–1.5 m (3–5 ft) thick surface layer of black to gray mottled silty clay. The average unit weight was 1,760 kg/m^3 (110 lb/ft^3), and the shear strength was estimated as 79–91 kPa (1.65–1.90 kip/ft^2).
- Stratum B was a 2.6–3.4 m (8½–11 ft) thick layer of gray and tan silty clay. Shrinkage cracks were observed in the top 1.5–2.4 m (5–8 ft) of this layer. The average unit weight of this soil was also 1,760 kg/m^3 (110 lb/ft^3), and the shear strength was estimated as 43–86 kPa (0.90–1.80 kip/ft^2).
- Stratum C consisted of 0.6–1.8 m (2–6 ft) of interbedded sand, silt, and clay. Shear strength could not be measured directly, but an average friction angle of 25° based on penetration test *N*-values of 4–12 was assumed. This estimation provides a shear strength of 30 kPa (0.63 kip/ft^2).
- Stratum D was a thick deposit of dark gray clay with occasional pebbles. The average unit weight was 1,520 kg/m^3 (95 lb/ft^3), and the shear strength was estimated as 57–81 kPa (1.20–1.70 kip/ft^2) (Nordlund and Deere 1970).

To use equations 7-1 and 7-2 to calculate bearing capacity, several assumptions should be satisfied:

- the ratio D/B should be less than 2.5;
- the shear strength should be averaged over a depth of $\tfrac{2}{3} B$ below the raft—in this case, a depth of 10.6 m (34.6 ft), which includes all four strata;
- the foundation should be loaded concentrically;
- the soil considered should be entirely cohesive; and
- the shear strength within the layers considered should not vary by more than 50% from average (Nordlund and Deere 1970).

The latter two requirements were not satisfied because layer C was essentially cohesionless and significantly weaker than the other layers. Nevertheless, a comparison is still useful. Using the dimensions of the Fargo Grain Elevator in meters and feet, respectively:

$$N_e = 5\left(1 + \frac{16}{5 \times 67}\right)\left(1 + \frac{1.8}{5 \times 16}\right) = 5\left(1 + \frac{52}{5 \times 218}\right)\left(1 + \frac{6}{5 \times 52}\right) = 5.3.$$

Nordlund and Deere (1970, p. 600) estimated an average shear strength over $\tfrac{2}{3} B$ of 37–58 kPa (0.77–1.22 kip/ft^2). Multiplying these by $N_e = 5.35$, the bearing strength becomes 197–312 kPa (4.11–6.52 kip/ft^2). The ratio between calculated and observed strength ranged from 0.80 to 1.37. These numbers are reasonably consistent with a factor of safety equal to 1, indicating a foundation loaded to failure.

It was also observed that for a two-dimensional raft foundation, the failure surface is roughly elliptical and much deeper and wider at the middle than at the ends. This result contradicts the assumption of a cylindrical failure surface used in two-dimensional analyses. The increase in settlement occurred at a load of about 153 kPa (3.2 kip/ft^2), or about 65% of the failure pressure, indicating plastic movement of part of the soil (Nordlund and Deere 1970).

The method of failure, settlement and loading records, and subsequent analysis show that the collapse of this grain elevator was a classic example of a full-scale bearing capacity failure. Even the most unsophisticated testing program and computation would have revealed that a net working pressure of [239 kPa] 5 ksf [kip/ft^2] was courting failure. A soil investigation limited to unconfined compression tests on untrimmed samples would have been adequate. Using these test results, a net failure pressure of [197 kPa] 4.11 ksf [kip/ft^2] would have been calculated. The minimum factor of safety is 1.5 so that a maximum working pressure of [131 kPa] 2.74 ksf [kip/ft^2] would have been allowed . . . a simple plot of load versus settlement for any one of the bench marks would have shown the elevator to be in imminent danger of collapse. Prompt unloading would have saved it. Why this data was not analyzed is a mystery. (Nordlund and Deere 1970, p. 605)

Lessons Learned

The two grain elevator failures occurred 42 years apart, and the landmark papers analyzing the failures were written almost two decades apart. In each case, the authors went into great detail concerning their field sampling and laboratory testing procedures, as well as the methods they developed to estimate the shear strength of the soils. Readers interested in those details should obtain and read the original references.

The key point is that engineering theories, by and large, do a pretty good job of predicting failure, provided that reasonable input data are obtained and analyzed with a critical eye. Of course, in each of these cases, the authors of these landmark papers had the advantage of knowing in advance the ultimate failure loads on the foundations. Designers do not have this advantage but can gain confidence from these two full-scale verifications of bearing capacity calculations.

Essential Reading

The key references on these cases are by Peck and Bryant (1953) on the Transcona Elevator and by Nordlund and Deere (1970) on the Fargo Elevator. Morley (1996) discusses the significance of the Transcona and other cases in the development of geotechnical and foundation engineering as a discipline.

Other Cases

Johnstown Flood

The Johnstown Flood of 1889 in Pennsylvania was another earthen dam failure, with much greater loss of life than the Teton Dam. The generally accepted total number of deaths is 2,209. This case study is reviewed in

Chapter 8. Although the original construction of the South Fork Dam was relatively sound, the later repairs of the dam, using low-quality fill materials including horse manure, made the failure inevitable.

Austin Concrete Dam Failure

The Austin, Pennsylvania, dam failure of 1911 occurred north of Johnstown and killed between 50 and 149 people. The case study is provided in Chapter 9. The dam was a concrete gravity type, but one element of the failure was a weak shear plane at the foundation, further weakened by water pressures caused by seepage under the dam.

Schoharie Creek Bridge

The 1987 Schoharie Creek Bridge failure may be considered a structural, foundation, or hydraulic (scour) failure, depending on one's point of view. It is discussed in Chapter 8, which covers fluid mechanics and hydraulics. The foundation of the bridge pier was washed away, causing the bridge superstructure spans to fall. The material under the piers was vulnerable to scour and should have been protected by riprap.

Agricultural Product Warehouse Failures

Two elements of the agricultural product warehouse failures, discussed in Chapter 4, pertain to geotechnical engineering. The pressures of cottonseed or other agricultural products are analogous to lateral earth pressure, making the warehouse walls, in effect, underdesigned walls. Also, the connection between the steel structure and the concrete foundation was flawed, leading to cracking of the foundation. The foundation was underdesigned and had insufficient reinforcement.

New Orleans Hurricane Katrina Levee Failures

The Hurricane Katrina levee failures of 2005 could have been discussed either in this chapter or in Chapter 8, Fluid Mechanics and Hydraulics. The levee failures were due, in large part, to poor selection of soil fill materials and poor foundation design. However, because the overall purpose of the system was flood control, the case is discussed in Chapter 8.

8

Fluid Mechanics and Hydraulics

FLUID MECHANICS BUILDS TO SOME DEGREE ON CONCEPTS from statics and dynamics. Fluid mechanics courses may be taught by faculty in other engineering disciplines, such as mechanical or chemical engineering. In contrast, hydraulic engineering courses are specific to civil engineering. A hydraulic engineering course may begin with a review of fluid mechanics.

Johnstown Flood

Johnstown, Pennsylvania, was a thriving community with a strong economy based on the coal and steel industries. The community was essentially wiped out by the historic Johnstown Flood of May 31, 1889, along with six other villages in the Conemaugh River Valley. The South Fork Dam, 22 km (14 mi) upstream of the town, burst and sent a wall of water approximately 12 m (40 ft) high at a speed of 32 km/h (20 mi/h) down the valley. The debris from the flood was trapped against a surviving railroad bridge below the town, and then it caught fire. About 2,200–3,000 people died in the flood and fire, and 3,500 people were left homeless. This brief introduction is summarized from Levy and Salvadori (1992, pp. 162–166).

Construction of the South Fork Dam began in 1839 and was interrupted for 10 years for lack of money. It was finally finished in 1853. The original purpose of the dam was to provide water for the Pennsylvania Canal, which runs from Philadelphia to Pittsburgh. The original design included a spillway 45 m (150 ft) wide to allow water up to 3 m (10 ft) deep to pass safely over the dam.

Railroads became increasingly more important for transportation than canals, and four years after completion the dam was sold to the Pennsylvania Railroad. The railroad had little use for it, and the dam fell into disrepair. In 1880, the dam was sold to the South Fork Hunting and Fishing Club of Pittsburgh. Members of the club included Andrew Carnegie and Henry Clay Frick.

The new owners made dangerous modifications to the dam. The outlet pipes were removed because there was no longer a need to feed the canal. The dam was lowered 0.6 m (2 ft), and a trestle bridge for a road was installed across the spillway. A screen was placed in front of the trestle to keep fish from escaping from the reservoir if water overtopped the spillway. The spillway, a critical safety fuse to prevent the breach of any dam, was reduced to about a third of its capacity.

Heavy rain began on the evening of May 30. The club's resident engineer, John G. Parke, Jr., observed the rising water in the morning and rode by horse to South Fork village and sent a warning telegram to Johnstown. He returned around noon and found water already 2.25 m (7.5 ft) above normal lake level and cutting into the dam outer face. Parke ate lunch and then came back to the dam.

By this time, the water was washing away the outer face and cutting a large hole into the dam. The dam breached, and within 45 min the lake was fully drained. The rainfall was heavy—about 127 mm (5 in.) in 34 h—but the original spillway would probably have been sufficient if the club had not made the modifications.

Construction of the Dam

The South Fork Dam was 22 m (72 ft) high and more than 275 m (900 ft) long. The dam itself was steep and covered with loose rocks, wild grass, bushes, and saplings rooting within the structure.

The reservoir behind it was about 3 km (2 mi) long and went by various names—Western Reservoir, Old Reservoir, Three Mile Dam, and finally Lake Conemaugh. The lake covered about 182 hectares (450 acres) and was up to 21 m (70 ft) deep. The dam held back approximately 20 million metric tons (20 million tons or about 16,000 acre-ft) of water. The distance to Johnstown downstream was 24 km (15 mi), with a drop in elevation of 137 m (450 ft) (McCullough 1968, pp. 39–41).

When the dam was originally conceived, canals remained an important form of transportation in New York, Pennsylvania, and Ohio. In 1836, the Pennsylvania legislature appropriated $30,000 for a reservoir to supply water to a new canal system from Johnstown to Pittsburgh during dry months. For much of the summer, there was not enough water to use the canals. Another $240,000 proved necessary to complete the dam, which was obsolete two years after it was finished. The canal had been intended to compete with the Erie Canal further to the north (McCullough 1968, pp. 50–51).

The site for the dam was surveyed and selected by the head engineer for the Canal, Sylvester Welsh. The design for the canal, originally 259 m (850 ft) long by 19 m (62 ft) tall, was developed by state engineer William E. Morris. Because of finances, the project continued off and on for 15 years. When eventually finished, it was somewhat longer and higher than the original plan (McCullough 1968, pp. 51–52). "Considerable care went into the construction of the new dam. The valley floor was cleared down to the bare rock" (Frank 2007). Then the newly completed Pennsylvania Railroad put the canal out of business. In 1857, the state of Pennsylvania sold the canal to the railroad, which could then use the right of way. The sale included the dam and reservoir, but the railroad had no use for it and simply let it sit for more than two decades. The dam broke in 1862, but the lake was only half full, and a watchman at the dam opened the sluices to release the pressure. While the reservoir was owned by the railroad, it was kept almost empty and posed little danger. Congressman John Reilly bought the dam and reservoir and held it for four years, selling the cast iron sluice pipes for scrap (McCullough 1968, pp. 53–55).

Repairs and Modifications

The South Fork Hunting and Fishing Club was organized in Pittsburgh in 1879. The club's 16 prominent Pittsburgh industrialists all purchased shares in the property. One of the later leaders of the club was industrialist Andrew Carnegie, who had been coming to these mountains for years beforehand. The club was able to purchase the lake property from Reilly for a reasonable price. The dam had fallen into disrepair and would require work. The club was chartered in Pittsburgh, in Allegheny County, but not in Cambria County, where the dam and lake were located. The illustrious, and mostly wealthy, membership included Carnegie, Henry Clay Frick, Henry Phipps, Jr., and other important members of the Carnegie empire, Robert Pitcairn and Andrew Mellon (McCullough 1968, pp. 42–49, 57–58).

Benjamin F. Ruff, the president of the club, originally proposed rebuilding the dam to only 12 m (40 ft) high and cutting the spillway 6 m (20 ft) deeper to handle any overflow. However, this change would have cost more than repairing the old break and restoring the dam to its original height (McCullough 1968, p. 55). Ruff repaired the dam in a slipshod manner, "dumping in . . . just about everything at hand" (McCullough 1968, p. 55).

Overall, the repairs cost the club about $17,000, and heavy rains washed away much of the restoration work over the next few years. Nevertheless, the construction of the club buildings and stocking the lake with fish continued (McCullough 1968, p. 56).

The newly refilled dam caused considerable worry in Johnstown. Matters were not helped by the club's secretive business dealings. There was a scare on June 10, 1881, and two men from the Cambria Iron Company inspected the dam, pronouncing it safe although the water rose to only 0.6 m (2 ft) below the crest. The danger passed, perhaps creating a false sense of security. Still, many in Johnstown lived in fear. The valley had a history of floods. Flooding was getting worse because the hills had been stripped of timber for lumber and to clear space for land to build on (McCullough 1968, pp. 63–66).

Daniel Johnson Morrell, who ran the Cambria Iron Works, was concerned about the safety of the dam. He sent his engineer John Fulton up to meet with the club's representatives and inspect it in November 1880. Fulton expressed two serious misgivings about the dam—the lack of a discharge pipe to drain the dam for repairs or in case of danger, and the poor quality of the materials used for the repair, which were already leaking. Morrell sent a letter reporting these findings to Ruff, who claimed that there was no leak and Fulton's calculations of how much water the dam held back were incorrect. Morrell offered to help pay for installing a proper discharge system, but the offer was not accepted (McCullough 1968, pp. 72–75).

Four other important changes had happened to the dam. Overall, the dam had been lowered about 0.3 to 1 m (1 to 3 ft) to provide enough width for two carriages to pass along the crest of the dam. Because the spillway was not lowered also, this meant that there was little safety margin above the spillway before the water would overtop and destroy the dam. Second, a screen was placed along the spillway under a trestle bridge to keep fish from going over. If debris became trapped against the spillway and bridge, as it almost certainly would in a flood, the spillway would be blocked. The third change was that the middle of the dam sagged slightly, by perhaps only 0.3–0.6 m (1–2 ft) at the location of the repaired break. As a result, the spillway was only 1.2 m (4 ft) or so below the middle of the crest. Finally, the club raised the level of the lake by an additional 7.6 m (25 ft) (McCullough 1968, pp. 75–77).

In total, the changes made it inevitable that the dam would fail and increased the effects of the failure. More water was now held back, the sluice pipes were gone, and the spillway would clog almost immediately as the screen caught the debris.

The Failure of the Dam

The South Fork Hunting and Fishing Club's resident engineer, John G. Parke, Jr., had actually had three years of civil engineering education at the University of Pennsylvania, which was unusual for the day. He had taken the job about three months before, and supervised a work crew of about 20 laborers (McCullough 1968, p. 20).

The storm that struck that evening was the worst downpour recorded in western Pennsylvania, with approximately 150–200 mm (6–8 in.) of rain in 24 h. Some places on the mountain recorded 250 mm (10 in.). At the dam, the rain started at 11:00 P.M., and Parke slept through the entire storm. The total rainfall at the dam was probably about 125 mm (5 in.) (McCullough 1968, pp. 21–22).

At this time, the population of Johnstown and the other communities packed tightly against it in the narrow valley was almost 30,000. Many were immigrants, in large part from Germany and Wales. The main industries were the Cambria mills, the Gautier wire works, and the woolen mills (McCullough 1968, pp. 26–28).

By morning, the rivers were rising in Johnstown at a rate of approximately 0.3 m (1 ft) per hour. Some citizens headed for higher ground. The flood in the town was approximately 0.6–3 m (2–10 ft) deep (McCullough 1968, pp. 79–82).

The first of three warning telegraph messages arrived some time between noon and 1:00 P.M. Little or no effort was made to spread the alarm. The wording of the message was along the lines of "SOUTH FORK DAM IS LIABLE TO BREAK: NOTIFY THE PEOPLE OF JOHNSTOWN TO PREPARE FOR THE WORST" (McCullough 1968, p. 87).

Two more messages arrived during the afternoon, at approximately 1:52 P.M. and about a half hour later. One read, "THE WATER IS RUNNING OVER THE BREAST OF LAKE DAM, IN CENTER AND WEST SIDE AND IS BECOMING DANGEROUS" and the other read, "THE DAM IS BECOMING DANGEROUS AND MAY POSSIBLY GO" (McCullough 1968, p. 96).

At the dam, work had continued through the morning once the danger was noted. At 6:30, Parke woke up and observed that the lake was up 0.6 m (2 ft), or halfway to the crest. He surveyed the creeks feeding into the lake, noting the large quantity of water still coming in. By the time he returned to the dam, a crew of approximately 50 men was working on the dam, trying to raise it and cut a spillway. Colonel Elias J. Unger, the new club president, was present and directing the work. An onlooker told Unger to pull out the bridge and the screen impeding the spillway, but he refused until it was too late (McCullough 1968, pp. 90–92).

By 11:00 A.M., water was starting to flow across the crest, and leaks were appearing on the outer face of the dam near the base. At this point, Parke rode to the telegraph office to send a warning to Johnstown. Unfortunately, perhaps in part because of his youth and because of other warnings that had not come true in the past, there was reluctance to pass on his warning. As the day wore on, more messages were sent on, as quoted above. In Johnstown, there was disagreement as to how seriously to take the warnings, and at any rate it was difficult to send word through the flooded streets (McCullough 1968, pp. 93–95).

Water continued to flow across the dam, and Parke went to dinner. He had thought about cutting a new spillway across the dam but lacked confidence to make such a drastic decision that would almost certainly destroy the dam. When he returned, water was cutting away the dam and had made a hole 3 m (10 ft) wide and 1.2 m (4 ft) deep. At 10 to 3:00 in the afternoon, the first break came through the dam, and 20 minutes later, the dam failed in what was later described as one big push outward. Over the next 30–45 min, the reservoir emptied at a flow rate comparable to Niagara Falls. The wall of water had started down the valley (McCullough 1968, pp. 99–102).

It would take almost an hour for the flood to reach Johnstown. The Conamaugh River Valley winds through steep mountains, slowing the flood and causing it to change direction many times. Despite the twists and turns of the valley, the speed reached about 65 km/h (40 mi/h). Some observers estimated the height of the wall of water at about 30 m (100 ft), although in reality it was closer to 12 m (40 ft). When the valley narrowed, it rose to as much as 21–23 m (70–75 ft). A stone viaduct dammed the valley temporarily before it was swept away. As the flood rolled down the valley, it rolled over itself, causing a powerful downward force at the front of the wall of water (McCullough 1968, pp. 105–112).

The rail line ran through the valley. One locomotive driver tied his whistle down so it blew continually—a recognized sign of warning at that time—and headed down the valley ahead of the flood to warn stranded trains downstream. Many were able to flee, but much of the rolling stock was swept away (McCullough 1968, pp. 115–126).

As the flood approached Johnstown, it destroyed Woodvale, killing 314 people, or approximately a third of the population (McCullough 1968, pp. 126–128).

The wall of water then smashed into Johnstown itself, where the Little Conamaugh River fed into Stony Creek, at roughly 4:07 P.M. It washed up into the hills and up Stony Creek and then washed back again. The destruction of the city took about 10 min. Much of the debris became jammed at the Pennsylvania Railroad 7-m (23-ft) high stone bridge at the far end of the town. This bridge formed a dam and kept the bulk of the water and debris from continuing on downstream. Some who were not drowned were pinned and buried in the debris pile, and perhaps 80 people later died when the pile caught fire (McCullough 1968, pp. 145–173). Today, the stone bridge is still there, and is shown in Fig. 8-1.

The first reporters to the scene wired back wild estimates of 10,000 dead, but the final generally accepted total is 2,209. The violent nature of the wall of water would make an exact accounting impossible—some victims were driven deep into the mud, and many bodies were swept downstream past the stone bridge. A third of the bodies found could not be identified. The estimated population of the valley was 23,000 at the time, meaning about 1 in 10 people had been killed (McCullough 1968, pp. 193–196).

Figure 8-1. The stone bridge.

With so much contaminated water and the bodies of people and animals, the situation was favorable for an outbreak of disease in the valley. Fortunately, this additional suffering was avoided through the cleanup efforts of the survivors and others who came into the valley to help them, the Pennsylvania state militia, and the Red Cross under Clara Barton. This work represented the Red Cross's largest peacetime effort up until that point (McCullough 1968, pp. 228–234).

Repercussions

The members of the South Fork Hunting and Fishing Club were able to skirt responsibility for the failure of the dam. The club was liable, not the members individually, and the club itself had almost no assets. The question of responsibility came down to whether the disaster had occurred because of human error or because of an unprecedented natural event, the heavy rains.

Colonel Unger and John Parke went to Johnstown three days after the flood. They were understandably eager to promote the position that the flood had been an unavoidable natural calamity. They told the newspapers how they had worked to prevent the disaster and assured the papers that everyone at the club was alive and unharmed. Parke noted that he had sent a warning, and at any rate the danger should have been obvious to anyone in the valley (McCullough 1968, pp. 213–214).

For a while, some club members even expressed doubts that their dam had failed, but once reports reached the site, it became obvious. An angry crowd of men from Johnstown went up to the dam site, but by then Unger and the club members had left. There was resentment about how the club members had left so abruptly when there was such a need for able-bodied men in the valley and also resentment as local farmers told about the poor-quality materials that had been used to repair the dam. Newspaper reporters passed on the accounts about the club, including condemnations of the maintenance and repair of the dam. Strangely, although much abuse was heaped on the club and its members in the abstract, few newspapers mentioned the names of any of the members. In Johnstown, the prevailing opinion was that human error, and not the elements, had produced the disaster (McCullough 1968, pp. 241–252).

A lawsuit was brought against the South Fork Hunting and Fishing Club. The club's members continued to stay quiet, though, particularly Frick and Carnegie. Member James Reed was a prominent Pittsburgh attorney, who led the defense. To begin with, the club had only $35,000 and a $20,000 mortgage still outstanding. Reed claimed that he could see no grounds for a lawsuit. A lawsuit was also brought against the Pennsylvania Railroad because 10 barrels of whiskey had been looted after the flood. That lawsuit was the only one that succeeded against either the club or the railroad. In the other cases, the juries agreed with the defendants' claims of a "visitation of providence." These findings ignored the fact that several small dams near Johnstown had not failed despite the magnitude of the rainfall (McCullough 1968, pp. 255–260).

In the 19th century, the difference in legal and political influence between the powerful and the humble was great. It was not surprising, therefore, that lawsuits against the most powerful men of Pittsburgh had little success in a Pittsburgh courtroom of the day.

Lessons from the Case

One important lesson is the effect of land use on design and performance of engineered facilities. Two important changes that took place in the valley were the removal of timber from the hillsides, which dramatically increased runoff, and the tripling of population after the Civil War. Similar changes often take place with any dam project—the land becomes more valuable, population increases (including downstream of the site), forests are cleared, and land is paved. These changes increase the extent of flooding, as well as the potential property damage and loss of life.

The United States has a vast dam infrastructure, much of which is poorly maintained. Critical elements to maintain are the safety valves—the pipes and spillways.

> The Johnstown disaster illustrated a basic problem with dams in the United States: large numbers were privately owned, unregulated and uninspected. Many dams are by-products of industries such as mining or agriculture, industries that are often based in remote areas. A dam at Buffalo Creek in West Virginia was one such dam. In 1972, the dam, which enclosed vast waste water from a coal mine, collapsed, killing 125 people. In 1980, the Federal Emergency Management Agency (FEMA) completed a huge survey and concluded that the majority of non-federal and privately owned dams in the United States were unsafe. They were "poorly engineered, badly constructed, and improperly maintained."
> (Wearne 2000, p. 200)

Of course, with changing land use patterns, remote areas may become much less remote over time. Changes in facility use need to be carefully considered so that the safety systems are left in place and maintained.

Runoff predictions also change over time, often adjusted upward as more data and better models become available. In the case of the South Fork Dam, significant rain had fallen before the big storm, and the ground was saturated. This antecedent moisture condition greatly increased runoff.

Engineers have an important part to play in emergency planning. Although such planning is a public policy matter, the policy should be based on sound engineering information. It is necessary to have a communications and evacuation plan to move people away from the hazard. False alarms have to be handled carefully, so as to keep the populace from losing faith in the predictions. No formal system existed in Johnstown to provide warning, and the nearly annual predictions of the dam failure—which, in hindsight, had considerable merit—served in large part to ensure that when the warning finally came it would be ignored.

On May 8, 2007, Lawrence H. Roth testified to congressional subcommittees on dam and levee safety.

> "There are now more than three thousand three hundred unsafe dams nationwide—an alarming number," Roth said. . . . According to the National Performance of Dams Program, which collects information on performance from dam owners and state and federal regulatory agencies, more than 30 dam failures have occurred in the past four years. . . . "The problem of hazardous dams is enormous," Roth went on to say. "Although catastrophic failures are rare, there were over one thousand dam safety incidents—including one hundred twenty-nine failures—between 1999 and 2006. The number of high-hazard dams—dams whose failure would cause loss of human life—is increasing dramatically, largely because of downstream development. By 2005, the number of high-hazard-potential dams totaled more than eleven thousand nationally." (Fitzgerald 2007)

Essential Reading and Other Resources

The most important book on the disaster is *The Johnstown Flood* by David McCullough (1968). McCullough includes extensive accounts of the individuals involved and was even able to interview some of the survivors decades after the disaster. Many of the survivors had left detailed personal accounts. This case is discussed by Levy and Salvadori (1992, pp. 162–166).

The Johnstown Flood is one of the few disasters of this type to be commemorated by a national memorial operation by the National Park Service. The Johnstown Flood National Memorial (http://www.nps.gov/archive/jofl/

home.htm) is located at the site of the failure of the South Fork Dam. The memorial website states, "We encourage everyone who views this web site to travel to the place where this event began, the South Fork Dam. The Johnstown Flood National Memorial preserves the remains of this dam and works with many area partners in order to tell you this story." The Johnstown Area Historical Association (http://www.jaha.org/) also maintains a Johnstown Flood Museum in Johnstown. Some photos from a visit to the South Fork Dam are shown in Figs. 8-2 and 8-3.

Malpasset Dam

The sudden and unexpected collapse of the Malpasset Dam in France led to considerable loss of life.

> The necessity of water for irrigation and drinking, joined with the wish to interrupt regular inundations by the Reyran River, are the reasons that led the French agriculture ministry to approve the building of the

Figure 8-2. The basin of the former reservoir.

Figure 8-3. Remains of the South Fork Dam.

Malpasset Dam in August 1950. Construction was started in April 1952 and was finished in December 1954. The dam was located in a narrow gorge of the Reyran River Valley, about 12 km [7.5 mi] upstream of Fréjus, a French town upstream of the Côte d'Azur area.

The dam was 66.5 m [218 ft] high with a variable thickness between 1.5 and 6.77 m [5 and 22 ft]. The upper crest was 223 m [732 ft] long. The dam formed a reservoir of 55×10^6 m^3 [14.5 billion gal or about 44,000 acre-ft].

The reservoir was filled slowly in the five years following the building, but the intense rainfall of November 1959 caused a rapid increase of the water level that reached 7 m [23 ft] below the dam crest. At that time, a water leak was observed along the right bank. From November 19 to December 2, 500 mm [20 in.] of rain fell on the basin of the Reyran River, the water level in the reservoir reached the crest of the dam, and water overflowed. Permission to open the bottom outlet gate was given only at 6 P.M. on December 2. The delay was due to the wish not to damage the highway bridge piers under construction downstream from the dam.

At 9:14 P.M. on December 2, 1959, the Malpasset Dam failed explosively, giving rise to a flooding wave more than 40 m [130 ft] high. The wave reached the Fréjus Gulf about 21 min after the dam break, and the valley morphology was drastically changed; some very large concrete blocks were carried very far from the dam. A large portion of the Esterel Freeway was destroyed. Only a little portion of the arch of the dam still remains in its original position. 421 casualties were reported.

Several inquiries were conducted in order to highlight the cause of the failure. The result was that the main causes of the disaster were the exceptional rainfall and a lack of detail in the geological survey. (Valiani et al. 2002, p. 465)

The valley and the extent of the flooding are shown in Fig. 8-4.

Design and Construction

The Malpasset Dam was a continuation of the work of the famed designer André Coyne, who had slowly been raising the allowable compressive stresses within arch dams. At Malpasset, the allowable compressive stress was increased from 2.4 MPa (355 lb/in.^{2}) to 4.7 MPa (680 lb/in.^{2}). When it was completed, Malpasset was the thinnest arch dam in the world (Ross 1984, p. 127). A plan view of the dam is shown in Fig. 8-5.

A thin curved arch structure such as the Malpasset Dam is efficient and requires much less material than a gravity dam. The structural behavior is similar to that of a half dome turned on its side. The structural integrity of arches and domes relies in large part on the strength of the supports. In the case of an arch dam, the supports are the valley walls. These walls carry the horizontal thrust from the arch. Therefore, a thorough geological investigation is necessary to locate any potential weak areas at the abutments. The right abutment of the Malpasset Dam was a rock face, and the left abutment used a wing wall to close the gap between the dam and the canyon wall (Levy and Salvadori 1992, pp. 166–172).

Failure of the Dam

The French investigation report noted,

Toward mid-November the water level stood at elevation 95.20 [m, 312 ft], the valve operator noticed certain leaks on the right bank approximately 20 m [66 ft] downstream of the dam: upon examination, these appeared to be due either to certain deep cracks of the dam or to the rising water level connected to the rains.

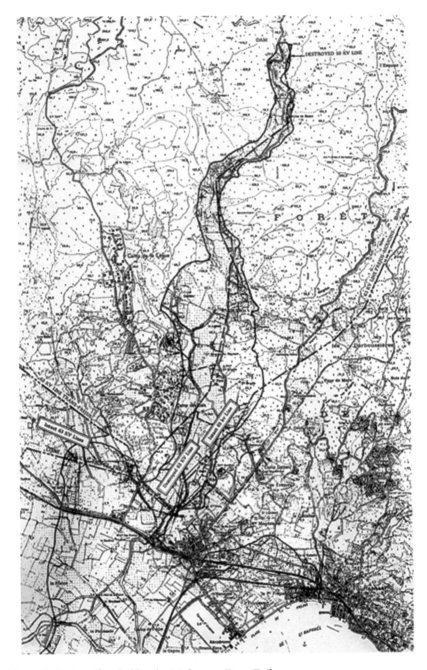

Figure 8-4. Area flooded by the Malpasset Dam Failure.
Source: Ministère de l'Agriculture (1960).

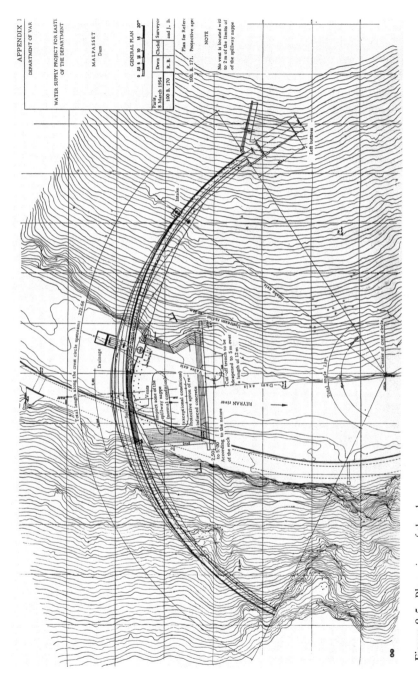

Figure 8-5. Plan view of the dam.
Source: Ministère de l'Agriculture (1960).

On the 27 November the outflows were clearly noticeable; on the 28, heavy rains fell in the region. On the 29 the water level reached elevation 95.75 [m, 314 ft]. It was clear that the seepage of water on the right bank did not proceed from the contact point of the bedrock with the dam, but was due either to deep infiltrations or to the rains. Renewed heavy rains descended on the night of the 29 to the 30 November. The reservoir filled up, elevation 96.30 [m, 316 ft] was reached on the morning of the 30 November and elevation 97 [m, 318 ft], around 6 P.M. Seepage on the right bank increased. On 1 December, rain rendered communication precarious: elevation 98 [m, 322 ft] was passed around noon of that day; at midnight elevation 99 [m, 325 ft] was reached, and at noon of 2 December elevation 100 [m, 328 ft] was reached. (Ministère de l'Agriculture 1960, pp. 5–6)

Normally, the valves would have been opened to discharge water from the dam at 98.50 m (323 ft) to reserve room for absorption of floods, with opening automatic at elevation 100 m (328 ft). However, work was in progress on a bridge below the dam, and there were concerns about safely discharging the water.

After a conference held near the dam on the afternoon of the 2 December, the order was given to the dam watchman to open the valve at 6 P.M. This operation was carried out without normal vibration, the level of the reservoir was 100.12 [m, 328 ft]. The discharge of the valve surpassed expectation; the valve operator noted a 3 cm [1.2 in.] lowering of the reservoir around 7:30 P.M. At 8:45 P.M. the dam watchman left the structure.

Around 9:10 P.M. the watchman who had returned to his home on the right river slope of the valley approximately 1,500 m [4,900 ft] downstream of the dam, heard successive crackings, a violent blast opened doors and windows, a brilliant flash appeared, and the electricity went out. The dam broke in an instant. (Ministère de l'Agriculture 1960, p. 6)

A plan view of the dam after the failure is shown in Fig. 8-6.

Investigations and Repercussions

French authorities launched an investigation, supervised by the minister of agriculture. Sabotage, earthquake, meteorites, explosions, downstream scour, and erosion were ruled out. Observers noted that the dam

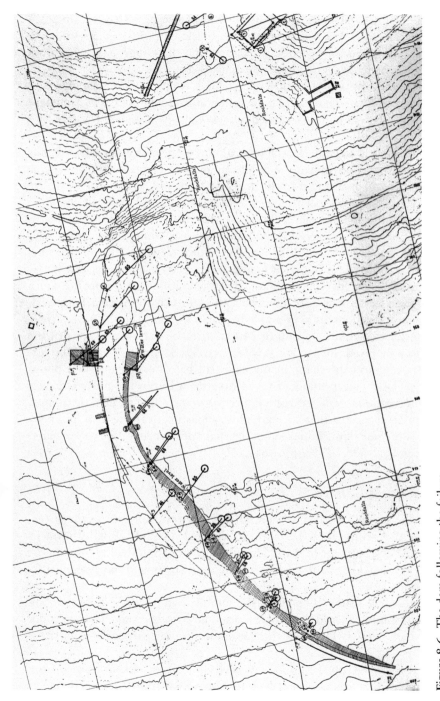

Figure 8-6. The dam following the failure.
Source: Ministère de l'Agriculture (1960).

began breaking apart near the center. This evidence suggested that an abutment had shifted. At the right abutment, blocks of concrete were found to be still bonded to the bedrock, indicating that the failure had not started there (Ministère de l'Agriculture 1960, pp. 6–11).

Overall, the quality of the concrete and the grout were adequate. The concrete and grout were found to be well bonded to the rock. The arch dam design calculations had been carried out correctly (Ministère de l'Agriculture 1960, pp. 11–24).

This result suggested a fault or weakness of the rock at the left embankment. The panel attributed the failure to an unpredictable shift of rock at that embankment (Levy and Salvadori 1992, pp. 166–172).

The 69-year-old engineer of the Malpasset Dam, André Coyne, died 6 months after the disaster. In France, unlike the United States, failures of engineering structures may lead to criminal prosecution. The engineer who accepted the dam on behalf of the Agriculture Department, Jacques Dargeou, was indicted for negligent homicide. Dargeou demanded a new investigation. The new investigation faulted Dargeou for not exploring the dam's foundation better but also implicated the late Coyne. After a long trial, Dargeou was acquitted (Levy and Salvadori 1992, pp. 166–172).

After this event, a group of 240 relatives of victims sued four of the engineers involved. Interesting new facts emerged. A month before the collapse, changes in the shape of the dam had been documented in a photo survey. The engineer who had taken over from Dargeou did not consider this information significant but had been slowly starting an investigation. A watchman suggested that strong highway blasting near the dam had been far over the safe limit. At the end of the trial, the four engineers were cleared (Levy and Salvadori 1992, pp. 166–172).

Lessons Learned

The technical cause of the failure was revealed during the investigations. A thin, clay-filled seam in the rock behind the left abutment allowed the abutment to shift. This displacement and the loss of support led to the cracking at the center of the dam (Levy and Salvadori 1992, pp. 166–172).

The clay seam was only 25–50 mm (1–2 in.) wide. The prosecutor, after the failure, alleged that this seam should have been taken into account in the dam's design. However, the clay seam was only part of the story (Ross 1984, pp. 127–129).

Ross (1984, pp. 127–130) reviewed some of the lessons learned from this failure that were discussed in an *Engineering News Record* letter to the editor published in the November 2, 1967, issue. The letter was written by P. Londe of Coyne & Bellier, the firm that had designed the dam. While acknowledging that the weak clay seam led to the failure, he stated that the problem was more complex and that such seams are often encountered in dam foundations and successfully dealt with. By this time, several years of rock mechanics research had been carried out. Londe stated that the explosive failure occurred because of the pressure of water percolating through the rock abutment. The three conditions required to trigger the failure were:

- upstream geological discontinuities,
- a downstream shear fault, and
- watertight rock under compression.

The last condition was unusually intense at the Malpasset Dam. Therefore, the dam did not merely slip on a clay seam. It was safe under the dam thrust alone, but the support was weakened by the unprecedented water pressure within the rock at the abutment.

Because the speed and depth of the flood caused by the Malpasset Dam failure were well documented, this case study has been used to develop and verify hydraulic models to improve future flood predictions (Valiani et al. 2002).

The velocity of the water after the dam break may be estimated using the Bernoulli equation between two points (1 and 2):

$$\frac{p_1}{\gamma} + \frac{v_1^2}{2g} + z_1 = \frac{p_2}{\gamma} + \frac{v_2^2}{2g} + z_2 = H \tag{8-1}$$

where p = pressure, γ = unit weight, v = velocity, g = gravity, and z = elevation above some reference point, at points 1 and 2, respectively. H is the Bernoulli constant.

For a dam break, point 1 represents the water just behind the dam as it breaks and point 2 represents the water just downstream. Both points are on the surface, so p_1 and p_2 are 0. The water velocity just before the dam break v_1 is assumed to be 0. Therefore, the velocity after the dam break v_2 may be determined from the difference in elevations $(z_1 - z_2)$ equal to the height of the dam. Therefore,

$$v_2 = \sqrt{2g(z_1 - z_2)} = \sqrt{2g(\Delta z)} \tag{8-2}$$

Note that this is the same equation found in Chapter 2 using the work-energy principle from dynamics. Therefore, for the Malpasset Dam with a height of 59 m (197 ft) and $g = 9.81$ m/s² (32.2 ft/s²), the final velocity would be

$$v_2 = \sqrt{2g(\Delta z)} = \sqrt{2(9.81)(59)} = 34 \text{ m/s } (112 \text{ ft/s})$$

Conclusions

The panel report concluded,

> The disaster had no witnesses and the description of the process of failure is thus necessarily hypothetical. However, one thing is certain: failure of the arch could be caused only by the supports giving way. In fact, an arch well designed and well built, as was the case, will not collapse unless the supports fail.
>
> How can we explain that failure? In the last analysis, these supports were all constituted by the foundation terrain. However, along the top of the left bank the arch had no direct contact with the ground except by the intermediary of a buttress. It may therefore be asked if this buttress which moved approximately 2 m [6.1 ft] was not the cause of the accident. (Ministère de l'Agriculture 1960, p. 31)

The report ends with these conclusions:

1. The disposition of the structure was correct; the calculations were exact and in conformity with standard practice.
2. Execution was very good, especially concerning the quality of the concrete and its bond with the bedrock.
3. The cause of the break must therefore be exclusively attributed to the ground below the foundations.
4. The most probable cause of the disaster should be attributed either to the presence of a slip plane or to the higher upstream fault described in the report, which for a considerable distance ran almost parallel with, and close to, the foundations of the arch in the upper part of the right bank. The already high deformability of the foundations was locally increased by the presence of this slip plane. The structure was unable to adapt itself to this increased deformability.
5. The absence of a deep grout curtain has no connection with the disaster.

As the result of these investigations, the Commission can affirm that the catastrophe of Malpasset should not diminish the confidence of engineers in dams of the arch type, the safety of which is ensured as long as the entire supporting structure is capable of permanently carrying the loads transmitted by such a structure. (Ministère de l'Agriculture 1960, p. 39)

Essential Reading

This case study is covered in pp. 166–172 of Levy and Salvadori (1992). The French Ministry of Agriculture *Final Report of the Investigating Committee of the Malpasset Dam* (Ministère de l'Agriculture 1960) was published in English translation, but it may be difficult to find.

Schoharie Creek Bridge

The Schoharie Creek Bridge in New York State collapsed on the morning of April 5, 1987, after three decades of service. The collapse of Pier 3 caused two spans to fall into the flooded creek. Five vehicles fell into the river, and 10 occupants died.

Bridges across waterways must be designed structurally not only to carry their own weight and traffic loads, but also to resist the hydraulic forces imposed by rivers and other bodies of water. Moreover, the construction of the bridge abutments and piers alters the river's flow and may lead to new patterns of erosion and deposition. The collapse of the Schoharie Creek Bridge illustrates the importance of designing bridge piers to resist scour. The case also illustrates the importance of the inspection and maintenance of bridges.

Design and Construction

The Schoharie Creek Bridge was one of several bridges constructed by the New York State Thruway Authority (NYSTA) for a 900-km (559-mi) superhighway across New York State in the early 1950s. The bridge was situated northwest of Albany in the Mohawk Valley (WJE and MRCE 1987).

The 1949 edition of the American Association of State Highway Officials (AASHO—now the American Association of State Highway and Transportation Officials, AASHTO) "Standard Specifications for Highway Bridges" (AASHO 1949) was used for the design of the Schoharie Creek Bridge. The preliminary design for the bridge was contracted out to Madigan-Hyland Consulting Engineers (WJE and MRCE 1987).

The New York State Department of Public Works (DPW), later named the New York State Department of Transportation (NYSDOT), approved the 165-m (540-ft) bridge for the crossing of the Schoharie Creek. The final design was submitted in January 1952 and consisted of five simply supported spans with nominal lengths of 30.5, 33.5, 36.6, 33.5, and 30.5 m (100, 110, 120, 110, and 100 ft). Concrete pier frames supported the bridge spans along with abutments at each end (Fig. 8-7).

The pier frames were constructed of two slightly tapered columns and tie beams. The columns were fixed within a lightly reinforced plinth, which was positioned on a shallow reinforced spread footing. The spread footing was to be protected by a layer of dry riprap. The superstructure was made up of two longitudinal main girders with transverse floor beams. The skeleton of the 200-mm (8-in.) thick bridge deck was made of steel stringers (WJE and MRCE 1987). A plan of the bridge spans is shown in Fig. 8-8.

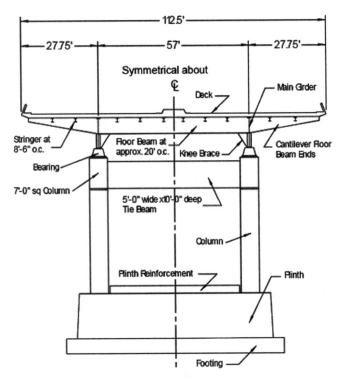

Figure 8-7. Schoharie Creek Bridge pier frames.

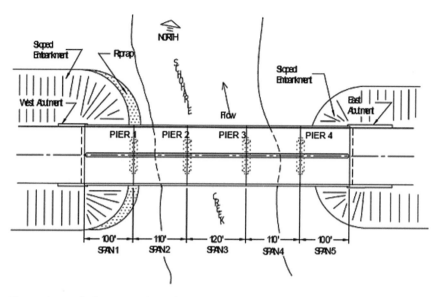

Figure 8-8. Schoharie Creek Bridge plan view.

The construction contract for the bridge was awarded to B. Perini and Sons, Inc., on February 11, 1953, and construction began shortly thereafter. Madigan-Hyland Consulting Engineers performed construction inspection for the bridge, in conjunction with DPW. The majority of the construction was completed and the bridge was opened to partial traffic during the summer of 1954. The Schoharie Creek Bridge was fully completed in October 1954 (WJE and MRCE 1987). Almost a year later, the bridge successfully survived a 100-year flood, but the damage from the flood of October 16, 1955, may have had a bearing on the collapse three decades later (NTSB 1988).

The as-built plans did not reflect the true condition of the bridge. They showed that sheet piling had been left in place to protect the piers. However, the sheet piling had been removed after construction (WJE and MRCE 1987, NTSB 1988).

Pier Modifications

Shortly after construction was completed, in the spring and summer of 1955, the Schoharie Creek Bridge pier plinths (shown in Fig. 8-7) began to form vertical cracks. The cracks ranged from 3 to 5 mm (⅛ to 3/16 in.) in width, and the locations of the cracks varied from pier to pier (WJE and MRCE 1987). The cracks occurred because of the high tensile stresses in the concrete plinth. The plinth could not resist the bending stresses between the two columns. The original designs called for reinforcement to be placed in the bottom portion of the plinth only because designers had confidence that the concrete in tension could resist the bending stresses without reinforcement.

It was later determined that the upper portion of the pier plinths had a tensile stress of 1.4 MPa (200 lb/in.^{2}) and there should have been "upwards of [39,000 mm^2] 60 square inches of steel in the upper face" of the plinth (WJE and MRCE 1987). In 1957, plinth reinforcement was added to each of the four piers to correct the problem of vertical cracking.

The plinth may be seen as an upside-down uniformly loaded beam, with the soil bearing pressure providing the uniform loading and the two columns acting as supports. It becomes obvious that the top of the plinth represents the tension face of the beam and requires reinforcement. However, to be properly anchored, the tension reinforcement must be extended past the supports—in this case, into the columns (NTSB 1988). This design is illustrated in Chapter 5's Fig. 5-15.

Obviously, this extension was not done, and it would have been difficult to extend the reinforcement through the columns without replacing the columns. Ironically, because the added plinth reinforcement was not adequately anchored, it may have contributed to the brittle and sudden nature of the subsequent collapse by supporting the plinth until most of it had been undermined (WJE and MRCE 1987, Thornton-Tomasetti 1987).

Several other problems occurred shortly after the completion of the bridge. Inspectors noticed that the expansion bearings were out of plumb, the roadway approach slabs had settled, the roadway drainage was poor, and the supporting material for the west embankment dry stone pavement was deficient. All of the problems mentioned and other minor problems were corrected by the fall of 1957 (WJE and MRCE 1987).

Collapse

The Schoharie Creek Bridge collapsed on the morning of April 5, 1987, during the spring flood (Boorstin 1987, Thornton et al. 1988). Rainfall totaling 150 mm (6 in.) combined with snowmelt to produce an estimated 50-year flood (WJE and MRCE 1987).

The collapse was initiated by the toppling of Pier 3, which caused the progressive collapse of Spans 3 and 4 into the flooded creek. The piers and spans are shown in Fig. 8-9. One car and one tractor-semitrailer were on the

Figure 8-9. Schoharie Creek Bridge collapsed spans.
Courtesy Howard F. Greenspan.

bridge when it collapsed. Before the road could be blocked off, three more cars fell into the gap. The drivers of the other vehicles were probably too close to the bridge to stop in time when it fell. Over the next three weeks, nine bodies were recovered. One of the victims was never found (NTSB 1988).

Pier 2 and Span 2 fell 90 min after Span 3 dropped, and Pier 1 and Span 1 shifted 2 h after that (Thornton-Tomasetti 1987). The National Transportation Safety Board suggested that Pier 2 collapsed because the wreckage of Pier 3 and the two spans partially blocked the river, redirecting the water and increasing the stream velocity to undermine Pier 2 (NTSB 1988). The failed plinth is shown in Fig. 8-10.

Six days later, a large section of the Mill Point Bridge located about 5 km (3 mi) upstream of the Schoharie Creek Bridge collapsed. Fortunately, the bridge had been closed since the flood because NYSDOT feared that its foundation had also been eroded (WJE and MRCE 1987).

Causes of Failure

Two teams investigated the Schoharie Creek Bridge failure—Wiss, Janney, Elstner (WJE) Associates, Inc., with Mueser Rutledge Consulting

282   BEYOND FAILURE

Figure 8-10. Schoharie Creek Bridge failed plinth.
Courtesy Howard F. Greenspan.

Engineers investigated for the NYSTA, and Thornton-Tomasetti, P.C., investigated for the New York State Disaster Preparedness Commission. The teams cooperated, and the chief role of Thornton-Tomasetti, P.C., was to review WJE's work. A number of other firms assisted in the investigation. A cofferdam was constructed around the failed piers, and the site was dewatered and excavated, to aid both the investigation and the construction of the replacement bridge (WJE and MRCE 1987).

Each of the teams prepared a report as to the cause of the failure, and they similarly concluded that the Schoharie Creek Bridge collapsed because of the extensive scour under Pier 3. *Scour* is defined as "the removal of sediment from a streambed caused by erosive action of flowing water" (Palmer and Turkiyyah 1999). The vulnerability of scouring under Pier 3 was affected by four important factors (Thornton-Tomasetti 1987):

- "The shallow footings used, bearing on soil, could be undermined." Therefore, the depth on which the footings stood was not enough to take them below the probable limit of scour.
- The foundation of Pier 3 was bearing on erodible soil. "Layers of gravel, sand and silt, interbedded with folded and tilted till," allowed high-velocity floodwaters to penetrate the "bearing stratum."

- "The as-built footing excavations and backfill could not resist scour." The area left around the footing for excavation was backfilled with erodible soil and topped off with dry riprap, "rather than being backfilled with riprap stone" to the entire depth of the excavation, as design plans specified.
- "Riprap protection, inspection and maintenance were inadequate."

The process of scouring under the piers began shortly after the bridge was built. In 1955, the bridge footings experienced floodwater flows unanticipated in the design of the bridge, a 100-year flood, and it is believed that the majority of the scouring energy was dissipated into moving the original riprap layer from around the footings. Once the backfill had been exposed, the years of peak flows removed the backfill material, and the backfill material in turn was replaced by sediment settling into the scoured holes (WJE and MRCE 1987).

The 1955 flood had an estimated flow of 2.17 million L/s (76,500 ft^3/s). The 1987 flood had an estimated flow of 1.8 million L/s (63,000 ft^3/s) and an estimated velocity of 4.6 m/s (15 ft/s). However, after the 1955 flood, berms were constructed upstream, and the velocity at the bridge may have been the same as that in the 1955 flood. Furthermore, the riprap placed at construction had probably been washed away during the 1955 flood and had not been replaced (WJE and MRCE 1987).

This process continued until so much material was removed that there was a loss of support capacity (Shepherd and Frost 1995). The upstream end of Pier 3 fell into a scour hole approximately 3 m (9 ft) deep (NTSB 1988). It was estimated that approximately 7.5–9 m (25–30 ft) along the length of the pier was undermined (WJE and MRCE 1987).

Sheet piles were used to keep water out of the excavation area during construction. The bridge design originally called for leaving the sheet piles around the piers (Levy and Salvadori 1992). The specified riprap would then fill the area left between the pier footings and the sheeting.

Unfortunately, the sheet piles were not left in place. Possibly, they could have prevented the scour altogether. Levy and Salvadori (1992, p. 146) note that "in the contract issued in 1980 for maintenance work, all reference to new stone riprap had been deleted by a nonengineer state employee who decided, after viewing the site from shore, that it was unnecessary." The NTSB also noted this incident (NTSB 1988).

The riprap was also too light. The specification called for riprap with 50% of the stones heaver than 1.3 kN (300 lb) and the remainder between 0.44 and 1.3 kN (100 and 300 lb). However, the investigators found that riprap weights of 4.4–6.7 kN (1,000–1,500 lb) should have been specified (WJE and MRCE 1987).

Although the main cause of the bridge failure was scour, several other items were considered during the investigation of the collapse. These items include the design of the superstructure, the quality of materials and construction, the inspection and maintenance of the superstructure, the inspection and maintenance of piers above the streambed, and the inspections performed using the guidelines available at the time of inspections. These items did not contribute to the collapse of the Schoharie Creek Bridge (Thornton-Tomasetti 1987).

Thornton-Tomasetti found six items that aggravated the tendency for scour (Thornton-Tomasetti 1987, pp. 3–5):

- The flood was greater than that anticipated by the designers and followed the 1955 flood and others that had disturbed the riprap.
- A curve in the river upstream of the bridge directed a higher velocity flow toward Pier 3.
- Drift material caught against the piers directed water downward at the base of Pier 3.
- Berms built in 1963 directed floodwaters under the bridge.
- An embankment west of the creek channel increased flood velocities.
- The Mohawk River Dam downstream was set for winter conditions and was 3 m (10 ft) lower than in the 1955 flood, increasing the hydraulic gradient.

Furthermore, Thornton-Tomasetti found a number of other factors that contributed to the severity of the collapse (Thornton-Tomasetti 1987, pp. 5–6):

- The bridge bearings allowed the spans to lift or slide off of the concrete piers.
- The simple spans were not redundant.
- The lightly reinforced concrete piers did not have enough ductility to permit frame action.
- The plinth reinforcement stopped the hinge action of the plinth cracks. Therefore, instead of dropping slowly into the scour hole, the plinth cracked suddenly.

The first two elements were common practice when the bridge was designed in the 1950s.

The National Transportation Safety Board conducted its own investigation and concluded that the probable cause of the accident was failure to maintain riprap. As contributing factors, they pointed to ambiguous construction plans and specifications, an inadequate NYSTA bridge inspection program, and inadequate oversight by the NYSDOT and FHWA (NTSB 1988).

Scour and Countermeasures

Scour removes material through three mechanisms (Thornton et al. 1988, Palmer and Turkiyyah 1999):

1. Long-term aggradation or degradation, the change in channel bottom elevation that occurs through normal erosion and deposition of material in the riverbed. Clearly, the degradation is of concern from a bridge safety standpoint.
2. Contraction scour when the stream width is narrowed by natural processes or bridge abutment and pier construction. By the continuity equation $Q = Av$ (where Q = flowrate, A = cross-sectional area, and v = velocity), if the channel cross-sectional area decreases, then the velocity must increase, resulting in an additional lowering of the channel bottom elevation.
3. "Local scour occurs when flow is obstructed by a pier or abutment placed in the floodplain. Vortexes that form at the pier or abutment remove stream bed material" (Huber 1991, p. 62).

Clearly, at a bridge pier, all three processes occur and are additive. Because the last two mechanisms occur only after the bridge has been constructed, the extent of potential scour may be difficult to estimate in advance.

Scour may be countered by riprap, by supporting piers on piles, by providing cofferdams around piers, and through other measures. The key is an adequate prediction of the hydraulic forces that occur during powerful floods. The analysis should also include the effects of potential land-use changes upstream, such as increases in runoff from development and associated paving.

Technical Aspects

The NTSB suggested a number of ways that the disaster could have been prevented or that the loss of life could have been reduced. The bridge, like other bridges on the Schoharie Creek, could have been supported on piles, which would have resisted scour. The AASHO 1949 provisions were unclear on whether piles were required for this bridge. In the absence of the piles, leaving the sheet piling in place and providing enough riprap would have helped protect the pier. However, the quantity estimates provided to the contractor by the design engineer did not have enough sheet piling or riprap for pier protection. The use of continuous spans, rather than simple spans, would have provided redundancy once Pier 3 failed and perhaps would have allowed for the redistribution of forces between the spans. Also, the plinth reinforcement added after the bridge construction was not anchored in the columns (NTSB 1988). The reinforcement does not cross the crack in the plinth of Pier 3.

However, the key lesson pointed out by the NTSB was operational, not technical. It is important for bridge owners to identify the critical features that can lead to the collapse of a bridge and to ensure that those critical features are inspected frequently and adequately (NTSB 1988).

The WJE Associates and Mueser Rutledge Consulting Engineers report notes that bridges must be designed for hydraulic, geotechnical, and structural effects. Of the three, only the geotechnical design, relying on the support strength of the glacial till, was satisfactory (WJE and MRCE 1987).

Bridge inspections play a major role in evaluating the superstructure and substructure for deterioration to determine if maintenance is required. Although the Schoharie Creek Bridge had been inspected annually or biennially since 1968, an underwater inspection of the pier footings had never been performed. The bridge was scheduled for an underwater inspection in 1987, but the bridge collapsed before the inspection took place (NTSB 1988).

The Thornton-Tomasetti report notes, "where riprap is used to prevent scour, inspection and restoration of protective riprap should be performed after every significant flood to avoid . . . progressive damage, and the replacement stones used should be heavier than those which were observed to shift" (Thornton-Tomasetti 1987, p. 20).

Because of the collapse of the Schoharie Creek Bridge and other bridges failing in a similar manner, bridge inspectors were further trained to recognize scour potential by examining and comparing any changes in the conditions from previous inspections (Huber 1991). Scouring failures also sparked the much-needed research for detecting scour potential.

Changes to Engineering and Management Practices

When the bridge was built, the tools of the day were not adequate for predicting scour. While the bridge was in service, the inspection procedures used were not sufficient to detect the scour. Since the collapse of the Schoharie Creek Bridge, important advances have been made. A study conducted in 1989 revealed that 494 bridges failed between 1951 and 1988 as a result of hydraulic conditions, primarily because of scouring (Huber 1991).

Summary and Conclusions

The collapse of the Schoharie Creek Bridge was an important event in the development of bridge design and inspection procedures. It is important to predict accurately the effects of scour and to design bridges to resist those effects. Lessons learned include the following:

1. proper selection of a critical storm for the design of bridges crossing water;
2. the need for regular inspections of the superstructure, substructure, and underwater features of the bridge; and
3. the importance of adequate erosion protection around piers and abutments susceptible to scour.

Similar Failure of the Hatchie Bridge

Two years after Schoharie, the 55-year-old Hatchie Bridge near Covington, Tennessee, also collapsed because of scour. A 9-m (28-ft) section of the bridge dropped 8 m (25 ft) into the water as a furniture van and several cars were crossing. The furniture van hit an additional bridge support, expanding the collapse, and the van and some cars were buried under the debris. The Hatchie Bridge had a timber pile foundation. However, over time the river had meandered, spreading to unprotected bridge piers that had originally not been in the main channel. Earlier field inspectors had recommended adding scour protection to these piers. At the time of the failure, the river was about 1 m (3 ft) above flood stage (Levy and Salvadori 1992, pp. 147–148).

Essential Reading

One important, comprehensive reference is the National Transportation Safety Board's (NTSB 1988) highway accident report "Collapse of New York Thruway (I-90) Bridge over the Schoharie Creek, near Amsterdam, New York, April 5, 1987." The case study is discussed in Levy and Salvadori (1992, pp. 143–147). This case study is featured on the History Channel's Modern Marvels *Engineering Disasters 9* videotape and DVD.

New Orleans Hurricane Katrina Levee Failures

The failure of the levees and the flooding of New Orleans during Hurricane Katrina on August 29, 2005, represent the first time in history that an engineering failure has brought about the destruction or near-destruction of a major U.S. city. The ASCE Hurricane Katrina External Review Panel stated that,

> The catastrophic failure of New Orleans's hurricane protection system represents one of the nation's worst disasters ever. . . . A storm of Hurricane Katrina's strength and intensity is expected to cause major flooding and damage. A large proportion of the destruction from Hurricane Katrina was caused not only by the storm itself, however, but also by the storm's exposure of engineering and engineering-related policy failures. (ASCE Review Panel 2007, p. v)

The Hurricane Protection System

The port of New Orleans has been vital to the national economy since before the founding of the United States. It is the gateway for imports and exports to the Mississippi Valley and the Midwest. Land throughout the area is formed from sediments carried down the Mississippi River, with substantial amounts of organic material. The city was founded in the early 1700s near the mouth of the Mississippi River on high ground, but as the city grew, adjacent low-lying marshlands were developed. The first levees were built around swampland that was then drained and developed. The area also contained bayous, many of which became canals. Over time, a system of flood walls and levees was gradually developed (ASCE Review Panel 2007, pp. 5–8).

The problem of protecting the city and the surrounding area is complicated by the fact that the underlying soft sand, silt, and clay are gradually sinking. Much of the city and its surrounding parishes is now below sea level. In the past, as the earth sank, new material was deposited by the river, but the flood control system now channels the sediments out into the Gulf of Mexico. Groundwater withdrawals and petroleum production have also increased the rate of subsidence. The rate of subsidence is approximately 4–5 mm (0.15–0.2 in.) per year overall and as much as 25 mm (1 in.) per year in some areas (ASCE Review Panel 2007, p. 8).

The first major hurricane protection project was authorized by Congress in 1946. Additional projects were authorized over the years. Congress gave the primary responsibility for the design and construction of the system to the U.S. Army Corps of Engineers (USACE). Some of the levees and pumping stations are owned and operated by other agencies. The overall USACE flood protection strategy was to build levees and flood walls around different segments of New Orleans, grouped into three main units. Based on guidance from Congress, USACE developed a "standard project hurricane" or SPH to use for design, based on storm records from 1900 to

1956. It was recognized that a hurricane stronger than the SPH was possible (ASCE Review Panel 2007, pp. 17–20).

The three types of flood protection structures designed and built by USACE were I-walls, T-walls, and levees. The levees were earthen dams made of compacted material or hydraulic fill, sloped at a rate of 1 vertical to 3 horizontal. As a result, a levee would need to be more than 6 times as wide as it was tall. Raising a levee, then, would require purchasing private property and demolishing buildings. Therefore, I-walls and T-walls were built on top of low existing levees to raise their height without widening them. I-walls were built from Z-type steel sheet piling, and T-walls were reinforced concrete structures shaped like an inverted T, supported by battered pilings. T-walls are much stronger than I-walls but also more expensive (ASCE Review Panel 2007, pp. 20–22). Figure 8-11 shows these flood wall and levee configurations.

One common error in the design and construction of the levees and flood walls was the use of an incorrect datum. The structures were intended to be built referenced to mean sea level (MSL). However, they were actually built to a land-based datum, which was incorrectly assumed to have a constant offset from MSL (ASCE Review Panel 2007, p. 22).

Because of the high annual rainfall of 1.5 m (60 in.) in the New Orleans area, pump stations are used to remove water. The levees surround a series

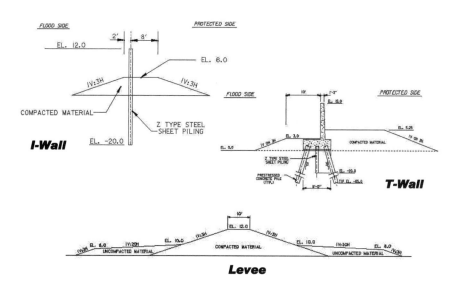

Figure 8-11. New Orleans floodwall and levee configurations.
Source: IPET (2007).

of bowls below sea level. The pumps were designed to handle storm water runoff but not water from levee and flood wall overtopping and breaches. The pumps remove water to canals and eventually to the surrounding lakes. The New Orleans system is one of the largest pumping systems in the world and consists of more than 100 pumping stations, some of which are almost a century old. The system can handle approximately 225 mm (9 in.) of water over a 24-h period. Power for the pumps is provided by diesel engines or the electrical grid with backup generators. Some of the older stations use 25 Hz power rather than the more common 60 Hz, requiring frequency changers (ASCE Review Panel 2007, pp. 22–23).

Failure of the System

Hurricane Katrina made landfall in the early morning of August 29, 2005, in southeast Louisiana to the east of New Orleans. Throughout the area, levees and flood walls failed or were breached in more than 50 locations. Eighty percent of the city of New Orleans was flooded, to a depth of more than 3 m (10 ft) in some neighborhoods. The extent of the destruction made it difficult to account for the victims, but the toll a year later was listed as 1,118 dead people and 135 missing and presumed dead. More than 400,000 citizens fled the city, many never to return. Property damage reached tens of billions of dollars (ASCE Review Panel 2007, p. 1).

Wind and storm surge are the damaging agents of a hurricane, storm surge at the coast and wind away from the coast. Storm surge is a combination of wind-induced water motion, reduced atmospheric pressure in the storm, and possibly high tide. Hurricane Katrina, unfortunately, came ashore at high tide, and the storm surge in Plaquemines Parish reached as much as 6.1 m (20 ft) above sea level. In Lake Pontchartrain, directly to the north of New Orleans, wind from the north piled water up as high as 3.7 m (12 ft) above sea level. The hurricane also brought heavy rainfall, increasing the probability of flooding (ASCE Review Panel 2007, pp. 13–16).

> The Lake Pontchartrain and Vicinity Hurricane Protection Project system experienced the worst damage during and after Hurricane Katrina and resulted in the most serious consequences to the city and people of New Orleans. The massive, destructive flooding of New Orleans was caused by ruptures at approximately 50 locations in the city's hurricane protection system. Of the [457 km] 284 miles of federal levees and floodwalls—there are approximately [563 km] 350 miles in total—[272 km] 169 miles were damaged. (ASCE Review Panel 2007, p. 25)

Failures of the system began even before Hurricane Katrina made landfall, with overtopping of the Mississippi River–Gulf Outlet levees and flooding of parts of St. Bernard Parish. Shortly after landfall, at 6:30 A.M., levees on the south side of the New Orleans East neighborhood were also overtopped and breached. Shortly thereafter, waves reached 1.2 m (4 ft) in the Industrial Canal, causing more overtopping and flooding. Four I-walls also breached, between about 5:00 and 8:00 A.M., even before the water rose high enough to overtop them (ASCE Review Panel 2007, pp. 25–27).

With all of the breaches, some neighborhoods flooded to the rooftops in minutes. Even where the flooding was slower, further from the sites of the breaches, the water rose approximately 0.3 m (1 ft) every 10 min. The deadliest breaches were in the Industrial Canal and the London Avenue Canal. These canals extended south from Lake Pontchartrain into the heart of the city, adding to the rapidity of the flooding (ASCE Review Panel 2007, pp. 28–31).

As the hurricane moved north that morning, the storm surge receded, but the damage had been done. Once the I-walls failed, the city continued to flood until the water level was equal to that of Lake Pontchartrain. By September 1, more than 80% of the city was flooded, much of it 2–3 m (6–10 ft) deep. The pump stations were no longer working, and in any case the water couldn't be pumped out until the levee breaches were repaired (ASCE Review Panel 2007, pp. 31–32).

The consequences of the failure are discussed in detail by the ASCE Review Panel (2007, pp. 33–46). In essence, the city and its economy were destroyed, and much of the population moved away permanently. A year and a half later, much of the city remained almost uninhabited and uninhabitable. The failures also, understandably, shook the public's faith in the civil engineering profession.

Investigations

After the disaster, the chief of engineers of the U.S. Army Corps of Engineers, Lieutenant General Carl A. Strock, ordered a thorough investigation. The investigation team was termed the Interagency Performance Evaluation Taskforce (IPET). He also requested that ASCE form an external review panel to assess IPET's work. Many other teams of investigators also assessed the performance of the levee system.

The IPET was asked to answer the following four questions:

1. What were the storm surges and waves generated by Hurricane Katrina, and did overtopping occur?

2. How did the flood walls, levees, and drainage canals, acting as an integrated system, perform and breach during and after Hurricane Katrina?
3. How did the pumping stations, canal gates, and road closures, acting as an integrated system, operate in preventing and evacuating the flooding due to Hurricane Katrina?
4. What was and what is the condition of the hurricane protection system before and after Hurricane Katrina and, as a result, is the New Orleans protection system more susceptible to flooding from future hurricanes and tropical storms? (ASCE Review Panel 2007, p. 2)

Seven of the major failures were breaches of the I-walls, which were devastating because these walls had been intended to protect heavy residential development. These failures included the 17th Street Canal Breach, the London Avenue Canal South Breach, the London Avenue Canal North Breach, and several breaches of the Industrial Canal (ASCE Review Panel 2007, pp. 47–58).

The 17th Street Canal Breach

The 17th Street Canal Breach occurred at about 6:30 A.M., with the failure of a 137-m (450-ft) section of I-wall. This failure flooded the Lakeview neighborhood. The water was about 1.5 m (5 ft) below the top of the wall when it failed, well below the design water level (ASCE Review Panel 2007, p. 47).

The levee and flood wall were built on a layer of organic marsh soil, on top of a layer of soft clay. The strength of the weak soil was badly overestimated by the designers. Soil data were taken from borings along the centerline over the entire 2,400-m (8,000-ft) length of the levee. The strength was highly variable, and in some sections the actual strength of about 12 kPa (0.13 ton/ft^2) was less than the design strength of about 18 kPa (0.19 ton/ft^2). Also, the soil along the centerline had been consolidated and strengthened by the weight of the levee, but the soil away from the centerline was weaker. The problem of weak soil was exacerbated by the choice of a low factor of safety of 1.3, in conflict with USACE guidance requiring a factor of safety of at least 1.4 or 1.5. The margin for error was simply too small (ASCE Review Panel 2007, pp. 47–50).

Another error in the design of the I-wall was that the effect of a water-filled gap was not taken into consideration. This error turned out to be a common feature of I-wall failures throughout the area. As the I-wall is pushed back by water, a gap opens adjacent to the wall in the levee and fills with water. Instead of a sliding failure surface that encompasses the entire base of the levee, the I-wall can now fail along a sliding surface extending up and away from the bottom of the water-filled gap. The effect of the gap is to reduce the overall sliding resistance by about 30%, effectively reducing the 1.3 factor of safety to 1.0. Full-scale tests carried out by USACE on an I-wall two decades before had shown 75 mm (3 in.) of deflection at the ground surface. As this behavior was verified by further research, however, the existing walls were not checked (ASCE Review Panel 2007, pp. 50–52).

The London Avenue Canal South Breach

Some time around 6:00 to 7:00 A.M., the London Avenue Canal South was breached. As with the 17th Street Canal, failure occurred with the water still about 1.5 m (5 ft) below the top of the wall. The failure mechanism, however, was different. In this area, the soil below the marsh was sand, not clay. Water seeping through the highly permeable sand caused upward pressures that tended to lift the levee. In essence, the levee and marsh layer floated and then broke apart as the sand was eroded by the rushing water (ASCE Review Panel 2007, pp. 52–53).

The failure of this levee, therefore, was brought about by seepage and subsurface erosion of the underlying sand layer. The importance of seepage was well known to USACE, and the designers of the levee had considered it. However, a review of the design documents after the failure did not show proper consideration of the marsh layer (ASCE Review Panel 2007, pp. 53–54).

Measures such as deeper sheet pile walls or relief wells would have reduced the uplift pressures or provided an additional line of defense against failure, but they were not used. The information available at the time of design was sufficient to develop a satisfactory system, if a more rigorous analysis had been used. A water-filled gap also probably contributed to this failure or its severity (ASCE Review Panel 2007, p. 54).

The London Avenue Canal North Breach

The breach on the west side of the London Avenue Canal North occurred around 7:00 or 8:00 A.M. This levee and its I-wall were underlain by about 4.6 m (15 ft) of marsh over a thick sand layer. The sand under the London Avenue Canal North levee was much looser and weaker than that under the London Avenue Canal South levee. The failure mode was probably similar to that of the 17th Street Canal. Underseepage through the sand layer may have also played a role in the sliding failure, by reducing sliding resistance. A water-filled gap reduced the factor of safety of this levee from about 2.0 to about 1.0 (ASCE Review Panel 2007, p. 55).

Breaches of the Industrial Canal

Several breaches of levees with and without I-walls occurred at various locations in the industrial canal. The source of the earliest flooding, at around 5:00 A.M., was probably a failure of the Industrial Canal East Bank north I-wall. In this area, the soil was marsh over soft clay over sand. This levee probably failed in much the same way as the 17th Street Canal, as a sliding failure exacerbated by a water-filled gap (ASCE Review Panel 2007, p. 56).

The Industrial Canal East Bank south and West Bank I-walls were overtopped by the hurricane floodwaters, with a peak water level about 0.5 m (1.7 ft) above the tops. As the water flowed over the I-walls, it scoured and eroded soil from the land side, undermining the sheet piles until the wall was pushed over by a lack of foundation support (ASCE Review Panel 2007, p. 56).

Levees without I-walls around New Orleans, including the Industrial Canal West Bank south, failed because of overtopping by the storm surge. Most of the 50 estimated levee breaches occurred in this manner. Once overtopping occurred, the earth fill levees simply washed away. Levees built of well-compacted clay with grass cover performed better than those made with high silt and sand content or hydraulic fill. Some of the weaker levees washed away completely (ASCE Review Panel 2007, pp. 57–59).

Breakdown of the Pumping System

With multiple levee and flood wall breaches, the city was doomed to flood until the water level inside the city rose to that of Lake Pontchartrain. Although it is difficult to envision a pumping system that could have kept up with the inflow, for various reasons the system in place turned out to be useless during and directly after Hurricane Katrina.

> The pumping stations throughout the New Orleans area could have been—but were not—an integral part of the hurricane protection system. Most hurricanes and tropical storms bring heavy rainfall, but the pumping stations were designed only to remove storm water runoff and routine seepage water from the interior drainage system and pump it into Lake Pontchartrain or other nearby bodies of water. (ASCE Review Panel 2007, p. 59)

The panel found the following reasons for the ineffectiveness of the pumping system:

- During Hurricane Katrina, total rainfall was more than 300 mm (12 in.). A few pump stations operated, but their output was only 16% of pumping capacity. The operating pump stations were unable to lower floodwaters significantly because they were not in the areas of the worst flooding.
- Many of the pump station operators evacuated areas in Jefferson and St. Bernard Parishes because the stations had not been designed to resist hurricane forces.
- Electrical power failed early in the storm, idling the many pump stations that were electrically powered.
- Many pump stations were flooded and damaged as the city flooded.
- Buildings housing some of the older pump stations could not withstand the hurricane wind and water forces.
- The pump stations discharged into the canals and waterways that had been breached, so even if they had been able to keep up with the inflow, they would have simply recirculated the water.
- Some pumps in Jefferson Parish lacked automatic backflow prevention devices. Therefore, water flowed through some idle pumps into the city (ASCE Review Panel 2007, pp. 59–60).

Contributing Factors

In addition to the direct causes of flooding discussed above, there were a number of factors that contributed to the large-scale failure of the system. The system was overwhelmed by Hurricane Katrina.

Lack of Full Appreciation of Risk

With the scale and complexity of the hurricane protection system, it was difficult to assess either the probability or the potential consequences of failure. A major dam would typically be designed for a likelihood of failure of once in 100,000 or 1 million years of operation. In contrast, the risk of a failure of the hurricane protection system resulting in 1,000 deaths was once in 40 years. Moreover, it is generally easier to release water pressure behind a dam and to evacuate people from downstream than it would be to evacuate an area enclosed by a levee. The risks of levee failure were never quantified before Hurricane Katrina and were greatly underestimated. If the true risk had been known, it is likely that an evacuation would have been ordered sooner. It is estimated that the probability of a hurricane on the scale of Katrina occurring in any given year is between 1 in 50 and 1 in 500 (ASCE Review Panel 2007, pp. 61–63).

Piecemeal Construction of Hurricane Protection System

The piecemeal construction of the system made it more vulnerable to failure. Rather than a system, it was in fact a collection of parts cobbled together, with elements built at different times and to different standards. The pump stations near the south end of the 17th Street, Orleans, and London Avenue Canals are masonry structures, almost a century old, that have never been upgraded. They could not resist the water-pressure loads from the storm surge. There were gaps for pump stations and hundreds of penetrations through the levees for roads, railways, and pipelines. The closures that were supposed to seal the gaps were often missing or inoperable. The segmented construction of the levees led to abrupt discontinuities (ASCE Review Panel 2007, pp. 63–64).

Under the attack of the storm surge, the system could only be as strong as its weakest link. There were many weak links, and many failed.

System Underdesigned

The standard project hurricane (SPH) used to design the system was much weaker than the probable maximum hurricane (PMH), as defined by the National Weather Service (NWS). The SPH was at the low end of the 162–177 km/h (101–110 mi/h) expected winds determined by the forerunner of the NWS in 1959 for New Orleans. The hurricane protection system was not evaluated for a PMH. The SPH was not updated when the NWS revised its predictions in 1979. When design criteria were updated, the previously designed and constructed elements of the system were not improved to the new standard. Hurricane Katrina's maximum wind speed in the Gulf of Mexico was roughly equal to the 259 km/h (161 mi/h) predicted for the PMH. Overall, if the system had been designed and built to withstand the PMH rather than the SPH, it would probably have performed better. Analyzing the existing system for the PMH would have identified the segments of the system that needed upgrading (ASCE Review Panel 2007, pp. 65–66).

Many Levees Not High Enough

The error in the datum used to set the height of the levees has already been mentioned. Given the known rate of subsidence, the levees should have been built higher so that the levees and flood walls would still be high enough even after sinking over time. Subsidence plus the use of an incorrect datum left many of the levees and flood walls up to 0.9 m (3 ft) lower than designed. Because the peak storm surges were only 0.3–0.9 m (1–3 ft) above the tops, many of these walls would not have been breached if they had been at the correct elevations (ASCE Review Panel 2007, pp. 66–68).

No Single Entity in Charge

The large number of agencies responsible for parts of the hurricane protection system led to "chaotic and dysfunctional" management of the system. Ten federal, state, and local agencies had overlapping and poorly defined responsibilities. There was no single agency empowered to provide system-wide oversight. Because of overlapping agency responsibilities, the floodgates at Frances Road in Orleans Parish were not closed. Separate agencies controlled design and maintenance. As a result, parts of the system were poorly maintained. Major gaps also occurred in the emergency response. Continuous operation of pumping stations might have reduced the flooding in the east bank of Jefferson Parish. In some cases, local boards successfully defeated improvements proposed by USACE. The local boards were also concerned with local development projects, such as parks and casinos, in addition to the hurricane protection system (ASCE Review Panel 2007, pp. 68–71). The system, as a whole, suffered from division of focus and responsibility.

Lack of External Peer Review of Design

Large, complicated engineering projects, particularly those that involve a substantial risk to public safety, are often subject to review by outside experts. Peer reviewers assess the overall direction of the project, the validity of assumptions, the appropriateness of analytical methods used, and the overall design approach. Before Hurricane Katrina, the USACE peer review requirements were weak and inconsistent. Peer review might have uncovered some of the faulty assumptions and systemic design errors (ASCE Review Panel 2007, p. 71). Ironically, after the failure of the system, the work has been subjected to an extraordinarily thorough peer review.

Lack of Stable Funding

The hurricane protection system funding provided to USACE was through federal appropriations, with some costs borne by local agencies. Funding was irregular at best, forcing delays to projects or reductions in scope of work. Rather than a system funded to protect the public, the scale of the construction was reduced to what could be paid for. Funds were saved, for example, by not armoring the land sides of the levees in high-risk areas. Overall, the funds provided by Congress were not adequate for the system, and USACE and the local agencies did not argue vigorously enough for adequate funding (ASCE Review Panel 2007, p. 72).

Conclusions and Recommendations

The hurricane protection system for New Orleans was and remains badly flawed. Moreover, loss of public confidence in the system has seriously hampered the reconstruction of the city. People remain reluctant to move back and invest.

According to the ASCE Review Panel, "we must place the protection of public safety, health, and welfare at the forefront of our nation's priorities" (ASCE Review Panel 2007, p. 73). The specific recommendations made by the panel, with 10 specific calls to action classified under four recommended changes in thought and approach, were the following:

- Understand risk and embrace safety
  - Keep safety at the forefront of public priorities
  - Quantify the risks
  - Communicate the risks to the public and decide how much risk is acceptable
- Re-evaluate and fix the hurricane protection system
  - Rethink the whole system, including land use in New Orleans
  - Correct the deficiencies
- Revamp the management of the hurricane protection system
  - Put someone in charge
  - Improve inter-agency coordination
- Demand engineering quality
  - Upgrade engineering design procedures
  - Bring in independent experts
  - Place safety first (ASCE Review Panel 2007, pp. 73–82).

Some specific deficiencies that need to be corrected, listed under call to action 5, "correct the deficiencies," were to establish mechanisms to incorporate changing information, to make the levees functional even if overtopped, to strengthen or upgrade the flood walls and levees, and to upgrade the pumping stations (ASCE Review Panel 2007, p. 78).

Essential Reading

An important report was published by the American Society of Civil Engineers Hurricane Katrina External Review Panel, entitled *The New*

Orleans Hurricane Protection System: What Went Wrong and Why (ASCE Review Panel 2007). The various reports prepared by the Interagency Performance Evaluation Task Force (IPET), entitled *Performance Evaluation of the New Orleans and Southeast Louisiana Hurricane Protection System*, are being published on the IPET website as they are released and revised, https://ipet.wes.army.mil/. There are a total of eight volumes. The Interim Final Volume I—Executive Summary and Overview (IPET 2007) is 147 pages long.

Other Cases

Vaiont Dam Reservoir Slope Stability Failure

The Vaiont Dam failure case study in provided in Chapter 7. The landslide that pushed a wave of water over the Vaiont Dam led to considerable flooding downstream and loss of life. The water was channeled by the valley walls, thus increasing the height of the flood. Ironically, the concrete dam itself did not fail, but the reservoir was filled by the landslide, rendering it useless.

Austin Concrete Dam Failure

The failure of a dam in Austin, Pennsylvania, in 1911 is discussed in Chapter 9. This concrete gravity dam slid and broke apart, releasing a flood that wiped out Austin and other villages.

9

Construction Materials

FOR A SUCCESSFUL PROJECT, THE DESIGN, MATERIALS, AND construction must all be satisfactory. The project may be envisioned as a three-link chain, governed by the weakest link. A sound design with quality materials may become a failure if construction is poor. Also, poor materials selection may decrease the safety or durability of a project.

Undergraduate courses in construction materials typically cover the properties and testing of common materials used in civil infrastructure. Usually, these courses have a laboratory component so that students can observe materials properties directly. The course may be primarily laboratory based, with little theory.

There is a useful distinction to be made between materials that are furnished in the final state, e.g., steel and wood, and those that are essentially manufactured on the job site, e.g., concrete, masonry, and asphalt. For materials manufactured on-site, the engineer may bear a much greater responsibility for ensuring that tests are performed and interpreted correctly. It is often necessary to accept or reject materials on the basis of limited test information.

For many students, this course may be the only exposure to wood and masonry they will have. Universities generally teach structural design of steel and concrete—and then the students graduate, and the first project they work on may be a masonry building with a wood-frame roof.

New materials, such as fast-setting repair materials, fiber-reinforced polymers (FRPs), and epoxies, are increasingly being used in construction. Advantages may include higher strength or stiffness or increased speed of construction. However, there may be long-term disadvantages to the use of the materials, and some of these disadvantages may not be known until the new materials have been in service for many years. For new materials, manufacturers may use different test protocols for key properties, such as creep and cracking resistance, making it difficult to compare the engineering characteristic properties of different products.

Newman (1986, p. 38) has observed that "it is a truism that good quality reinforced concrete is made from cement, sand, aggregates, water, and reinforcement—and so is bad quality concrete. So what has happened?" In his article, he goes on to suggest some of the causes found and remedies applied in Europe.

Corrosion of metals is also an important issue. In 1850, an early wire cable suspension bridge across the Angers River collapsed as about 500 soldiers were marching across it in a storm, and almost half of them died. The cause was corrosion of the cable wires within the anchorages. The sealed anchorage system, intended to protect the cables, had instead hastened their corrosion. Unfortunately, by this time there were hundreds of bridges of this type, and the anchorage systems were difficult to inspect (Scott 2001, p. 6).

Corrosion also factored in to the collapse of the Berlin Congress Hall. The structure was a donation from the U.S. government to the people of Berlin, and it was designed by a U.S. architect and built in 1957. It had a large, hyperbolic, parabolic, prestressed concrete roof. The arch of the roof was stiffened by a ring beam 400 mm (1.3 ft) long and 8.3 m (27 ft) wide.

On May 21, 1980, the southern overhang of the roof collapsed, killing one person and injuring several more. A forensic investigation was carried out by German engineers. It was found that the structural system was sensitive to unplanned deformations and would not have permanently resisted the applied stresses. Insufficient mortar in the tendon ducts and subsequent corrosion of the tendons combined to cause the failure. Both conditions were necessary for the collapse (Buchhardt et al. 1984).

Over the 23-year life of the structure, stress fluctuations led to cracking in the concrete. The cracks diminished the concrete's protection of the reinforcing steel tendons. This condition led, in turn, to severe corrosion and breaking of the wires that made up the tendons. A contributing factor was the fact that the roof concrete was of poor quality and was uneven and porous (Feld and Carper 1997, pp. 320–321).

Austin Concrete Dam Failure

Quality of concrete construction is important, particularly in cold weather. The September 30, 1911, failure of a concrete gravity dam in Austin, Pennsylvania, claimed between 50 and 149 lives. Greene and Christ (1998) put the death toll at 78. It occurred fewer than 150 miles north of Johnstown, Pennsylvania, 22 years after its more widely known disaster.

Dam Construction

The Bayless Pulp and Paper Mill decided to replace a smaller dam with a larger dam in 1909 to increase production. The new dam was a gravity dam, designed to hold back the water by its mass and friction against the foundation. Cost and time constraints led to pressures to complete the dam before winter. The dam blocked the Freeman Run Valley and was 161 m (530 ft) long at the base and 166 m (544 ft) at the crest. The dam was 15 m (49 ft) high at the crest, 13 m (42 ft) high at the spillway, 9.8 m (32 ft) thick at the base, and 0.76 m (2.5 ft) thick at the crest. The reservoir filled half of the valley and was designed to hold 750,000 m^3 (200 million gal or 613 acre-ft) of water (Freiman and Schlager 1995b, pp. 205–206).

The dam was built from 12,200 m^3 (431,000 ft^3) of cyclopean concrete. The cyclopean concrete consisted of boulder-sized pieces of native sandstone rock, bound with conventional cement and concrete. Twisted steel rods 32 mm (1¼ in.) in diameter and 7.6 m (25 ft) long, were spaced 0.8 m (2.7 ft) apart to reinforce the dam. An earth fill embankment 8.2 m (27 ft) high and sloped 3 horizontal to 1 vertical was rolled against the upstream face of the dam (Freiman and Schlager 1995b, p. 207).

The dam's foundation was the rock of the valley floor, which consisted of thin layers of shale between narrow layers of sandstone. About 6,000 m^3 (212,000 ft^3) of soil was excavated at the foundation, and loose material was washed and cemented together with grout. A 1.2 × 1.2 m (4 × 4 ft) footing extended into the bedrock, and the abutments were also cut 6.1 m (20 ft) into bedrock (Freiman and Schlager 1995b, p. 207).

The dam was not yet finished as winter approached, and the last hurried concrete construction was carried out with temperatures below freezing. By December 1, 1909, when the dam was completed, a large crack passed completely through the dam, fracturing some of the cyclopean boulders. A second crack appeared later that month. With the dam empty, and no visible settlement of the foundation, it appeared that the cause was contraction of the concrete (Freiman and Schlager 1995b, pp. 207–208).

The Near-Disaster

The dam reservoir was filled within about six weeks of completion. On January 17, 1910, a combination of heavy rain and snowmelt caused by warm temperatures put floodwater over the spillway. Heavy seepage near the toe of the dam and cracks on the downstream face of the dam were observed. To lower the reservoir, Bayless used dynamite to blast two notches in the dam crest. Superficial repairs were made to the dam after the water level dropped, and then the water level was allowed to rise to normal pool again (Freiman and Schlager 1995b, p. 208).

During this event, the dam developed six prominent vertical cracks, and the top center of the dam bowed 0.76 m (30 in.) downstream. The dam had been built without construction joints, so the cracks divided the dam into seven separate segments. The blasting had been necessary because the dam had no spillways or gates to allow controlled water release (Greene and Christ 1998, p. 9).

Dam Failure

The dam failed on the afternoon of Saturday, September 30, 1911. A hole opened at the west abutment, and the force of the water pushed out the blocks of concrete, which slid downstream. The water channeled as a large wave between the high vertical valley walls, and it smashed into the town of Austin (Freiman and Schlager 1995b, p. 208). The remains of the failed dam are shown in Figs. 9-1 and 9-2.

As the water destroyed a lumber mill, it picked up the logs and lumber as debris. The mill's whistle blew, but for many the warning would be too late. Most of the buildings in the town were destroyed (Greene and Christ 1998, p. 10). Figure 9-3 shows the extent of the damage in Austin.

The fast-moving water ruptured a large gas main and set it on fire, burning much of the wreckage left by the flood. The city's fire protection, of course, had been destroyed in the flood. And 5 km (3 mi) down Freeman Run, the wall of water completely destroyed the town of Costello. The town of Wharton, 13 km (8 mi) past Costello, was heavily damaged, but no one was killed there (Freiman and Schlager 1995b, pp. 209–210). The pattern of a flood followed by devastating fire in Austin was reminiscent of the bridge fire in Johnstown more than 20 years before.

Causes and Investigations

Two separate inquiries were launched, one by the Pennsylvania State Water Commission and one by the District Attorney of Potter County,

CONSTRUCTION MATERIALS 305

Figure 9-1. The failed Austin Dam.

Figure 9-2. The failed Austin Dam.

Figure 9-3. Damage from the Austin Dam failure.
Courtesy Library of Congress.

where the town of Austin was located. The dam engineer, Chalkley Hatton, asserted that his design had been sound, but that the foundation had been undermined by seepage. He claimed to have told Bayless to build a concrete cutoff trench upstream, but his recommendation was not followed. Hatton also blamed poor construction practices—placing concrete at temperatures below freezing and not allowing enough time for the concrete to cure before the dam was put into service. Even by the standards of 1909, the dam foundation investigation was inadequate and the construction methods were poor. The cyclopean sandstone boulders were weak themselves, as shown by the fractures through them when the dam first cracked (Freiman and Schlager 1995b, pp. 210–211).

Investigations also showed that the cracks corresponded to cold joints between concrete cast on different days. The faces of all of the cracks but two were discolored, indicating that they had occurred some time before the dam failure. Foundation excavations showed that considerable water seepage had undermined the foundation. A block of underlying rock had slid downstream along with the dam itself. Because of the seepage, the hydrostatic force lifting up the gravity dam, and hence reducing friction, turned out to be much greater than that assumed in the design (Freiman and Schlager 1995b, p. 210).

The *Engineering News,* forerunner of *Engineering News Record,* dispatched an editor, F. E. Schmitt, to investigate the disaster. He surveyed the wreckage in the field, mapping the final location of the blocks. Schmitt identified two main breaches in the dam, one 30 m (100 ft) from the left abutment and one the same distance from the right abutment. Two blocks, one at each breach, had been pushed about 9 and 30 m (30 and 100 ft) downstream, respectively. He concluded that "the whole appearance of the wreck points directly to a sliding failure" (Greene and Christ 1998, pp. 10–12).

The site geology and the in situ materials explain how such a sliding failure could have occurred. Schmitt identified "remarkably upfolded and crumpled masses of laminated shale strata suggesting a great downstream pressure" between two of the failed concrete sections. The site geology consisted of a thin layer of sandstone between two layers of shale. The dam foundation rested on top of the sandstone. The sliding plane developed in the lower shale layer, and the upper sandstone layer slid along with the concrete blocks. The sandstone layer, which the designer had thought made an appropriate foundation, was in fact only 0.6 m (2 ft) thick (Greene and Christ 1998, pp. 12–13).

Greene and Christ state,

> Geology played a major part in the disaster at Austin. Had the dam's foundation been taken deeper into bedrock, it may have prevented the underseepage that ultimately brought about the high uplift condition that caused the structure to become unstable. Uplift pressure reduces the effective weight, and therefore the stability, of a concrete dam. When the effective weight of a dam is reduced, the horizontal force of the water load is able to move the structure. In addition, had the foundation been taken below the shale layer (where the sliding plane developed), the dam's stability would have been significantly enhanced. In an article written by the project design engineer, Mr. T. Chalkley Hatton, it is stated, "The failure of this dam was not the result of poor workmanship, but poor judgment upon my part in determining its foundation. I should have sought the advice of a man more skilled in determining foundations for dams than myself." Hatton went on to say, "The great mistake I made in building this dam was trusting the rock foundation to be impervious." (1998, p. 13)

Hatton made these admissions in *Engineering News* in 1912. In Hatton's defense, Greene and Christ (1998, p. 13) point out that the Bayless

Mill imposed severe constraints on his design. He had designed a foundation cutoff wall that would extend far enough into bedrock to prevent seepage, but he was overruled because of the additional expense.

Operation of the dam was also clearly at fault. Once cracks appeared in the dam, stronger measures should have been taken to ensure the safety of those downstream. The first crack was more than adequate warning, but construction continued without change. The implications of the second crack were also ignored.

> Blasting holes in the dam, which they did to bring down the water level, was an astounding, absurd blunder that probably weakened whatever remained of the structure's integrity. When . . . Bayless . . . disregarded the engineer's recommended repairs, the dam's fate was sealed. Although the reservoir should never have been filled after cracks were observed in December 1909, it definitely should have been abandoned completely after the near disaster of January 1910. (Freiman and Schlager 1995b, p. 211)

Aftermath

Pennsylvania had suffered two dam failures with substantial loss of life, Johnstown and Austin, just over two decades apart.

> Not surprisingly, very soon after the failure of Austin Dam, legislation was enacted in Pennsylvania calling for state supervision of the design, construction, and operation of dams and reservoirs within the commonwealth. A comprehensive dam safety program was thus initiated in Pennsylvania to prevent the recurrences of disasters such as those suffered at Johnstown and Austin. The failure of the dam at Austin not only raised awareness of dam safety but also taught designers of dams worldwide valuable lessons as to what is needed to prevent this type of tragedy in the future. (Greene and Christ 1998, p. 13)

Lessons Learned

There were clearly two major problems with the Austin Dam; one was the inadequate foundation design, and the other, the shoddy concrete construction. The mode of failure clearly shows that the dam slid in separate blocks. This evidence suggests that the most important factor was the low friction against the dam base.

It is by no means clear that a gravity dam resting on shale would have resisted water pressure indefinitely, even had it not cracked. However, the poor quality of the concrete construction probably also played some part, at least in the extent of the disaster. The cracks and weakness of the concrete divided the dam into the blocks, and once the weakest block began to slide, the breaches would widen. Cracks within a dam allow hydrostatic pressures to work to break the dam apart.

If it had held together, the dam would have been safer, although perhaps not safe enough. The small amount of reinforcing steel provided clearly had no effect on the cracking or structural integrity of the dam. A broader discussion of the statics of concrete gravity dams is provided at the end of Chapter 2. The discussion includes the calculation of factors of safety against overturning and sliding, as well as the effect of hydrostatic uplift forces.

As with some other projects discussed in this book, such as the Quebec Bridge in Chapter 3, time and economic pressures compromised safety. The decision to omit the foundation cutoff wall from the original design, combined with the refusal to pay for the concrete cutoff trench upstream, doomed the dam to a sliding failure. The rush to complete the dam before the winter led to the freezing of the concrete. Early freezing considerably reduces the strength of concrete.

Concrete shrinks. If the shrinkage is restrained, as with a long gravity dam, it will crack. As a result, control joints or "designed cracks" are generally put in concrete structures. In concrete dams, control joints are provided with water stops to keep water from penetrating into the dam and weakening it. The cracks that divided the Austin Dam occurred pretty much where control joints should have been provided, dividing the failed dam into blocks of roughly equal size.

The accounts referenced do not describe any extraordinary event on September 30, 1911, that triggered the dam failure, unlike the record rains that triggered the Johnstown flood. Perhaps the dam simply got tired of holding back the water. Like the South Fork Dam responsible for the Johnstown Flood, the Austin Dam did not have pipes or any other means to rapidly reduce the water level in case of danger. The remains of the dam are still visible today along Pennsylvania Route 872, north of the town of Austin.

Essential Reading

This case is discussed by Freiman and Schlager (1995b, pp. 205–212). Another source is Greene and Christ's "Mistakes of Man: The Austin Dam Disaster of 1911," published in *Pennsylvania Geology* (1998).

Liberty Ship Hull Failures

The Liberty ships of World War II were an engineering and industrial solution to a specific military and political problem. The problem was that is was necessary to build merchant ships faster than the German Navy submarines could sink them. These ships were desperately needed to supply the United Kingdom and sustain its war effort. The problem was complicated by the fact that the Great Depression had greatly reduced U.S. shipbuilding capacity.

Between 1930 and 1937, U.S. shipyards built only 71 merchant ships. However, 5,777 were built between 1939 and 1945. This surge was made possible by changes to construction methods, as well as the standardized Liberty freighter and similar T2 tanker designs. One shipyard built a Liberty ship in five days. The massive increase in production was possible in large part because of a change from riveted to electric arc-welded construction. Shipyards were organized for maximum efficiency around the welding process. New welding equipment was developed for heavy steel ship plate. The success of the German pocket battleships had pointed out the efficiencies of welded ship hulls. By saving a thousand tons of weight in the hull, they could carry that much more armament (Tassava 2003, pp. 90–93).

Tassava notes,

> Many experts understood that compared to a riveted ship, a welded ship could be stronger and faster; welding eliminated the considerable weight of thousands of rivets and decreased the resistance created by overlapped plates and rivet heads. In the shipyard, too, welding seemed superior to riveting. When properly organized, welders could work more quickly and efficiently than riveters, and thus make possible higher rates of production. By eliminating the need to create overlaps through which rivets could be driven, welding allowed a shipbuilder to purchase less steel, saving money and driving down the cost of a ship. (2003, pp. 93–94)

Despite these desirable features, welding was considered an unproven technique for shipbuilding. Stresses due to "warpage and shrinkage" from the welding temperature changes were difficult to understand. There were few experienced welders available, and the best welders in the country were hired by the shipbuilders to train and supervise other welders. Many women and African-Americans learned welding through accelerated courses at the shipyards (Tassava 2003, pp. 93–94).

The training programs were effective in turning out large numbers of productive welders. The training emphasized skills, not theory. Some of the new welders became skilled enough to pass the difficult American Bureau of Shipping certification exam. Overall, however, the volume of work was too great for the available number of certified welders. Stress tests and X rays were used to spot-check weld quality (Tassava 2003, pp. 95–96).

Ship Failures

One of the best known Liberty ship failures was the *S.S. Schenectady*:

> At 10:30 P.M. on January 16, 1943, an explosive boom shattered the cold night that had settled over the Swan Island shipyard outside Portland, Oregon. Rushing to the fitting-out docks where the yard's T2 tanker vessels were completed, graveyard shift workers discovered that the yard's very first ship, the *S.S. Schenectady*, had cracked in half amidships. The ship's deck and side shells had fractured completely; only the plates running along the bottom of the [162-m] 532-foot long hull held the fore and aft sections together. (Tassava 2003, p. 87)

Although this wasn't the first cargo ship failure, it was highly visible, as the first ship built in the new yard. In March, another tanker, the *Esso Manhattan*, split in half while entering New York harbor. In January 1943, about 20 ship failures occurred, with 4 or 5 suffering Class I damage or total hull failure like the *Schenectady*. Twenty Class I failures occurred in January 1944, with 120 failures the next March. Many failures occurred in the open ocean. Because the *Schenectady* and *Esso Manhattan* failures occurred in port, the ships could be put in drydock, repaired, and put back into service (Tassava 2003, pp. 88–90).

Given the rapid expansion of the welding workforce, suspicion fell immediately on the quality of the welders. Given the prejudices of the day, the high proportion of women and African-Americans did not help. Within 3 months of the *Schenectady* incident, an investigating committee of engineers and supervisors identified eight relevant factors, four of which related to poor workmanship. As the problems continued, the welding engineers continued to blame the welders. There were, perhaps, two motives. One was to advance the war effort. The other was for the welding engineers to build their status as professionals (Tassava 2003, pp. 96–97).

In contrast, Admiral Emory Land, the chairman of the Maritime Commission, which had jurisdiction over shipbuilding, "identified several production errors that could restrict the three-dimensional flow of a steel plate as a welding arc heated it . . ." (Tassava 2003, p. 98). Any of these could lead to residual stresses, making the ship more susceptible to fracture. Cold waters could also increase fracture susceptibility, and some ships had been lost in cold Arctic waters on the supply runs to the northern ports of the Soviet Union.

Admiral Land argued that the incidence of cracking in the Liberty ships was in fact rather low, affecting only 74, or 3% of the 2,246 ships built through 1943. Overall failure rates varied by shipyard, from a low of 0.75% for the Bethlehem Steel Fairfield shipyard outside Baltimore, to the relatively high 5.95% for Kaiser's Oregonship shipyard facilities (which included the *Schenectady*). The higher and lower failure rates were found in both old and new shipyards, suggesting that the rapid expansion of new shipyards was not necessarily the problem. He also pointed out that harbor collisions had taken more freighters out of service than cracking problems, and that 90% of the failed ships had been repaired and returned to service. In comparison, he noted, the sturdy B-17 heavy bomber had a much higher 13% structural failure rate (Tassava 2003, pp. 98–99).

Solutions

The solutions fell into three categories—improvements to shipyard practice, changes to the design, and retrofits of the completed ships.

IMPROVEMENTS TO SHIPYARD PRACTICE

In the rush to build the ships as quickly as possible, welders had been deployed in large numbers without regard to equalizing temperature stresses. Now, single welders or teams worked symmetrically on opposite sides of a subassembly or plate. Different yards used either semiautomatic or manual welding machines. The Maritime Commission hired the Lincoln Electric Company, one of the largest welding equipment manufacturers, to evaluate work practices in the yards. Gradual improvements were made to training methods. Radiographic testing with X rays proved particularly useful for finding defects, such as tiny bubbles or hollow spots in welds. One shipyard that used radiographic testing extensively had a low failure rate (Tassava 2003, pp. 99–101).

CHANGES TO THE DESIGN

Changes were also made to the standard ship designs. "Investigations revealed that many Liberty ship fractures began in the square corners of cargo-hold openings" (Tassava 2003, p. 101).

These findings suggested that, by and large, the welders were not to blame. If there had been an epidemic of bad welding, why would the failures all occur at the same place? Bad welds would be randomly distributed throughout the hulls, and thus the failures could start pretty much anywhere.

Two antifracture design changes were made. The first was to redraw the Liberty ship plans to round off the troublesome hatch corners. The second was to install crack-arresting devices to stop cracks from propagating through the hulls. It had been observed that fractures were generally stopped by perpendicular barriers, such as riveted or especially strong seams (Tassava 2003, p. 101). In the early Class I failures, such as the *Schenectady* and *Esso Manhattan*, the cracks had been able to travel completely through the hull.

Retrofits of the Completed Ships

Of course, by the time the ships were redesigned, there were already several thousand in service. These ships needed to be retrofitted. Rounded fillets were welded to the hatch corners to reduce the stress concentrations, and crack arresters were installed. Ironically, the largest crack arrestor had to be riveted on. A massive 87-m (287-ft) gunwale bar was riveted along each side of the middle of a 134-m (441-ft) long ship. The bars were installed in 9.1-m (30-ft) sections, with 6,200 rivet holes required per ship (Tassava 2003, pp. 101–102).

Technical Aspects

The Liberty ship fractures occurred because of the combination of stress concentrations and fatigue. These two phenomena are fairly important by themselves, but the combination of the two may be lethal.

Figure 9-4 shows a typical curve for determining stress concentration factors for fillets. The stress concentration factor K is a function of the radius of the fillet. The nominal average stress is multiplied by K to get the maximum stress:

$$\sigma_{max} = K\sigma_{ave} \qquad (9\text{-}1)$$

Note that as the radius approaches zero, for a sharp corner, the stress concentration factor approaches infinity. The solution represented by this curve is based purely on elastic considerations and does not account for material yielding. The stress concentration factor remains the same whether the material is steel, concrete, wood, or silly putty.

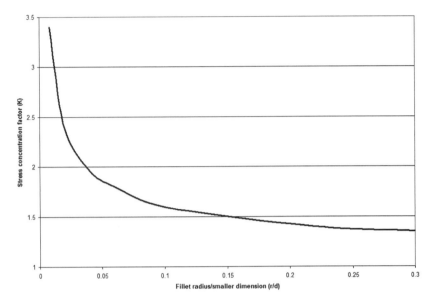

Figure 9-4. Stress concentration factors (Wider dimension D 10% larger than smaller dimension d).

With ductile materials, such as mild steel, stress concentrations are relieved as the steel yields. The theoretical maximum stress, shown in Fig. 9-4, does not occur because the steel at the stress concentration yields, and the stresses are redistributed.

Different materials have different S–N curves describing the relationship between applied stress and number of cycles to failure. Figure 9-5 shows a typical S–N curve for structural steel. S is the stress level, and N is the number of cycles to failure, which increases as S decreases. When $S = 220$ MPa (32 kip/in.^{2}), the material fails immediately with the first loading cycle. If S is sufficiently small, the material may theoretically sustain infinite load applications, and the stress is said to be below the material's endurance limit. If the approximate number of load cycles N can be predicted, the designer needs to merely ensure that all stress ratios are low enough so as to remain below the S–N curve.

When stress concentrations and fatigue are combined, ductile materials, such as mild steel, can fail in a sudden, brittle manner. With the stress concentration, K increases, S increases, and N decreases. The sharp hatch corners of the Liberty ships, coupled with locked-in residual stresses from welding, meant that a few load cycles from wave action or other sources could lead to the brittle fracture.

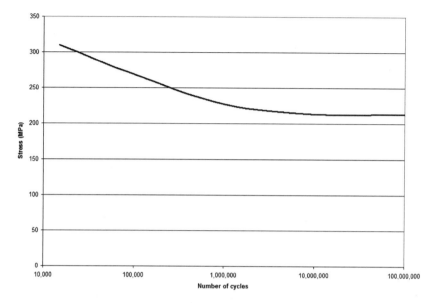

Figure 9-5. Fatigue S–N diagram (1 MPa = 145 psi).

Conclusions

The gunwale bars and the other retrofits were successful. No ship with gunwale bars ever failed in service. Through the end of the war, only 127 of the 4,694 Liberty ships and T2 tankers ever suffered a Class I fracture, and no ships with antifracture devices failed. Two years after the war,

> An official board of investigation determined that faulty workmanship caused exactly 25% of the 2,504 fractures which had occurred up to August 1, 1945, that a combination of inferior workmanship and inadequate design caused another 20%, and design so poor that "perfect workmanship would have done little to prevent the failures" caused a stunning 55%. (Tassava 2003, p. 103)

It seemed, in retrospect, that the blame placed on the welders had been misdirected.

Petroski (1994) commented on the change from riveted to welded shipbuilding.

> The riveted joints of overlapping plates had served as effective arresters of any cracks that might begin to grow in the hull, and which could

be repaired in due course. Welded construction not only removed such obstacles from the paths of any cracks that might develop, but also the very process of welding embrittled the adjacent steel and made it behave much like a ship of glass. (pp. 56–57)

Essential Reading

A useful review of the Liberty ships, as well as the problems and solutions of welding, was written by Christopher James Tassava, "Weak Seams: Controversy over Welding Theory and Practice in American Shipyards, 1938–1946" in *History and Technology* (Tassava 2003). Tassava is a historian, not an engineer, but he does a good job of reviewing the technical issues associated with the failures. This case study is featured on the History Channel's Modern Marvels *More Engineering Disasters* videotape and DVD. However, the History Channel version seems oversimplified.

Willow Island Cooling Tower Collapse

Willow Island was the worst construction accident in U.S. history.

Design and Construction

Two natural-draught hyperbolic cooling towers were to be built for the Pleasants Power Station, a coal-fired power station in Willow Island, West Virginia. These towers were designed to be large, chimneylike towers with a distorted hourglass shape, which would allow air to circulate without fans because the warmer air inside would rise naturally. The base diameter was 109 m (358 ft), but both diameter and shell thickness would change with the height of the tower (Schlager 1994). The cooling tower dimensions are shown in Fig. 9-6, and the two towers are shown in Fig. 9-7.

These two towers were designed and built by the Hamon Cooling Tower Division of Research-Cottrell, Inc. (R-C), Bound Brook, New Jersey. R-C was an environmental control company that designed and built air and water pollution control systems for utility companies (Peterson 1978a). Research-Cottrell held a $12 million subcontract on the pair of 131-m (430-ft) tall towers (Ross 1984). The general contractor of the project was United Engineers and Constructors.

The towers were built using a patented lift-form technique that had been successfully used for construction of 36 cooling towers. The lift form scaffolding was made up of five basic components: jumpform beams, anchor

CONSTRUCTION MATERIALS 317

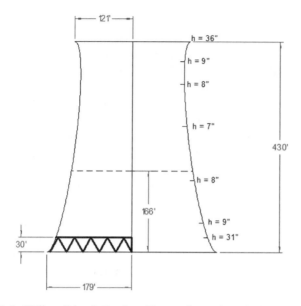

Figure 9-6. Willow Island Cooling Tower dimensions (1 in. = 25.4 mm, 1 ft = 0.305 m).

Figure 9-7. Willow Island Cooling Towers.
Courtesy National Bureau of Standards/National Institute of Standards and Technology.

318 BEYOND FAILURE

assemblies, jacking frames, formwork, and scaffolding platforms (Velivasakis 1997). Some of these components can be seen in Fig. 9-8.

A four-level-high system of scaffolding was used, and working platforms were suspended from the inside and outside jacking frames. At the top level, the construction materials were received by the hoisting system, steel reinforcement was distributed, and concrete was delivered. The second level was used only during the formwork adjustment process. Levels three and four of the scaffolding system provided access to the jumpform beams, and final surface preparation such as patching and grouting was done from these levels (Lew 1979). The scaffolding system was entirely supported by the previously completed portion of the tower. Each day, a 1.5-m (5-ft) lift was completed, and the entire scaffolding system moved with the jacking frame to the new elevation.

The daily routine to prepare for concrete placement consisted of four procedures. First, workers loosened the forms from the last concrete lift by removing the wedging from the formwork. Next, the forms were adjusted to

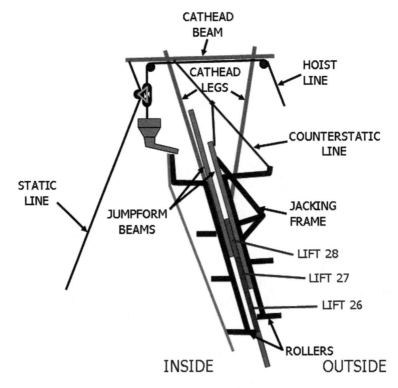

Figure 9-8. Willow Island Cooling Tower jump form system.

accommodate the changing diameter of the shell, and the jacking of the entire formwork and scaffolding system took place at the next elevation. Then, the lowest jumpform beam was unbolted and moved to its new location at the top. Finally, the formwork was wedged into position (Velivasakis 1997).

After the tenth lift was completed, the concrete and construction materials were carried to the working platforms by an elaborate hoisting system. Six cathead gantry cranes powered by twin drum hoists delivered the materials (Lew 1979). The legs of each cathead gantry were attached to the aluminum jumpform beams, which were then attached to the wall at rib locations approximately 3.7 m (12 ft) apart (Lew 1980).

The catheads moved up the lift form scaffolding as the construction advanced. A static line guided all of the materials as they were hoisted to the working platforms. The static line was attached to the slide plate at the interior end of the cathead at one end and secured to an anchor point on the ground at the other end (Velivasakis 1997). During construction, because of the changing geometry of the tower, both the catheads and static line had to be adjusted periodically.

Failure

The first cooling tower was completed in August 1977, but on April 27, 1978, during construction of the second tower, the formwork system failed. While concrete was being placed for lift 29, 52 m (170 ft) above the ground, lift 28 collapsed, and 51 workers fell to their deaths. The scene of the collapse is shown in Fig. 9-9.

On the day of the failure, cathead gantries Nos. 4 and 5 were being used to deliver reinforcement and concrete. Failure at cathead gantry No. 4 initiated when the third bucket of concrete was being delivered to the platform, and the tower began to collapse inward (Lew 1980). The platforms, formwork, fresh concrete, and most of the day-old lift all collapsed.

Schlager (1994) states,

> Those on the ground said that the interior work platform peeled away "like someone opening a can" and it, along with the exterior platform, fifty-one workers, and tons of concrete, were pulled inside the tower and collapsed into a heap of jumbled debris. The formwork ripped loose simultaneously in both circular directions and met directly across from the starting point. (p. 291)

Another worker made sure that the *New York Times* let the families know that, "It was all over in 20 or 30 seconds. It went so fast and none of those poor men could have suffered too much" (Peterson 1978b, p. 1).

320 BEYOND FAILURE

Figure 9-9a, b. Willow Island Cooling Tower collapse.
Courtesy National Bureau of Standards/National Institute of Standards and Technology.

Causes of Failure

Immediately after the collapse, the National Bureau of Standards (NBS) began an investigation on behalf of the Occupational Safety and Health Administration (OSHA) to determine the cause. The entire system was divided into three sections to be examined. First was the hoisting system, second was the scaffolding system, and third was the tower itself (Lew 1980).

The hoisting system was investigated because the failure initiated where cathead gantry No. 4 was located. Once all of the parts were collected from the debris, they were examined to see which of the components failed first. Lab tests were performed on the hoisting cable, static line, chain hoist, and anchor device of the system. These results, along with the field observations, indicated that the hoisting system did not initiate the failure.

The next section of the tower examined was the scaffolding system. According to the NBS investigation, the scaffolding system was not the most probable cause of failure. They determined in lab tests that bolt failure was not a cause because the hoisting system could not produce a large enough force to cause bolt failure (Lew 1980).

Next, the NBS looked at the tower itself as the most probable cause of the collapse. To do this, the strength and other mechanical properties of the concrete in lift 28 had to be determined. Concrete specimens were made using the materials supplied by the same company that supplied concrete to the construction site. These specimens were then cured in a chamber that simulated the temperature at the site for the 24 h before the collapse. The weather that week had been cold and rainy, with temperatures just above 16 °C (in the 60s °F) during the day and just above 0 °C (in the 30s °F) at night (Ross 1984). Estimating the collapse to have taken place approximately 20 h after the completion of the tower section located near cathead gantry No. 4 and curing the concrete at 4.4 °C (40 °F), the compressive strength of this section was approximately 1,500 kPa (220 lb/in.^{2}) (Lew 1980). It was concluded that this section on the tower did not have adequate strength to resist the applied construction loads.

Overall, the NBS investigation concluded the following (Lew 1979, pp. 47–48):

1. At the time of failure, the concrete bucket was in transit from the base of the tower to cathead gantry No. 4. Eyewitness accounts and measurements indicate that the bucket was approximately 60 ft (18 m) below the cathead beam. Therefore, it is not believed that the concrete bucket hit the cathead, causing it to fail.

2. The cables for catheads Nos. 4 and 5 were broken after the onset of the collapse. Therefore, the breakage of the cables did not initiate the failure.
3. Field and lab tests show that the collapse did not initiate because of a component failure of the hoisting, scaffolding, or formwork systems.
4. The compressive strength of the concrete near cathead gantry No. 4 was estimated to be 1,500 kPa (220 lb/in.^{2}) at the time of the collapse.
5. Analysis showed that the resultant stresses at several points along lift 28 equaled or exceeded the shell strength in compression, bending, and shear. Failure at any of these points would have propagated, causing the overall collapse.
6. The most probable cause of the collapse was determined to be the imposition of construction loads on the shell before the concrete of lift 28 had gained adequate strength to support these loads.

While the NBS was investigating for OSHA, Lev Zetlin Associates (LZA) was performing another investigation, on behalf of the general contractor. LZA's findings disagreed with those of the NBS investigation. LZA claimed that the most probable cause of failure was the early removal of anchor bolts and cones from the lower portion of lift 27. LZA believed that if the anchor bolts had been left in place and stayed attached to the jumpform beams, the collapse would have never occurred (Velivasakis 1997).

Because of the contradictions found in these investigations, H. S. Lew and S. G. Fattal of NBS analyzed three remaining questions after the investigations were complete. The first question was what should have been the strength of the concrete in lift 28 so that the shell could have resisted the applied construction loads. The second question dealt with the bolts that should have been in lift 27. The third question dealt with the location of the static line at the time of the collapse (Fattal and Lew 1980).

After analyzing the hoisting loads, including dynamic effects that were inherent in the hoisting system, the stress resultants in the shell were determined. These values were then compared to the resistance of the shell throughout each shell profile. The results showed that crushing failure in lift 28 would initiate if the strength of the concrete was 6,900 kPa (1,000 lb/in.^{2}) or less (Fattal and Lew 1980).

To answer the second question, models were analyzed with and without the lower bolts in place. They found that because "... of the additional bolts in lift 27 the magnitudes of the critical stress resultants were reduced . . ., but were still well beyond the ultimate capacity of the concrete section in lift 28" (Fattal and Lew 1980, p. 18).

For the answer to the third question, the analysis was performed with the ground anchor point of the static line at different distances from the shell of the tower. It was determined that if the base of the static line was not moved nearer to the center of the tower (its location at the time of the collapse) the critical stress resultants would have been less than the ultimate strength of the concrete in lift 28. Therefore, the NBS investigators concluded, "that the concrete in lift 28 would probably not have failed and consequently, collapse would not have occurred" (Fattal and Lew 1980, pp. 18–19) if the base of the static line was not moved.

Legal Repercussions

OSHA issued 10 willful citations and six serious citations against R-C (Ross 1984). Five of the willful citations appeared to be directly related to the collapse. The charges included failing to test field-cured concrete specimens before removing the formwork; not properly anchoring the scaffolding and formwork and catheads; not having key personnel, design specifications, and erection instructions on site; and not adequately training employees (Morrison 1980). The other five citations were considered unrelated to the accident. Criminal charges were dropped, but R-C paid substantial amounts in damages and fines imposed by OSHA.

Also, OSHA issued two serious citations each to Pittsburgh Testing Laboratory, the company that performed concrete testing for R-C, and the Criss Concrete Co., which supplied the concrete for the tower. These companies also faced lawsuits from the victim's families, as well as from R-C (Morrison 1980).

Professional and Procedural Aspects

The three companies that faced legal repercussions were not the only ones under fire after the disaster. OSHA was criticized for lax enforcement of regulations. At the time of the accident, the number of qualified federal safety inspectors for construction projects in the entire state of West Virginia was seven (Ross 1984). This amount was considered "absurdly low" by the state commission.

As a result of this disaster, OSHA adopted new guidelines. One large step that OSHA took toward protecting future construction workers was making changes in the U.S. Construction Safety Act. One of these changes shifted more responsibility from the engineer to the contractor for formwork decisions. Another major change was that OSHA removed a table that provided a schedule for formwork removal. Now, the act requires that the concrete specimens be tested before removal of formwork or any system that relies on the strength of the concrete (Schlager 1994).

Other guidelines that OSHA adopted after the accident included having a specialist review construction plans for cooling towers and requiring that a detailed safety manual be developed as part of the construction plan. Also, OSHA improved their inspection procedures by increasing the items checked at cooling towers and adding an inspection for compliance with the construction plan (Lew 1980).

Conclusions

This case study shows the importance of safety standards during the construction process. A process that had been used successfully 36 times, including once at the same site, drastically failed, causing the death of 51 workers. The men were paid for only eight hours a day, regardless of how long it took to do a day's work, which provided an incentive to rush the work. OSHA now has added many requirements, such as a construction plan, and has increased safety inspections to make sure that another disaster like this does not take place.

Another lesson that was learned by this accident was not to tie together all of the formwork. Not long after the Willow Island cooling tower collapsed, at a cooling tower in Satsop, Washington, two workers were killed and a third seriously injured when a steel form pulled away from the concrete (Ross 1984). If these forms had all been tied together, the loss of life would have been much greater.

The *Charleston* [West Virginia] *Daily Mail* discussed the disaster in a retrospective published on February 12, 1999:

> On Thursday, April 27, 1978, crews of laborers showed up on the site of the latest project at the plant, the erection of twin cooling towers. By 10 A.M., the first [680-kg] 1,500-pound bucket of concrete had been hoisted to the men working on scaffolds at the top of the first tower, 168 feet in the air. In what was described as "slow motion," one section of the scaffolding holding the workers collapsed in on the tower, which started a chain reaction collapse around the rest of the scaffolding. . . .
>
> Workers, who were allowed to leave work as soon as they completed the day's pour, had not given each section sufficient time to cure. One family, the Steeles, lost its father and four sons in the disaster, and seem-

ingly each of the nearly 8,000 residents of tiny Pleasants County was touched personally. (*Charleston Daily Mail* 1999)

Essential Reading

Key references are the two NBS reports: Lew (1979), *Investigation of Construction Failure of Reinforced Concrete Cooling Tower at Willow Island, West Virginia*, and Fattal and Lew (1980) *Analysis of Construction Conditions Affecting the Structural Response of the Cooling Tower at Willow Island, West Virginia*. Another point of view is provided by E. E. Velivasakis (1997), "The Willow Island Cooling Tower Scaffold Collapse: America's Worst Construction Accident," in *Forensic Engineering*. This case study is also featured in the History Channel's Modern Marvels *Engineering Disasters 6* videotape and DVD.

Boston's Big Dig Tunnel Collapse

Two principles emerge again and again in construction materials and structural engineering. The first is that new and innovative materials, such as epoxies, often have properties that are poorly understood when they are introduced or used in new applications. They may, in fact, introduce failure modes that have not previously been considered because they did not govern the performance of the materials used earlier. The second principle is that structural systems, when they fail, often fail at the connections.

The Boston Central Artery, or Big Dig, was one of the most monumental transportation efforts undertaken in the United States. It was a matter of planning and discussion for decades—I remember talking about it in class at MIT in the fall of 1984. It became notorious across the United States for cost overruns and construction delays.

When I was a professor in Alabama in the 1990s, state transportation engineers routinely referred to the "Big Dig" as an excuse not to tackle the tough congestion problems on Highway 280 coming into Birmingham. They also may have resented the large fraction of the federal transportation budget that had to be devoted to the Boston project because those funds were not available for projects in other states.

In defense of the project, it is not a simple thing to build a new interstate-quality highway through the heart of a congested city, through a century's worth of forgotten underground infrastructure. The project was complicated by Boston's legendary traffic, as well as the tendency for Boston Harbor to try to work its way into any tunnels in the area.

The Tunnel Collapse

There was, therefore, great relief among engineers when the Big Dig finally opened, with only a few loose ends. Then a heavy tunnel ceiling panel fell on a car, killing a woman. Because the collapse occurred in a tunnel, the incident fell under the jurisdiction of the National Transportation Safety Board (NTSB).

The NTSB Investigation

The *Cleveland Plain Dealer* reported on July 11, 2007, that the investigation had been completed:

> The fatal Big Dig tunnel collapse in Boston could have been avoided if authorities had considered that the epoxy securing tons of ceiling panels could pull away slowly. The National Transportation Safety Board approved a report saying the likely cause of the accident that killed a woman was "use of an epoxy anchor adhesive with poor creep resistance." (*Cleveland Plain Dealer* 2007, p. A9)

In a public meeting on July 10, 2007, one year after the accident, the NTSB released a synopsis of the highway accident report that would soon be published on the case. The report synopsis blamed the engineer and the contractor, with errors by the epoxy supplier and the Massachusetts Turnpike Authority also cited:

> The safety issues identified during this investigation are as follows:
>
> - Insufficient understanding among designers and builders of the nature of adhesive anchoring systems;
> - Lack of standards for the testing of adhesive anchors in sustained tensile-load applications;
> - Inadequate regulatory requirements for tunnel inspections; and
> - Lack of national standards for the design of tunnel finishes. (NTSB 2007)

The investigation also found that there had been an incident involving the same epoxy seven years before. Anchor displacement had been observed in the high-occupancy tunnel in 1999, but the engineers and contractors did not continue to monitor the performance of the epoxy.

Some of the report's major conclusions, as provided in the synopsis, are summarized or quoted below:

- By July 2006, many of the adhesive anchors supporting the D Street portal ceilings had displaced enough to put them in imminent danger of failure.
- Improper or deficient anchor performance could not, by itself, account for the anchor failure. The design calculations were consistent with the actual in-service loads.
- The selection of an adhesive anchoring system to support the ceiling panels was appropriate, if the proper material had been used.
- The engineers "failed to account for the fact that polymer adhesives are susceptible to deformation (creep) under sustained load, with the result that they made no provision for ensuring the long-term, safe performance of the ceiling support anchoring system." Therefore, the adhesive specification did not require creep resistance.
- The adhesive product supplied and used to support the anchors, Power-Fast Fast Set epoxy, had poor creep resistance. The manufacturer's information for the product was inadequate and misleading and left out the fact that testing had shown the epoxy to be subject to creep under sustained loading.
- "As shown by the displaced anchors in the D Street portal, the maximum load capacity of an adhesive anchor, which relates to short-term loading, does not indicate that the anchor will be able to support even lighter loads over time, and thus a larger design safety factor cannot compensate for an adhesive material that is susceptible to creep."
- "After unexplained anchor displacement was found in the Interstate 90 connector tunnel in 1999 and 2001, Bechtel/Parsons Brinckerhoff and Modern Continental Construction Company should have instituted a program to monitor anchor performance to ensure that the actions taken in response to the displacement were effective. Had these organizations taken such action, they likely would have found that anchor creep was occurring, and they might have taken measures that would have prevented this accident. Powers Fasteners, Inc.'s, response to the anchor displacements that occurred in 1999 in the high-occupancy tunnel of the D Street portal was deficient in that the company did not identify the source of the failures as creep in the Fast Set epoxy adhesive and took no followup action to ascertain why its product had not performed in accordance with the users' expectations."

- The Massachusetts Turnpike Authority should have regularly inspected the area above the suspended ceilings. If it had, the anchor creep would have been detected.
- Ultimate load tests should have been conducted on the adhesive anchors before installation.
- Installing adhesive anchors in an overhead application is difficult. This type of work makes it likely that voids will be introduced into the adhesive, which will reduce holding power and reliability.
- "The circumstances of this accident demonstrate a general lack of knowledge and understanding among design and construction engineers and builders of the complex nature of epoxies and similar polymer adhesives, and in particular, the potential for those materials to deform (creep) under sustained tension loads."
- Test protocols and standards are needed to determine the capacity and reliability of adhesive anchors in sustained tensile load applications.

The NTSB determined that "the probable cause . . . was the use of an epoxy anchor adhesive with poor creep resistance, that is, an epoxy formulation that was not capable of sustaining long-term loads" (NTSB 2007).

Creep to Failure

Creep is continued deformation under a sustained load. It is a time-dependent phenomenon. When a load is applied to a material that creeps, the resulting deformation has two components. The first deformation component is elastic, remains constant as long as the load remains the same, and is removed when the load is removed. The second component, creep, increases gradually over time, and when the load is removed, deformation (permanent set) remains.

Structural steel does not creep, except at high temperatures. Concrete, wood, and masonry creep, and the amount of creep deformation is in the range of two to three times the instantaneous elastic deformation. It is necessary to take creep into account for predicting long-term deformations, but the behavior is well understood and engineering solutions are well established.

On the other hand, new materials, such as epoxies and plastics, may have much higher creep deformations. In a tensile anchor application, the anchor may pull partway out of the hole, reducing the bond strength of the connection. It is also possible for an anchor to pull completely out. As the anchors creep, the loads may shift to other anchors. As this failure suggests, the long-term behavior of certain epoxy materials in tension is not yet well understood.

Recommended Changes to Practice

The NTSB suggested the following corrective actions:

- The Federal Highway Administration (FHWA) and the American Association of State Highway and Transportation Officials (AASHTO) should develop test protocols and standards to determine the capacity and reliability of adhesive anchors in sustained tensile load applications. Until the standards have been developed, FHWA should prohibit the use of adhesive anchors in these applications where failure would result in risk to the public.
- Legislation should establish and implement a tunnel inspection program similar to the National Bridge Inspection Program (which was established after the failure of the Point Pleasant Bridge discussed in Chapter 3). This legislation should be supported by FHWA and AASHTO's cooperative development of design, construction, and inspection guidance for tunnel finishes.
- State departments of transportation should prohibit the use of adhesive anchors in these applications where failure would result in risk to the public and should identify existing facilities where such a risk exists. Those sites should be inspected and repaired.
- The International Code Council (ICC) should require creep testing for qualification of all anchor adhesives and disqualify for use in sustained tensile loading any adhesives that have not been creep tested or that have failed creep testing. ICC building codes and evaluation reports should be revised in light of this failure.
- Powers Fasteners, Inc., and Sika Corporation should revise their packaging and product literature and packaging to state that the relevant products should be used for short-term loading only.
- AASHTO should "Use the circumstances of the July 10, 2006, accident in Boston, Massachusetts to emphasize to your members through your publications, Web site, and conferences, as appropriate, the risks associated with using adhesive anchors in sustained tensile-load applications where failure of the adhesive would result in a risk to the public" (NTSB 2007). The American Concrete Institute, the American Society of Civil Engineers, and the Associated General Contractors of America were urged to undertake similar educational efforts.

Essential Reading

The NTSB highway accident report (2007), "Ceiling Collapse in the Interstate 90 Connector Tunnel, Boston, Massachusetts, July 10, 2006," is available online at http://www.ntsb.gov/publictn/2007/HAR0702.pdf.

High-Alumina Cement

High-alumina cement is a specially manufactured high early strength cement. However, its use led to some well-publicized structural failures in the United Kingdom, and it has been banned for structural use in that country. It differs in composition and properties from conventional Portland cement and is manufactured from bauxite and limestone. It has also been called fondu cement, aluminous cement, or calcium aluminate cement. It has the advantage of being highly resistant to sulfate attack and was developed for this purpose. It also gains strength quickly. Unfortunately, the main hydration product, which provides strength and durability, is unstable and vulnerable to a process known as conversion. Conversion may occur at either room temperature or elevated temperatures. As the cement undergoes conversion, it may lose up to half its strength, which is obviously unacceptable for structural purposes. It was used in England for the manufacture of prestressed concrete units. After failures in the United Kingdom, most other countries banned the use of high-alumina cement in structures, although some failures occurred in Spain as late as the 1990s. In structural applications, there is really no way to guarantee that conversion will not occur. High-alumina cement continues to be used in refractory applications because it can withstand temperatures as high as 1,350 °C (2,460 °F) (Neville 1995, pp. 92–103). In these applications, the resistance to high temperatures more than makes up for any loss of strength.

The case of high-alumina cement illustrates the difficulty of making appropriate use of new materials. At first glance, it seems ideal for use in prestressed structural applications because high early strength is needed for tendon release. The conversion problem did not become widely known until later, although engineer Adam Neville had argued against the structural use of high-alumina cement from the beginning. Before new materials are used in structural applications, it is necessary to test how they will perform over the long term.

Other Cases

For many of the cases discussed in other chapters of this book, material properties and behavior were involved in the failure.

2000 Commonwealth Avenue

The collapse of 2000 Commonwealth Avenue in Boston, discussed in Chapter 5, occurred primarily because of overloaded floor slabs and premature removal of formwork, combined with poor construction practices throughout the project. There was, however, a role played by the quality and the strength development of the concrete.

After the collapse, the necessary records of concrete strength tests proved to be difficult to find. This, by itself, is a strong indicator of poor construction management and quality control. Subsequent tests showed that the concrete was of poor quality.

Also, it was not protected during cold weather construction. As a result, it would reach the desired strength either late or not at all.

Skyline Plaza in Bailey's Crossroads

At Skyline Plaza in Bailey's Crossroads in northern Virginia, as at 2000 Commonwealth Avenue, a large portion of a structure collapsed when formwork was removed too early. The case is also discussed in Chapter 5. Although the concrete was probably of adequate quality, the collapse occurred in cold weather before the concrete had had the chance to achieve adequate strength.

Point Pleasant Bridge

The Point Pleasant Bridge collapse of 1967 was blamed on fatigue and stress-induced corrosion of high-strength steel. The full case study is presented in Chapter 3. Higher strength steel has less ductility and is potentially more susceptible to fatigue failure than mild steel.

Comet Jet Aircraft Crashes

The full case study is presented in Chapter 3. Unlike the Liberty ships and the Point Pleasant Bridge, which were made of steel, the Comets were made of aluminum. Aluminum does not have an endurance limit—there is no stress level for which aluminum may safely undergo an unlimited number of stress cycles.

Sampoong Superstore

The collapse of the Sampoong Superstore in Seoul, South Korea, is discussed in Chapter 10. This failure involved widespread corruption on the part of engineers, builders, and building officials. Low concrete strength played a small role in the collapse.

Cold-Formed Steel Beam Construction Failure

The collapse of cold-formed steel beams supporting a concrete placement is reviewed in Chapter 6. The properties of cold-formed steel sections are significantly different from those of hot-rolled shapes, although the material itself is similar. Therefore, those who design with this material need to know and use the proper codes and standards.

10

Management, Ethics, and Professional Issues

ENGINEERING PROGRAMS, PARTICULARLY CIVIL ENGINEERING programs, are required to cover management, ethics, and other professional issues in the curriculum. Depending on how the curriculum is structured, these topics may be addressed in a separate course or integrated in other courses. The cases reviewed in this chapter are designed to support either model. The ASCE Code of Ethics, a fundamental foundation for civil engineering practice, is provided in Appendix B.

Unlike Chapters 2 through 9, this chapter does not provide a separate section on Other Cases. Virtually all of the cases reviewed in those chapters have important management and ethics lessons, so such a section would be extensive as well as redundant. Instructors looking for additional case studies are encouraged to review those other chapters.

Citicorp Tower

This is a story of a collapse that never happened. If the Citicorp Tower had failed under wind pressure in midtown Manhattan, thousands, if not tens of thousands of people, would have been killed. Fortunately, the structural flaw was identified and fixed.

The case of William LeMessurier and the emergency repair of the Citicorp Center Tower in New York City has been used by some as an example of ethical behavior by an engineer. This point of view is described in Morganstern (1997). The abstract of the Morganstern paper states,

> What's an engineer's worst nightmare? To realize that the wind-bracing system he designed for a skyscraper like Citicorp Center is flawed—and hurricane season is approaching. In 1978, William J. LeMessurier, one of the nation's leading structural engineers, discovered after Citicorp Center was completed and occupied, conceptual errors pertaining to joint weakness, tension, and wind force. Alarmed by the magnitude of the errors and the danger they presented, LeMessurier acknowledged the flaws, immediately drew up new plans, and saw that all of the necessary changes to the braces were put into effect. LeMessurier's exemplary behavior—encompassing honesty, courage, adherence to ethics, and social responsibility—during the ordeal remains a testimony to the ideal meaning of the word, "professional."

The Citicorp Center skyscraper had been completed, and the engineer and architect had moved on to other projects. Morganstern (1997, p. 23) describes the incident that raised the question of the structural safety of the building:

> On a warm June day in 1978, William J. LeMessurier, one of the nation's leading structural engineers, received a phone call at his headquarters, in Cambridge, Massachusetts, from an engineering student in New Jersey. The young man, whose name has been lost in the swirl of subsequent events, said that his professor had assigned him to write a paper on the Citicorp Tower, the slash-topped silver skyscraper that had become, in its completion in Manhattan the year before, the seventh tallest building in the world. . . .
>
> The student wondered about the columns—there are four—that held the building up. According to his professor, LeMessurier had put them in the wrong place.
>
> "I was very nice to this young man," LeMessurier recalls. "But I said, 'Listen, I want you to tell your teacher that he doesn't know what the hell he's talking about, because he doesn't know the problem that had to be solved.' I promised to call back after my meeting and explain the whole thing."

Most buildings with a square floor plan, of course, have the main load-bearing columns at the corners. However, the northwest corner of the block for this building was occupied by the historic St. Peter's Church, which did not want to move although it would have been replaced at the same place with a new church. Therefore, LeMessurier and Hugh Stubbins, the architect, decided to place the main load-bearing columns at the center of each side. Figure 10-1 shows the church under the building, on the left side of the photograph.

This design put the entire structure on four massive nine-story-high stilts, with the skyscraper appearing to hover over the block. To carry the gravity loads at the corners of the building, multistory diagonal chevron braces framed into the main columns. A total of 48 braces was arranged in six tiers of eight each. Figure 10-2 shows one of the tower legs.

LeMessurier called the student back and explained that the position of the columns increased stability against quartering winds and suggested that he now had something interesting to tell his professor. He also taught architecture and engineering as an adjunct professor at Harvard and MIT. Therefore, he began to review the structure to prepare notes for a structural engineering course he was teaching to architecture students at Harvard.

He had previously determined the safety of the wind braces under wind loading acting perpendicular to the sides of the building. This is the

Figure 10-1. St. Peter's Church under the Citicorp Tower.
Courtesy Bob Pitt, University of Alabama.

Figure 10-2. Citicorp Tower legs.
Courtesy Bob Pitt, University of Alabama.

standard way in which structures are analyzed for wind forces. He decided now to analyze the forces in the braces under quartering winds—blowing at 45 degrees against two building faces simultaneously. For tall buildings, wind is usually the controlling structural load case.

The results of his analysis surprised him. With a more conventional structure, the reduced wind pressure effects acting over a larger area due to quartering winds cancel out, and the structural member forces remain the same. With the Citicorp Tower, however, the brace forces increase by 40% in four of the braces and cancel out in the other four.

The results were even more disturbing because LeMessurier had recently learned of a structural change that had been made to the braces. The original design called for welded connections. Bethlehem Steel had suggested that bolted joints would provide sufficient capacity at a much lower cost. He had not been told of the change and had been under the impression that the tower's joints had been welded.

Under the original structural analysis, the change would have made no difference because the building dead load would overcome any tension forces in the braces, and therefore the braces would always be in compression. However, with the revised calculation, the tension forces became larger and would place the braces and their connections into net tension. Morganstern (1997, p. 24) notes, "At any given level of the building, the compression figure remains constant; the wind may blow harder, but the structure doesn't get any heavier." Thus, the 40% increase in tension became a 160% increase in bolt forces.

Wind forces increase as the square of the wind velocity—when the wind speed doubles, the pressure and force quadruple. The critical event that could cause the building to fail, therefore, was a hurricane. Fortunately, hurricanes may be tracked long before they make landfall.

Unfortunately, failure under wind loading would be sudden and catastrophic. Frame-type buildings have considerable ductility and redundancy. If one element fails, as a general rule the forces can be redistributed to other members with reserve capacity.

A tension connection failure, such as at the Hyatt Regency walkways, gives no warning. There is no time to evacuate. Moreover, because the Citicorp Tower would have failed under lateral wind loading, it would have collapsed sideways and not straight down.

On the web (Case Western Reserve University 2007a), the case study about William LeMessurier superimposes a circle with a radius equal to the building height onto a map of midtown Manhattan, centered at the base of the building. It shows the dozens of blocks at risk from a collapse. The sideways failure, coupled with the lack of warning, would have caused a loss of life far exceeding that from the collapse of the two World Trade Center towers on September 11, 2001.

Another design decision reduced the factor of safety still further. The American Institute of Steel Construction specification in effect at the time provided different safety factors for columns, or compression members, and truss members. LeMessurier's engineers had used the lower safety factor for trusses, and therefore used a smaller number of bolts.

One mitigating factor was the tuned mass damper (TMD) that had been installed in the top of the tower. A 410-tonne (410-ton) block of concrete floated on a film of oil to reduce the motion of the building. It was not intended to be a safety device, although it reduced vibration. The TMD would have little effect on sustained wind pressure, which is treated essentially as a static load.

The degree of risk depended on the actual wind forces acting on the tower. The Citicorp Tower, like other skyscrapers, had been designed on the basis of wind tunnel tests. This work had been done at the Boundary Layer Wind Tunnel at the University of Western Ontario. Review of the test results did not help the situation—it was possible for a dynamic vibration response of the structure to increase some of the member forces. Because the steel tower was light and flexible, it was susceptible to crosswind vibration.

The Alternatives

LeMessurier took the information and analyzed the structure again. The weakest connection was at the 30th floor and could fail in a windstorm with a return period of roughly every 16 years. If the TMD worked, it would increase the return period to 55 years—but the TMD required electrical power, which could fail in a storm.

LeMessurier considered his options:

- silence because only his wind tunnel consultant knew of the problem, and he would probably not reveal the situation without LeMessurier's permission;
- suicide, similar to the above, without the difficulty of keeping the secret; or
- blowing the whistle on himself.

The short return period settled the issue. The risk to public safety was too clear to allow the problem to remain unsolved. He called the architect and the owner, Citicorp, and quickly met with their lawyers. Once they understood the problem, they offered full support for repairing the building and dealing with the public. Les Robertson, the structural engineer for the World Trade Center towers, was brought in to consult on the problem. Robertson brought disaster management as well as structural engineering expertise to the team.

The Plan

The obvious retrofit would be to weld the connections together. The problem was that the building was now occupied. LeMessurier proposed building a small plywood enclosure around each connection so that a welder would be able to work without causing too much disruption. It was necessary to locate enough steel plate and certified welders; 200 bolted joints had to be upgraded.

Because the TMD had been identified as important to the overall safety of the building, backup power generators were set up to keep it supplied with power under all conditions. Also, engineers from the company that had installed the TMD provided full-time, on-site technical support.

Strain gauges were also installed on critical members to provide structural monitoring. However, they were not installed by union electricians, and the wires were mysteriously snipped in the middle of the night.

Hurricane season was approaching. It was also necessary to set up weather monitoring and to make plans for evacuating the building and the surrounding area of midtown Manhattan. Robertson and Citicorp officials met with New York's emergency officials to prepare a plan.

Publicity

An obvious concern was how the public would react to the situation. Before work could start, the approval of the New York City building department would have to be obtained. At this point, it would become difficult to keep word from leaking out. Citicorp faced the difficult public relations task of explaining why a brand-new building had to be fixed and of articulating the risk to the public.

The Citicorp subsidiary Citibank's public affairs department issued a press release on Tuesday morning, August 8, 1978. The release said that some of the connections were being strengthened, on the advice of the engineers who had designed the building. The retrofit was attributed to new wind tunnel data, and the public was assured that there was no danger. The owner's representative presented the work as an abundance of caution, a "belt and suspender" approach. The press release was accurate, insofar as it went, but obviously left a great deal out.

The city's acting building commissioner was told the full truth. One concern was finding enough welders, so an agreement was struck to expedite testing and certification of additional welders for the project.

New York City has an aggressive press corps, and not all were satisfied by the news release and accompanying explanation. A reporter from the *New York Times* called LeMessurier's home in Cambridge, Massachusetts. When LeMessurier returned the call later in the day, he found that the *Times* and all the other New York newspapers had just gone on strike. The strike would continue until October, when the crisis would be over.

Executing the Repair

Work started right away. Much of the welding was done at night because no welding was allowed during office hours. The work continued seven days a week.

LeMessurier set the priority for which joints were fixed in what order, based on the degree of risk. As each joint was repaired, the return period of the critical wind increased.

On the first of September, a storm in the Atlantic strengthened to a hurricane and appeared to be headed for New York. The team carefully and nervously tracked the storm. By this point, the building could withstand a 200-year storm. Fortunately, the storm turned away.

The weather watch ended September 13, and the emergency measures were stood down. The tower was now strong enough to weather a 700-year storm, even without the dampers.

Repercussions

All the people had cooperated during the emergency because the most urgent thing was to get the building repaired. On September 13, Citicorp informed LeMessurier and Stubbins that the company wanted them to foot the bill for the work. The estimates ranged from $4.3 to $8 million for the structural repairs alone. After some discussion, LeMessurier offered the $2 million limit of his liability coverage. Eventually this was accepted, and Stubbins was held harmless.

According to Morganstern, there were no villains. All the people involved behaved well. LeMessurier's reputation was enhanced.

Morganstern (1997, p. 29) closes his paper the way LeMessurier concludes the story when he relates it to students.

> You have a social obligation. . . . In return for getting a license and being regarded with respect, you're supposed to be self-sacrificing and look beyond the interest of yourself and your client to society as a whole. And the most wonderful part of my story is that when I did nothing bad happened.

The story is told much the same way at: http://www.onlineethics.org/cms/8888.aspx.

An Alternative Point of View

Eugene Kremer (2002) takes a different view in his paper "(Re)Examining the Citicorp Case: Ethical Paragon or Chimera," which may be found on-line at http://www.crosscurrents.org/kremer2002.htm. Kremer first made his case at an ethics and architecture conference in New York City on April 6,

2002. His essay was published in *Cross Currents*, the magazine of the Association for Religious and Intellectual Life.

Kremer's paper narrates many of the same facts as Morganstern. He provides a more in-depth discussion of the dramatic architectural appearance of the tower and of the state of New York City's economy in the 1970s. The tower was a source of great pride.

The story of the retrofit of the tower was concealed for most of two decades by the parties involved, until Morganstern's 1995 *New Yorker* article. The case quickly became widely used in ethics courses for engineers and architects. One example is the Online Ethics website referred to earlier. The Illinois Institute of Technology Center for the Study of Ethics in the Professions (http://ethics.iit.edu/index.html) invited LeMessurier to relate his story for the center's 20th anniversary celebration. Kremer also cites several engineering ethics textbooks that use the case as an example of ethical behavior.

Kremer cites six aspects of the case worthy of further attention.

Analysis of Wind Loads

The Citicorp Tower should have been checked in the first place for quartering winds because building codes require consideration of the most severe loading case. Because this structural system departed so dramatically from convention, there was an accompanying duty to exercise more than ordinary caution.

It has been argued that the building code of the city of New York did not explicitly require analysis under quartering winds. Building codes, however, set minimum standards and do not reflect the state of the art. In fact, the designers of other tall buildings in New York and elsewhere had considered quartering winds. Two of LeMessurier's senior engineers stated that quartering winds were considered early in the design of the Citicorp Tower. The diagonal braces seemed simple and easy to understand, and it seemed obvious that if the building were safe for winds from any side it would also be safe against quartering winds.

The Change from Welded to Bolted Joints

The decision to change welded joints to bolted joints was an important change that should have been considered more carefully. The original design called for five full-penetration welded joints for each eight-story diagonal brace. The structural steel fabricator proposed a change to bolted joints and offered a $250,000 credit to Citicorp. LeMessurier's New York office approved the change. The decision was critical to the safety of the building and should have involved all of the key people on the project.

Professional Responsibility

Kremer suggests that by even considering suicide or silence as alternatives, LeMessurier failed to keep his focus on protecting the public, which was where it properly belonged.

Public Statements

LeMessurier had a duty to publicize the problem with the building. Kremer contends that the statement issued by Citibank was misleading and designed to obscure the threat to public health and safety. LeMessurier knew better, and in fact had supplied an essential part of the cover story. Kremer suggests that inaccurate statements were made to the *Wall Street Journal* and the *New York Daily News*, as well as to *Engineering News Record* by LeMessurier himself on August 17, 1978. This untruth violated Canon 3 of the ASCE Code of Ethics, which states that engineers shall issue public statements only in an objective and truthful manner.

Public Safety

The Red Cross estimated that a collapse of the Citicorp Tower could kill up to 200,000 people in the 156 city blocks in the neighborhood. The cover story denied the public the right to make a critical decision to evacuate for themselves. The desire to avoid public panic did not justify withholding critical safety information from these people. The first canon in ASCE Code of Ethics requires an engineer to hold public safety paramount.

Advancing Professional Knowledge

LeMessurier withheld the story for almost two decades. An address he gave as late as 1995 traced his experiences in wind engineering and discussed the Citicorp Tower in detail without mentioning the problem. Canon 7 of the ASCE Code of Ethics makes clear an engineer's obligation to advance knowledge of the profession. Instead, LeMessurier sought to hide the technical details of the problem and its solution. In the intervening two decades, another building might have been designed to the original standard, posing a risk to the public. Overall, Kremer's analysis offers an interesting counterpoint to the story of LeMessurier as an exemplar of professional engineering ethics.

Stanley Goldstein, who had worked in LeMessurier's New York City office and ran the project's later phases for LeMessurier, told a local chapter of the New York Society of Professional Engineers that the version related by Morganstern was inaccurate. He says that the Cambridge office was aware of the change from welds to bolts and provided the design forces for the connection. Goldstein points out that the *New Yorker* article

> doesn't state that the bolts were substituted for butt welds, which because they would have developed the full strength of the members rather than just what the calculated forces were, would have obviated the need for any reinforcing at all. (Korman 1995)

Robert J. McNamara, who was the project manager in Cambridge, claims that LeMessurier studied quartering winds early in the design process and was aware of the change from welds to bolts when it was made, not much later, as claimed in Morganstern's paper. In response to the comments, LeMessurier stated, "My mistake was in designing a structure that was innovative, and I didn't check and dog people carrying it through carefully enough" (Korman 1995).

The Intervening Two Decades

Is it true, as Kremer alleges, that LeMessurier buried his story for two decades? I'm able to offer an unusual perspective on this issue because I took his course Structural Design of Buildings during the fall 1985 semester at MIT, about seven years after the incident and a full decade before Morganstern's paper.

During the course, we discussed various wind load-resisting systems, including the Citicorp Tower. My notes from December 6, 1985, which may have been the last day of class, reflect an in-depth discussion of the case. Unfortunately, my notes are sketchy, which is probably more of a reflection of my own laziness late in the semester than any efforts to conceal on LeMessurier's part.

An extract from my notes for that day is provided as Fig. 10-3.

The notes reflect the following elements later related by Morganstern:

- A problem was found with the new building in 1978.
- Four connections are shown per diagonal. The connections were intended to be full penetration wells but were replaced by bolts at the suggestion of Bethlehem Steel.
- Calculations for the forces in the diagonals due to quartering winds are shown. For the eight braces per floor, four have essentially no wind force and four have approximately 1.4 times the force calculated from the two-dimensional analysis. The 2D analysis was a homework problem earlier in the course.

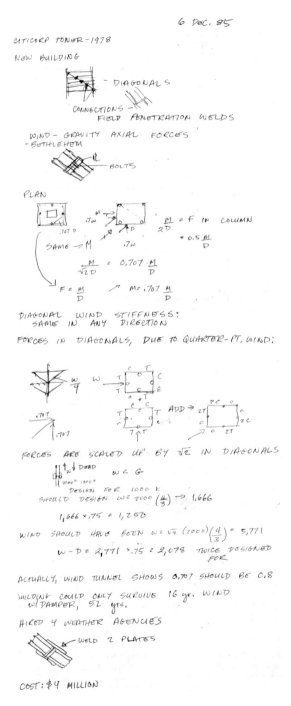

Figure 10-3. Extract from course notes.

- Connections should have been designed for wind force minus dead load ($W - D$), and the actual tension difference in the force was twice what was designed for.
- Wind tunnel results showed that the 0.707 pressure multiplier for quartering winds should in fact have been a slightly larger 0.8.
- The building could only survive a 16-year wind without the damper working or a 52-year wind with the damper.
- Four weather agencies were hired to monitor approaching storms.
- The retrofit required welding two plates per connection, at a cost of $4 million.

My notes partially answer the question of how forthcoming LeMessurier was to his students at Harvard and MIT. Certainly he went to the trouble to relate the main technical details. The notes are silent on ethical aspects, but this may simply have been poor notetaking on my part and may not reflect every detail of the lecture.

Essential Reading

LeMessurier's side of the story is told quite well in Morganstern's 1997 paper, which provides greater narrative depth than the account provided above. The paper, entitled "The Fifty-Nine Story Crisis," was published in the January 1997 issue of the ASCE *Journal of Professional Issues in Engineering Education and Practice*. However, this paper does not have any diagrams or graphics. The article was previously published in *The New Yorker* magazine on May 29, 1995.

The story is told in the "Moral Exemplars" section of the Online Ethics Center for Engineering and Science at Case Western Reserve University's website http://www.onlineethics.org/cms/8888.aspx. The narrative closely follows Morganstern's paper. Some graphics and a video clip have been added.

The Morganstern paper and the Online Ethics Center website tell the story from LeMessurier's point of view. Kremer's opposing view may be found at http://www.crosscurrents.org/kremer2002.htm.

Space Shuttle Challenger

Each space shuttle mission involves a large amount of safety engineering, whether measured in dollars or person-hours. Yet two of the space shuttles have exploded catastrophically, with a total loss of vehicle and crew. In contrast, the shuttle's predecessor, the Saturn V rocket, had a perfect safety record.

The shuttles were intended to make space travel routine and to reduce the costs of putting satellites into orbit and of building and maintaining the space station. They could be reused, unlike rockets such as the Saturn V. The shuttles have never fulfilled that promise, either in achieving the desired frequency of the flights or in reducing the cost per flight.

Engineering, particularly rocket engineering, requires a can-do spirit, optimism, and a willingness to overcome obstacles. Experimental aviation is inherently risky. It is difficult to advance space technology while maintaining an appropriate level of safety.

The *Challenger* exploded 73 seconds after launch on January 28, 1986. It was the 25th shuttle mission. The seven astronauts killed included a high school teacher, Christa McAuliffe, who was to broadcast lessons from space. Government investigations and redesign efforts grounded the shuttle program for the next two years (Freiman and Schlager 1995a, p. 161).

Space Shuttle Design and Operation

The reusable portions of the space shuttle consist of the orbiter, an external fuel tank, and the solid rocket boosters (SRBs). The orbiter flies into space and returns to land on a runway. The external tank and SRBs provide the thrust to lift the orbiter into space and are jettisoned into the ocean and recovered.

Each of the two SRBs is 45 m (149 ft) long and 3.7 m (12 ft) in diameter and weighs 8.9 MN (2 million lb) before ignition. Although solid rockets produce much more thrust per unit mass than liquid rockets, they cannot be turned off or controlled once ignited. Morton Thiokol won the contract to produce the SRBs in 1974 and used a scaled-up version of the Titan missile, which had an excellent performance record (Space Shuttle *Challenger* Disaster 2007).

Each SRB is made of seven hollow metal cylinders, which are built separately, shipped in connected pairs, and finally joined together at the Kennedy Space Center in Florida for launch. At Kennedy, they are connected with tang and clevis field joints, each with 177 clevis pins (Space Shuttle *Challenger* Disaster 2007).

Each of the field joints is sealed with two O-rings, a primary and a secondary ring. The Titan only had a single ring, but a second was added to the SRBs for redundancy and safety. The purpose of the O-rings is to keep hot combustion gases from escaping from the inside of the motor through the field joint. Heat-resistant putty is applied inside the inner O-ring to pro-

MANAGEMENT, ETHICS, AND PROFESSIONAL ISSUES 347

tect the ring itself from the gases. The gap between the tang and the clevis is adjusted with shims to reduce the gap and increase the compression of the O-ring (Space Shuttle *Challenger* Disaster 2007). The O-ring joint is shown in Fig. 10-4.

Several factors affect the size of the gap, including dimensional tolerances of metal cylinders and the tang and clevis assembly, the temperature, the O-ring diameter, the shims, and the loads applied to the joint. When the booster is ignited, internal pressure presses against the putty and forces the O-ring into the gap between the tang and clevis to form a seal (Space Shuttle *Challenger* Disaster 2007).

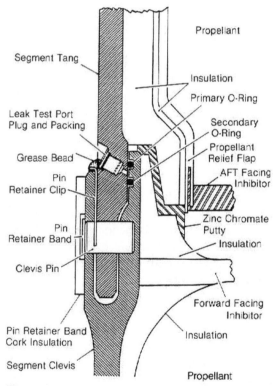

Figure 14
Solid Rocket Motor cross section shows positions of tang, clevis and O-rings. Putty lines the joint on the side toward the propellant.

Figure 10-4. Space Shuttle Challenger O-ring joint.
Courtesy National Aeronautics and Space Administration.

The pressure against the SRB walls also causes the cylinder to increase in diameter or balloon slightly. This, in turn, causes the gap in the tang and clevis joint to open, a phenomenon known as joint rotation. Morton Thiokol discovered the effect during testing in 1977 and implemented three changes to decrease joint rotation. The changes were improved dimensional tolerances of the metal joint and O-ring, an increase in O-ring diameter, and the use of shims. Morton Thiokol's testing also indicated that in some instances, the second O-ring might not seal at all, eliminating the redundancy of the design (Space Shuttle *Challenger* Disaster 2007).

From the beginning in 1958, NASA's space program had been built around a strong culture of safety. However, the shift from rockets to space shuttles was accompanied by a shift of emphasis to making the program fully operational and cost effective. By the early 1980s, NASA had lost much of its internal capability for overseeing technical quality control of the work done by outside contractors. National security instructions called for 24 shuttle flights per year, with 6 reserved for exclusive Air Force use. However, NASA found that each shuttle required at least 1,240 h between scheduled launches, not the 160 originally assumed. Still, 1986 was to be a breakthrough year with 15 missions using all four shuttles. There was considerable pressure throughout the organization to meet this ambitious schedule (Freiman and Schlager 1995a, pp. 161–165).

Leaks in the Primary Seal

A new problem emerged after the second shuttle mission in November 1981. Postflight examination of the SRB field joints showed that the O-rings were eroding during flight. Although the joints still sealed effectively, some hot gases got past the putty and damaged the O-rings. Morton Thiokol investigated different types of putty to improve joint performance (Space Shuttle *Challenger* Disaster 2007).

The onlineethics.org website uses the case of Roger Boisjoly as one of its "Moral Exemplar" studies (Case Western Reserve University 2007b). Boisjoly was an engineer with more than a quarter century of experience in the aerospace industry who became involved in trying to improve the O-rings used in the space shuttle solid rocket boosters.

Boisjoly observed a large amount of blackened grease between the two primary seals of the solid rocket boosters in January 1985 during a postflight hardware inspection of Flight 51C. This evidence suggested that hot combustion gases had compromised the integrity of the seals. He reported the findings and was asked to make a presentation at the Marshall Space Flight Center in Huntsville, Alabama. The solid rocket booster manufacturer, Morton Thiokol, was also asked to make a presentation. The presentations were part of the flight readiness review for Flight 51E, which was scheduled for April 1985. Boisjoly suggested that the lower than usual launch temperatures had interfered with the seating of the O-rings and were thus responsible for the amount of gas blow-by. NASA management insisted that Boisjoly soften his position for the final review board (Case Western Reserve University 2007b).

Boisjoly and another engineer, Arnie Thompson, discussed the low-temperature problem and proposed further testing. Resiliency testing on the O-ring material showed that the rings had problems seating at low temperatures. Another postflight inspection, that of a nozzle joint from flight 51B from April 29, 1985, showed seal erosion. Boisjoly thought that the primary seal might not have sealed at all during the two-minute flight, and he became more concerned because if the same thing happened to a field joint, the secondary seal could also fail (Case Western Reserve University 2007b).

During the Flight Readiness Review for mission 51F, scheduled for launch on July 1, 1985, Boisjoly presented his findings, including his test results from a few months earlier. The 7-mm (0.28-in.) diameter O-ring specimens were compressed to 1 mm (0.04 in.) with a decompression distance of 0.75 mm (0.03 in.) at a rate of 50 mm/min (2 in./min). At 38 °C (100 °F), the seals did not lose contact, at 24 °C (75 °F), they lost contact for 2.4 s, and at 10 °C (50 °F), they lost contact for 10 min. Increasing the seal diameter to 7.5 mm (0.295 in.) helped, but the temperature dependence was clear (Case Western Reserve University 2007b). During the period that the seals lost contact, hot gases could get by. Because no tests were performed at temperatures lower than 10 °C (50 °F), it was not possible to say what would happen at lower temperatures. However, the trend had been clearly established.

Boisjoly was concerned about the lack of attention to the problem. After overcoming initial resistance, he was able to form a solid rocket motor seal erosion team to address the problem. NASA headquarters asked Morton Thiokol to make a presentation on all of the booster seal problems. Boisjoly was told by NASA to emphasize the joint improvements but not the urgency of the problem. He did not get any response to his presentation, and the seal erosion team remained frustrated by lack of management support (Case Western Reserve University 2007b).

Morton Thiokol ordered new steel billets for a redesigned field joint in July 1985. However, because these billets would take many months to manufacture, they were not available by the time of the fatal *Challenger* launch (Space Shuttle *Challenger* Disaster 2007).

The Meeting Before the Launch

A final meeting was held the night before the *Challenger* launch. Overnight, the temperatures were forecast to fall to −3 °C (26 °F). Boisjoly presented his arguments about the temperature sensitivity of the seals and made his recommendation against the launch. NASA asked those present for their launch decisions. Joe Kilminster, the vice president of Space Booster Programs, and George Hardy of NASA did not recommend launching (Case Western Reserve University 2007b).

The engineering argument against launching was that there were no data to show that a launch below 12 °C (53 °F) was safe. NASA had specified a booster operation temperature as low as −1 °C (31 °F). Thiokol had understood this to be a storage temperature for the booster and that the launch temperature specification was 4 °C (40 °F). Bob Lund, Thiokol's engineering vice president, noted that the predicted temperatures at launch were outside of the database and recommended delaying the launch until temperatures rose above 12 °C (53 °F). Larry Mulloy, the Marshall Center's solid rocket booster project manager, countered that the data were inconclusive. Jerald Mason, a Thiokol senior executive, commented that a management decision was needed. Mason told Lund to "take off your engineering hat and put on your management hat." Kilminster wrote out a new recommendation to launch, which the engineers refused to sign. NASA approved the boosters for launch, although the predicted temperature was outside of the specifications (Space Shuttle *Challenger* Disaster 2007).

The Columbia mission before *Challenger* had been postponed seven times. *Challenger* had a tight launch window. It was necessary to fly in the morning because an afternoon launch would mean that a possible emergency landing in Casablanca on the west coast of Africa would have to be made at night on an unlit runway. The *Challenger* was also supposed to recover the Spartan–Halley research observatory in orbit, which meant that the launch had to be before January 31 (Freiman and Schlager 1995a, pp. 164–165).

The Shuttle Explosion

That night, the temperature fell to −8 °C (18 °F). Ice on the vehicle became a concern. Several times during the final countdown, key personnel had to override the low-temperature safety limitations. At launch, the coldest spot on the booster was one of the field joints, at −2 °C (28 °F). At ignition, the joint rotated, but the O-rings were too cold to seat. Hot gases at a temperature of 2,760 °C (5,000 °F) shot past the O-rings, and cameras spotted about nine smoke puffs during the launch. At 72 s into the flight, the shuttle *Challenger* exploded, killing all seven aboard (Space Shuttle *Challenger* Disaster 2007).

Boisjoly had continued to protest the launch decision, but watched the launch. After the explosion, he went directly to his office and sat in silence (Case Western Reserve University 2007b).

Investigations

The *Challenger* explosion was investigated by a presidential commission, as well as the U.S. Senate and House of Representatives. This disaster represented a rare engineering failure where a number of people knew the cause immediately.

The Rogers Commission, appointed by President Ronald Reagan, recorded 15,000 pages of testimony and reviewed 170,000 pages of documents and hundreds of photographs. More than 6,000 people were interviewed as part of the three-month investigation. The commission's findings were released on June 6, 1986, in a 256-page report. The House Committee on Science and Technology also conducted hearings and issued a report. After the incident, a number of astronauts and other key personnel resigned from NASA (Freiman and Schlager 1995a, pp. 169–171).

Whistle Blowing

The larger and more important the project, the greater the economic and political consequences of delay or failure. This situation makes whistle blowing on large projects particularly difficult, and events rarely work out well in the long term for whistleblowers. However, for civil engineers, whistle blowing is required by the ASCE Code of Ethics.

Essential Reading

The U.S. House of Representatives Committee on Science and Technology report (Committee on Science and Technology 1986), "Investigation of the Challenger Accident, October 29, 1986," is available online at http://history.nasa.gov/rogersrep/51lcover.htm and http://www.gpoaccess.gov/challenger/64_420.pdf. A summary is provided in Freiman and Schlager (1995a, pp. 161–172). Two ethics websites, onlineethics.org (Case Western Reserve University 2007b) and Texas A&M University (Space Shuttle *Challenger* Disaster 2007), feature this case study.

Sampoong Superstore, Korea

Whether we like it or not, civil engineering, like other engineering professions, is becoming increasingly international. Many large U.S. construction and design firms compete throughout the world. In some countries, bribery and corruption are epidemic. However, the ASCE Code of Ethics and other professional standards do not have an international exception, and engineers licensed in the United States are bound by those standards, as well as by U.S. laws, no matter where they practice or where the project is located. Bribes are not a tax-deductible business expense. In some cases, U.S. firms have a difficult time competing successfully in construction markets where bribery is rampant.

This statement is not to imply that there is no corruption in the U.S. construction industry. Although it is not addressed in any reports, it is difficult to understand how a mismanaged project like 2000 Commonwealth Avenue (discussed in Chapter 5) could proceed without a bribe or two changing hands.

It is useful to study failures outside the United States, particularly those failures in which shoddy construction, bribery, or corruption may play a part. Earthquakes overseas often expose broad patterns of poor construction. An earthquake that would kill a few dozen people in California may kill hundreds or thousands in the developing world. Buildings sometimes collapse because of unauthorized construction of additional stories without permits. The principles of physics and engineering do not respect international borders.

Failures outside of the United States present some difficulties for researchers because the relevant documents and reports may be unobtainable or may not be available in English. News reports are generally sketchy with respect to technical matters and may oversimplify the case.

Another significant difference with U.S. practice is that engineers and contractors may be charged in criminal court. Some of the engineers involved with the Vaiont and Malpasset dam failures (in Chapters 7 and 8, respectively) were tried and convicted.

The collapse of the Sampoong Superstore in Seoul, South Korea, represents an example of a structural collapse attributed in large part to corruption. The late 1980s were an exciting time in Seoul and the rest of South Korea. In 1988, Seoul hosted the summer Olympic Games. The games gave South Korea an opportunity to show off its technological advances, and the nation took full advantage. I was stationed in South Korea as an Army officer during this period, and it was fascinating to watch the process unfold.

In Korea, large family-dominated firms called *chaebols* are responsible for much of the economy. Names that would be familiar in the U.S. include Hyundai, Samsung, and Daewoo. The *chaebols* often cross many industries, such as automotive, construction, retail, and shipbuilding. It would not be unusual for a *chaebol* to own a department store as well as the construction company that built it. Some of these *chaebols* are active in the international construction market.

Design and Construction

The Sampoong department store opened in December 1989. It was a nine-story building with four basement floors and five above grade. The building was laid out in two wings (north and south) connected by an atrium lobby. By the mid-1990s the store's sales amounted to more than half a million U.S. dollars a day (Wearne 2000, pp. 99–100).

Unfortunately, the store had been built on a landfill site that was poorly suited to such a large structure. Woosung Construction built the foundation and basement and then passed the project on to Sampoong's in-house contractors. Woosung had apparently resisted some proposed changes to the building plans, such as the addition of the fifth floor (Wearne 2000, p. 100).

Sampoong made significant changes to the structure. The most important was the conversion of the original use as an office block to that of a department store. Other changes included changing the upper floor from a roller-skating rink to a traditional Korean restaurant. Stricter standards had to be met for fire, air conditioning, and evacuation. Although the structure apparently met all building code requirements, the revised design was radically different from the original (Wearne 2000, p. 100).

Collapse

The building was put into service.

> For five and a half years business thrived. In June 1995 the store passed a regular safety inspection. But within days there were signs something was seriously wrong: cracks spidering up the walls in the restaurant area; water pouring through crevices in the ceiling. On June 29 structural engineers were called in to examine the building. They declared it unsafe. Company executives who met that afternoon decided otherwise. They ordered the cracks on the fifth floor to be filled and instructed employees to move merchandise to the basement storage area. (Wearne 2000, p. 100)

Some employees heard rumors of the structural damage and impending collapse but remained in their departments to work. At 6:00 P.M. on June 29, the center of the building collapsed, similar to a controlled implosion, in about 10 s. The five-story north wing, about 91 m (300 ft) long, fell into the basement, leaving only the façade standing (Wearne 2000, pp. 100–102).

Customers were concentrated in the basement and in the fifth-floor restaurant. The customers and employees had no time to run. Some survivors were found in the wreckage, and one was brought out 17 days after the collapse. The overall death toll was 498 (Wearne 2000, pp. 100–107).

Investigations and Conclusions

The technical causes of the collapse seemed straightforward. "The investigating committee noted design errors, many construction faults, poor construction quality control, reduction in the cross-section of the columns supporting the fifth floor and roof and change in use of the fifth floor . . ." (Gardner et al. 2002, p. 523).

Two Korean professors, Lan Chung of Dan Kook University in Seoul, and Oan Chul Choi of Soongsil University, investigated. Their findings were summarized by Wearne (2000, pp. 107–111):

- The store was a flat slab structure, without cross beams supporting the slab. This made the structure inherently less redundant. This design explained why the building had collapsed so quickly and completely.
- Quality of the concrete was not the cause. Samples tested for compressive strength did not show extraordinary weakness.
- The foundations and basement built by Woosung Construction had survived the collapse, and the foundation rested on rock. Therefore, foundation problems could be ruled out as the cause of the collapse.
- When the building design had been converted from an office block, it had been necessary to cut holes for escalators in each floor slab and remove some supporting columns.
- The change of use also required installation of fire shutters. Large chunks of the concrete columns had been cut away to fit the fire shutters.
- The fifth-floor conversion to a restaurant had added considerable weight. In a traditional Korean restaurant, diners sit on the floor, which must be heated. The floor and embedded heating system was 0.9 m (3 ft) thick and made of concrete. Refrigerators also increased the dead load.

- Because the floor plan of the restaurant was not compatible with that of the lower floors, the placement of the support columns was irregular. Columns did not line up from floor to floor. Therefore, the slab between the fourth and fifth floors, not columns, transferred loads.
- The building's large, heavy water-cooling blocks for air conditioning had been installed on the roof, rather than on the ground. In summer, when the collapse occurred, they were full of water and weighed a lot. They had been placed on the roof to avoid noise complaints from neighbors.
- Rather than adding columns, the builders increased the thickness of the roof slab to accommodate the water-cooling system.
- The roof system had about a quarter of the capacity required to support the water-cooling system.
- The building owners had recently moved the water-cooling blocks from the back of the building to the front. Instead of lifting them with a crane, workers slid them across the roof, causing considerable structural damage. Cracks up to 25 mm (1 in.) wide were observed where the blocks had been moved.
- On the lower four floors, columns specified to be 890 mm (35 in.) thick were only 610 mm (24 in.) thick and had only 8 reinforcing bars rather than the 16 specified.
- Slab dead loads had been miscalculated, based on 100-mm (4-in.) thick slabs when some slabs were three or four times thicker.
- Some reinforcement had not been installed, and connections between slabs and walls were poor.
- To maximize sales space, the spans between the columns had been increased to almost 11 m (36 ft), which was much too large.

Professor Chung and his colleagues blamed the Sampoong department store collapse unequivocally on "human ignorance, negligence, and greed." The prime cause, they said, was the "illegal alteration of the architectural design and usage of the building." They cited the negligence of supervision of the planning authorities and the refusal to act on any of the indications of structural problems by the management as crucial contributing factors to the disaster. Cracks and leaks had been appearing in the building for more than five years.... (Wearne 2000, p. 111)

The concrete used was only 18 MPa (2,600 lb/in.^{2}) rather than the specified 21 MPa (3,000 lb/in.^{2}). Actual concrete strength from samples taken after the collapse ranged from 18.4 to 19.3 MPa (2,700–2,800 lb/in.^{2})

(Gardner et al. 2002, pp. 523–524). The originally specified value is rather low for structural concrete, and the further reduction in strength would reduce the slab punching shear capacity.

The effective slab depth for negative moment areas had been reduced from the specified 410 mm to 360 mm (16 to 14 in.) because the reinforcement was improperly placed. Also, the change in use had increased the dead load on the fifth floor by 35%. The lightweight concrete topping used on the roof had more than twice the dead load assumed in the design. Dead load of drop panels was also neglected (Gardner et al. 2002, pp. 523–525).

Punching shear of concrete slabs is discussed in Chapter 5. The factors that reduced the punching shear capacity of this structure included (Gardner et al. 2002, pp. 524–525):

- the reduction in concrete strength,
- the reduction in effective slab thickness,
- the reduction in column diameter, and
- omission of a drop panel at the top area of column line 4 and E, reducing slab thickness from 450 to 300 mm (18 to 12 in.).

Also, moment transfer to columns was only checked for the exterior columns, not the interior columns. The South Korean structural concrete building code requirements were identical to the U.S. ACI (1983) 318 code. The ACI 318 code now requires carrying some positive moment steel through the columns, which might have minimized the extent of the collapse (Gardner et al. 2002, pp. 525–529).

The final report was delivered by the Seoul District Prosecutors Office, entitled *The Final White Book of Finding Out the Real Truth of the Collapse of the Sampoong Department Store*. The public was outraged. In particular, the news that the senior executives had fled the building without warning others was disturbing. The report on the collapse, as well as earlier structural and construction failures, suggested a widespread pattern of corruption in the country's construction business. A government survey of high-rise structures found 14% were unsafe and needed to be rebuilt, 84% required repairs, and only 2% met standards. Joon Lee, the chairman of Sampoong, and his son Han-Sang Lee, were convicted and sent to prison for 10½- and 7-year terms, respectively. Twelve local building officials were found guilty of taking bribes of as much as $17,000 (U.S. equivalent) for approving changes and providing a provisional use certificate (Wearne 2000, pp. 111–112).

The cause of the Sampoong collapse, then, was not a technical issue as much as outright fraud. The Korean construction industry, protected by government regulation from outside competition, had become complacent. Bribes were used to get around the usual government checks and balances that serve to protect public safety.

It is difficult for a firm that insists on maintaining ethical standards to operate in such an environment. It is worth noting that Woosung Construction lost the project after balking at making the requested changes.

Essential Reading

This case study is discussed by Wearne (2000, pp. 99–113) in Chapter 5, entitled "Crooked Construction: Sampoong Superstore." A technical paper on the collapse entitled "Lessons from the Sampoong Department Store Collapse" (Gardner et al. 2002) was published in the *Cement & Concrete Composites* journal.

Misuse of the Professional Engineer License

Although there are other paths to licensure, the most common way to earn a professional engineer (P.E.) license today is to graduate from an Accreditation Board for Engineering and Technology (ABET) accredited engineering program, pass the Fundamentals of Engineering (FE) examination, practice engineering for four years under the supervision of a licensed P.E., and then pass the principles and practice examination.

There are more than 50 licensing jurisdictions in the U.S. states and territories, and as a general rule a separate license must be acquired for each jurisdiction. Licensing requirements may be different in the various jurisdictions but generally follow a national model law. Each jurisdiction has its own licensing board that enforces requirements and may suspend or revoke licenses. Land surveying has a separate LS or PS (Professional Surveyor) license, and in some jurisdictions structural engineering also has a separate examination and license. In the states where I hold licenses—Alabama, Ohio, and Virginia—there is no separate licensure for structural engineers.

Depending on jurisdiction, a P.E. may be referred to as a registered engineer, a licensed engineer, or a professional engineer. Within the jurisdiction, only a person with a P.E. in that jurisdiction may offer or perform engineering services. Professional engineers are also bound by the jurisdiction's code of ethics, which are usually similar to the ASCE Code of Ethics.

Beyond the licensing boards, engineering is a largely self-regulating profession. Engineers are supposed to restrict their practice to areas in which they are qualified by education or experience, but the P.E. license is the same for all. A person who earns a P.E. license through education and experience in electrical engineering, for example, is not directly prevented from practicing civil engineering. Obviously, this situation has the potential to create problems.

In the Harbour Cay Condominium collapse in Cocoa Beach, Florida, discussed in Chapter 5, the retired NASA engineers who designed the building had probably not been designing reinforced concrete rockets for NASA. Feld and Carper note that,

> The project was built at a time when a considerable amount of construction was under way in Florida, straining the ability of local building department staff to keep pace with developers. Five of the parties involved in the Harbour Cay project were charged with negligence by the Florida Department of Professional Registration. Both structural engineers surrendered their licenses and will never again practice in the state of Florida. The architect who designed the project was suspended from practicing in Florida for a period of 10 years, and two contractors were disciplined. This failure serves as a reminder that such tragic occurrences are still possible in uncomplicated low-rise projects, despite the availability of sufficient knowledge to prevent them. (1997, p. 274)

These two engineers were clearly practicing outside of their expertise and experience. Punching shear is critical for structures of this type, and the change in the shop drawings on the height of the chairs should never have been allowed. The ACI code is revised every three to four years, and often the revisions are significant. There is a substantial investment of time required to keep up with the code, but engineers who don't should not attempt to design these structures.

I've encountered similar problems in my own practice. In the agricultural product warehouses case, described in Chapter 4, the engineer who designed the foundation had graduated with a bachelor's degree in civil engineering about three or four decades before. His entire career had been spent in preparing and approving soil laboratory reports, and there was nothing in the record to suggest that he had designed reinforced concrete since taking an undergraduate course on the topic. When the building owner asked him to design the foundation, he should have referred the owner to a more qualified engineer. Instead, he produced a design that turned out to be inadequate. No reinforcement was provided to resist the tensile forces applied to the foundation.

Property Loss Investigations

Failures may also happen with residential structures because of a number of factors, such as wind, water damage, foundation settlement, and many others. Insurance companies hire engineers to perform property loss forensic investigations. The insurance company wants to know the cause of the property loss, whether the cause of the loss was covered, and possibly an estimate of the cost to repair or replace the structure. Often, the question arises as to whether a structure is economically repairable and how it could be repaired.

Generally, there is no loss of life or injury involved. These are not dramatic collapses like some of the larger cases discussed in this book, but they are of course important to the property owners involved.

I have had cases where homeowners have disputes with insurance companies as to whether the damage to their homes occurred because of foundation settlement or because of some other event (hurricane winds, tornadoes, or earthquakes). The distinction is important—in the case of foundation settlement, insurance does not pay for the damage. The insurance company hires an engineer to examine the damage and make a recommendation.

In three cases, I've been hired by homeowners' attorneys to inspect residences and to recommend whether they should challenge findings of settlement. In two of those cases, I reviewed reports written by the same investigator, both of which found that the damage occurred because of settlement. The expert had a degree in mechanical engineering and had taken some short courses in topics such as wood design. I'm not sure he had the necessary background to come to his conclusions. For example, he did not appear to have ever taken a course related to geotechnical engineering, which is rather useful for assessing foundation settlement.

For many small firms or solo practitioners, these residential and small structural investigations may be the bread and butter of the practice. I've lost track of exactly how many I've done, but it's several dozen at least.

There is an inherent bias in this system, particularly if the evidence is ambiguous. An engineer who often finds foundation settlement damage will probably continue to get work from the insurance company. Therefore, there is unspoken but subtle pressure to find in favor of settlement. In the wake of a large event, such as a hurricane or flood, this sort of work may be highly lucrative for a small engineering firm because of the high volume.

This bias becomes especially pronounced in the case of a large event, such as a hurricane. After a hurricane, a company that insures a large number of properties in the affected area faces massive payouts. If the insurance adjustors can deny or reduce claims, the overall payout may be greatly reduced.

Dietz and Preston (2007) documented widespread efforts by insurance companies to reduce claims in property loss situations. They also described pressure applied by an insurance company to an engineering firm to alter report conclusions and reduce claims after Hurricane Katrina. The dispute concerned whether losses were classified as wind damage, which was covered by insurance, or flood damage, which was not.

> An engineer involved in Katrina, Bob Kochan, CEO of Forensic Analysis & Engineering Corp., says State Farm asked him to redo his reports because the insurer disagreed with the engineers' conclusions. . . . Kochan says he complied so State Farm didn't cut its contract with his company. . . .
>
> Randy Down, an engineer at Raleigh, North Carolina-based Forensic, wrote this Oct. 18, 2005, e-mail response to Kochan: "I have serious concern about the ethics of this whole matter. I really question the ethics of someone who wants to fire us simply because our conclusions don't match theirs." (Dietz and Preston 2007)

This situation is not meant to suggest that performing these engineering investigations for insurance companies is an inherently unethical practice. If I thought it was unethical, I wouldn't do it. However, I suggest than an engineer or firm engaged in such work understand the ethical pressures involved and follow the ASCE Code of Ethics. In this matter, as in so many others discussed throughout the book, the Code of Ethics is an important guide.

Appendix A

Notes to the Professor

THERE IS A DOCUMENTED NEED FOR FAILURE AWARENESS in the undergraduate engineering curriculum. Engineering students can learn a lot from failures, and failures play an important role in engineering design. This need has been expressed in a number of papers and at a number of conferences over the past two decades. This book is a specific response to that need and will provide much-needed access to examples and a heightened appreciation of the role failure analysis knowledge can play in higher education and public safety.

Many of the key technical principles that civil engineering students should learn can be illustrated through case studies. For example, the author has discussed the Hyatt Regency walkway collapse, the Tacoma Narrows Bridge failure, and other well-known cases with students in courses on statics, mechanics of materials, and other courses. These cases help students:

- grasp difficult technical concepts and begin to acquire an intuitive feel for the behavior of systems and structures,
- understand how engineering science changes over time as structural performance is observed and lessons are learned,

- analyze the effects of engineering decisions on society, and
- appreciate the importance of ethical considerations in the engineering decision-making process.

The main obstacle to integrating case studies and lessons learned from failures into existing courses has been that many faculty do not have time to research and prepare case studies. Although there are many references available, they are difficult to translate into classroom lectures without considerable added effort on the part of the instructor.

There are three ways to introduce failure analysis and failure case studies into civil engineering education. A small number of colleges and universities offer courses in forensic engineering or failure case studies. Often, these courses are offered at institutions such as the University of Texas, Mississippi State University, or the University of Colorado at Denver, which have practicing forensic engineers on the faculty (Delatte and Rens 2002). Clearly, this approach depends on the availability of qualified and interested faculty.

Another method is to use case studies in capstone (senior) design projects (Delatte and Rens 2002). This method also depends on interested and qualified faculty, as well as on the availability of appropriate projects (which must be sufficiently free of liability concerns).

These two approaches offer great depth in the topic, but because of their inherent limitations, their application is likely to remain limited. As a result, even at colleges and universities where courses are offered in this area, few undergraduates are likely to be able to take them. Although some people might argue for a required stand-alone course in failure analysis for all undergraduate civil engineering students, the argument is likely to fall on deaf ears as programs shrink their credit-hour requirements. However, this book would be an excellent text for a civil engineering failure analysis course.

A more promising approach is to integrate failure case studies into courses throughout the curriculum. Many professors have done this on an informal basis for years. The author used this approach at the U.S. Military Academy while teaching two courses in engineering mechanics: statics and dynamics and also mechanics of materials (Delatte 1997). He continued the approach in engineering mechanics and civil engineering courses at the University of Alabama at Birmingham (Delatte 2000, Delatte and Rens 2002, Delatte 2003) and at Cleveland State University.

Why Study Failures?

In a survey conducted by the ASCE Technical Council on Forensic Engineering (TCFE) Education Committee in December 1989, about a third of the

87 civil engineering schools responding indicated a need for detailed, well-documented case studies. The University of Arizona said, "ASCE should provide such materials for educational purposes," and Swarthmore College suggested, "ASCE should provide funds for creating monographs on failures that have occurred in the past" (Rendon 1993a).

The ASCE TCFE conducted a second survey in 1998, which was sent to all Accreditation Board for Engineering and Technology (ABET) accredited engineering schools throughout the United States (Rens et al. 2000b). Similar to the 1989 survey, the lack of instructional materials was cited as a reason that failure analysis topics were not being taught. One of the unprompted written comments in that survey was this: "A selected bibliography is needed on the topic, which could be accessed via the Internet."

The use of case studies is also supported by the latest pedagogical research. *From Analysis to Action* refers to textbooks lacking in practical examples as an emerging weakness (Center for Science, Mathematics, and Engineering Education 1996, p. 2). Much of this document refers specifically to breadth of understanding, which may be achieved through case studies. Another issue addressed (Center for Science, Mathematics, and Engineering Education 1996, p. 19) is the need to "incorporate historical, social, and ethical issues into courses for engineering majors." The Committee on Undergraduate Science Education in *Transforming Undergraduate Education in Science, Mathematics, Engineering, and Technology* (Committee on Undergraduate Science Education 1999) proposes that as many undergraduate students as possible should undertake original, supervised research. *How People Learn: Brain, Mind, Experience, and School* (Bransford et al. 1999, p. 30) refers to the need to organize knowledge meaningfully to aid synthesis and develop expertise.

This work raises the question of whether failure analysis is merely tangential to, or is in fact fundamental to, civil engineering education. Put another way, are failure case studies simply interesting, or should they be an essential component of a civil engineering curriculum?

Failure Case Studies and Accreditation Requirements

ASCE TCFE Education Committee surveys of civil engineering departments reported in 1989 and 1998 (Bosela 1993, Rendon-Herrero 1993a and 1993b, and Rens et al. 2000a) found that many respondents indicated a need for detailed, well-documented case studies. Some of those replying felt strongly that incorporation of failure case studies should not become part of accreditation evaluations. However, unless something is specifically mandated by the Accreditation Board for Engineering and Technology (ABET), it is likely to be a low priority for inclusion in a curriculum.

There is certainly an argument to be made that failure analysis should be mandated by ABET. It may also be argued that, in a sense, it already is. Under ABET (2007) Criterion 3, Program Outcomes and Assessment,

> Engineering programs must demonstrate that their students attain:
>
> (a) an ability to apply knowledge of mathematics, science, and engineering
> (b) an ability to design and conduct experiments, as well as to analyze and interpret data
> (c) an ability to design a system, component, or process to meet desired needs within realistic constraints such as economic, environmental, social, political, ethical, health and safety, manufacturability, and sustainability
> (d) an ability to function on multi-disciplinary teams
> (e) an ability to identify, formulate, and solve engineering problems
> (f) an understanding of professional and ethical responsibility
> (g) an ability to communicate effectively
> (h) the broad education necessary to understand the impact of engineering solutions in a global, economic, environmental, and societal context
> (i) a recognition of the need for, and an ability to engage in life-long learning
> (j) a knowledge of contemporary issues
> (k) an ability to use the techniques, skills, and modern engineering tools necessary for engineering practice.

Programs often struggle with how to document the fact that their graduates understand the impact of engineering solutions in a global and societal context, engage in life-long learning, and demonstrate knowledge of contemporary issues (criteria h, i, and j, respectively). These outcomes can be difficult to demonstrate. One method of documenting these particular outcomes is to include case studies of failed engineering works in the curriculum. Many case studies show the direct societal effect of failures and demonstrate the need for life-long learning by highlighting the evolutionary nature of engineering design procedures.

Case studies also address the revised criterion c, design within realistic constraints. Case studies, and specifically failure case studies, illuminate how "economic, environmental, social, political, ethical, health and safety, manufacturability, and sustainability" affect design, behavior, and performance of engineered systems.

Criteria for civil engineering programs are more specific. Students must demonstrate

> An understanding of professional practice issues such as: procurement of work, bidding versus quality-based selection processes, how the design professionals and the construction professions interact to construct a project, the importance of professional licensure and continuing education, and/or other professional practice issues. (ABET 2007)

These professional issues are integral to many of the case studies addressed in this book. As an example, some project failures may be traced to poor interaction and communication between the designers and the builders.

Failure Case Studies and the Civil Engineering Body of Knowledge

The ASCE (2004) report *Civil Engineering Body of Knowledge for the 21st Century: Preparing the Civil Engineer for the Future,* prepared by the Body of Knowledge Committee of the Committee on Academic Prerequisites for Professional Practice, goes beyond ABET. The Body of Knowledge (BOK) defines 15 outcomes. The first 11 are identical to the ABET a–k. BOK outcomes 12–15 are the following:

12. an ability to apply knowledge in a specialized area related to civil engineering.
13. an understanding of the elements of project management, construction, and asset management.
14. an understanding of business and public policy and administration fundamentals.
15. an understanding of the role of the leader and leadership principles and attitudes.

For those failures with complex technical causes, failure case studies may be used to deepen understanding within specialized civil engineering areas (outcome 12). Failures can expose and highlight the subtleties of structural and system behavior that are the province of the specialist. Some specialties, such as earthquake and geotechnical engineering, have historically relied heavily on failure case studies to advance the state of the practice.

Outcomes 13, 14, and 15 may also be addressed through failure case studies. In many failures, the technical issues involved may not be particularly complex or unusual. Instead, breakdowns may come in the project management and construction processes or in the management of the facility by the owner (outcome 13). Pressures of business and public interests may encourage engineers to take short cuts, with harmful consequences (outcome 14). Some failures might have been averted with stronger leadership (outcome 15). A more thorough discussion of the relationship between failure case studies, ABET, and BOK outcomes is provided in Delatte (2008).

Pedagogical Benefits of Case Studies

Learning that occurs in multiple learning skills domains and exercises higher level learning skills is crucial to successful engineering education. This learning must, however, occur efficiently because engineering curricula are already overcrowded. This is one reason why failure case studies should be an essential part of engineering classes. The single activity of using a case study as part of a traditional course lesson plan simultaneously fosters learning in three different learning domains, thus making learning more efficient:

1. Affective: The failure is interesting and sometimes dramatic, thus increasing initial acquisition and permanent retention of knowledge from the learning exercise because of the emotional state of the student during the learning process.
2. Cognitive: The failure validates the science, showing that our engineering tools work and thus motivating the students to learn and retain more knowledge.
3. Social: Students discover or rediscover how engineering decisions affect individuals, communities, and society.

As a result of including case studies, students will demonstrate an ability to process failure analysis, apply ethics in engineering, and demonstrate an understanding of the engineer's role in and their value to society. Students will also demonstrate a greater depth of knowledge by developing intuition about expected behavior of engineered systems, understanding load paths, and better visualizing the interaction of components of engineered systems. Finally, students should experience a change in attitudes about quality engineering as a result of studying failures of engineered systems.

Use of Cases

Some of the ways to use case studies and a suggested format were reviewed in Delatte and Rens (2002). These methods include the following:

- Introductions to topics—Use the case to illustrate why a particular failure mode is important. Often the importance of a particular mode of failure only became widely known after a failure. Examples include the wind-induced oscillations of the Tacoma Narrows Bridge and the failure of Air Force warehouses in the mid-1950s that pointed out the need for shear reinforcement in reinforced concrete beams.
- Class discussions—Link technical issues to ethical and professional considerations. Add discussions of standard of care, responsibility, and communications to coverage of technical topics.
- Example problems and homework assignments—Calculate the forces acting on structural members and compare them to design criteria and accepted practice. This work can have the added benefit of requiring students to compare design assumptions to actual behavior in the field under service loads and overloads.
- Group and individual projects—Have students research the cases in depth and report back on them. This will also help built a database of cases for use in future classes. Students gain valuable research, synthesis, and communication skills.

Common Threads

The use of case studies as common threads through the curriculum can best be illustrated through an example. The 1907 collapse of the Quebec Bridge during construction, discussed in Chapter 3, represents a landmark of both engineering practice and forensic engineering. The Quebec Bridge was the longest cantilever structure attempted until that time. The bridge project was financially troubled from the beginning. This problem caused many setbacks in the design and construction. Construction began in October 1900. In August 1907, the bridge collapsed suddenly, and 75 workers were killed in the accident; there were only 11 survivors from the 86 workers on the span.

A distinguished panel was assembled to investigate the disaster. The panel's report found that the main cause of the bridge's failure was improper design of the latticing on the compression chords. The collapse was initiated by the buckling failure of Chord A9L, immediately followed by Chord A9R. Theodore Cooper had been the consulting engineer for the Quebec Bridge project, and most of the blame for the disaster fell on his shoulders. He mandated unusually high allowable stresses and failed to require recalculation of the bridge dead load when the span was lengthened.

This case study illustrates a number of important teaching points from engineering courses:

1. Statics—Truss analysis. The bridge was a cantilever truss. As the two arms of the bridge were built out from the pier, the moments on the truss arms increased, and the compressive stresses in the bottom chords of each arm also increased. Both the method of joints and the method of sections, traditionally taught in statics courses, may be used to analyze the compressive strut forces at the different stages of bridge construction. See Chapter 2.
2. Mechanics of materials—Allowable stresses. Mr. Cooper increased the allowable stresses for his bridge well beyond the limits of accepted engineering practice without experimental justification. He allowed compressive stresses that were considerably higher than that provided by the modern AISC code and were highly unconservative, given the state of knowledge at the time. The compression struts of the truss were too large to be tested by available machinery, so their capacity could not be precisely known. Development of engineering codes and standards requires tradeoffs between structural safety and economy, and there must be mechanisms for resolving disputes among competing criteria. See Chapter 3, which has this case study.
3. Mechanics of Materials—Structural deformation. The bending of the critical A9L member reached 57 mm (2¼ in.) and was increasing at the time of the collapse. The bending was discussed at the site and by Mr. Cooper, attempting to supervise the project from New York, but no action was taken. In fact, the bending showed that the member was slowly buckling.
4. Mechanics of materials—Buckling of columns and bars. The critical A9L compression member failed by buckling. It was a composite section, which meant that it required latticing to require the members to bend together. The moment of inertia and buckling capacity of the composite section may be compared to that of the individual truss members, showing the importance of the latticing system.
5. Structural analysis—Predicting, computing, and correcting dead loads. One critical error made in the design was that the dead load

was greatly underestimated. When material invoices showed that 17–30% more steel had gone into parts of the structure than had been planned for in the design, no attempt was made to analyze the bridge for the new loads. See Chapter 4.
6. Design of steel structures—Analysis and design of built-up members. This point follows from the discussion of buckling of columns and bars. Many existing steel bridges use built-up members, and engineers involved in assessing and rehabilitating such structures need to know how to evaluate member capacity and likely failure modes. See Chapter 6.
7. Engineering management—The requirement for the engineer of record to inspect the work on site. Mr. Cooper attempted to supervise the construction of a bridge in Quebec from his office in New York City. When problems arose, the problems were referred to him for a decision. The absence of an on-site engineer with authority to stop the work meant that there was no way to head off the impending collapse. A meeting was held to decide what to do, and the bridge collapsed just as the meeting was breaking up.
8. Engineering ethics—Professional responsibility. Mr. Cooper planned for the Quebec Bridge to be the crowning achievement of an illustrious career as a bridge engineer. However, by this time his health was poor and he was unable to travel to the site. He was also poorly compensated for his work. After the collapse, organizations such as ASCE began to define better the responsibility of the engineer of record. Unfortunately, the collapse of the Hyatt Regency walkways three quarters of a century later showed that much remains to be done.

As an example, the following problem statement may be used in a structural analysis, capstone design, or professionalism course, in conjunction with the Quebec Bridge collapse case study. The problem should be assigned before the discussion of the case study, probably as an overnight homework. After discussion of the case study, students should be better able to identify potential problems with an unusual construction technique:

You are the engineer for a cantilever truss bridge across a major river in North America. The bridge owner has asked you to prepare specifications, including allowable stresses, and has emphasized that they have a shaky financial situation. The bridge was initially intended to be 488 m (1,600 ft) long. To reduce the cost of the piers, they have been moved into shallower water and the bridge will now be 549 m (1,800 ft long). When completed, it will be the longest bridge of this type in the world.

Problem: List all of the things you can think of that can go wrong during this bridge construction project.

Once the collapse case has been discussed, the problem may be reassigned with the additional assignment to propose communication and quality control measures to ensure against collapse. Students should refer to the case study in formulating their answers.

Assessment

The case study materials developed so far have been well received by faculty across a wide range of civil engineering programs, as well as some other related programs. To date, however, the benefits identified have been anecdotal (although nevertheless impressive). There remains a need to identify, quantify, and assess the effect of case studies on teaching and learning.

Surveys have found widespread agreement that faculty consider failure case studies important and useful (Bosela 1993, Rendon-Herrero 1993a and 1993b, and Rens et al. 2000a). Several of the faculty participants in the failure case study workshops have reported back that the case studies have been excellent for motivating their students to learn. So far, the formal assessment of the effect of using case study materials in courses has been limited. Some assessment methods and results have been published by Delatte et al. (2007 and 2008).

Desired student learning outcomes are improved understanding of technical issues in civil engineering and engineering mechanics and improved understanding of ethical, professional, and procedural issues in civil engineering and engineering mechanics.

The primary assessment question is: "In what ways does the use of failure case studies improve students' ability to demonstrate competencies that prepare them to be better professional civil engineers?"

The assessment questions are as follows:

- Does the use of failure case studies improve students' ability to demonstrate competencies that better prepare them as professional engineers for the 21st century?
- How does the implementation of failure case studies encourage deep learning in civil engineering students?
- What has been the time commitment and value-added experience for faculty who integrate failure case studies in the curriculum that improves student learning of civil engineering concepts?

Typical Course Topics

As an aid to the instructor, some typical course topics that case studies may be used to address are summarized below by chapter.

Chapter 2: Statics and Dynamics

Typical topics in a statics course include:

- Determining force and moment resultants. This topic includes determining the resultants of distributed loads, including wind, soil, and static water pressures, as well as dead, live, and snow loads.
- Developing free-body diagrams. This topic includes classification of supports—e.g., fixed supports, pinned supports, rockers, and rollers. Support classifications, e.g., frictionless pins or fully fixed connections—may seem exact, but the behavior of actual supports can only approximate these idealized models. With a correct classification of supports, the free-body diagram is the basis for the problem solution. If it isn't correct, the solution won't be either.
- Using equations of equilibrium to solve for unknown forces and reactions.
- Classification of rigid bodies as statically determinate or indeterminate and properly or improperly constrained. Improperly constrained rigid bodies allow movement in at least one direction and will be unstable under certain loading conditions.
- Analysis of trusses. Trusses are idealized as assemblages of two force members, joined by frictionless pins. Truss members are in either pure tension or compression.
- Analysis of pin-connected frames and machines. This analysis introduces the concept of forces transferred between members at connections.
- Calculating internal forces in frame and machine members. This analysis may include developing shear and moment diagrams for beams, or this topic may be deferred until a course in mechanics of materials.
- Determining forces in cable and pulley systems, including suspension bridges.
- Analysis of dry friction, including wedges, screws, and other simple machines.
- Calculating centroids and area and mass moments of inertia for areas and volumes. Area and mass moments of inertia have limited

application in the statics course but are important in the subsequent dynamics and mechanics of materials courses.

Chapter 3: Mechanics of Materials

Typical topics in a mechanics of materials course include:

- normal and shearing stresses and strains;
- failure stresses, allowable stresses, and factors of safety;
- design of connections for normal, bearing, and shear stress;
- stress–strain relationships for materials, including the elastic-perfectly plastic (elastoplastic) model for yielding materials, such as structural steel;
- axial forces and deformations;
- use of compatibility relations to solve for reactions of statically indeterminate structures;
- stress concentrations due to discontinuities, such as fillets and holes in members;
- fatigue failure and relationships between stress range and number of cycles to failure;
- torsion stresses and deformations;
- bending stresses, shear and moment diagrams, and beam design;
- beam transverse shearing stresses;
- beam deflections;
- stress and strain transformation and principal stresses and strains; and
- column buckling and design.

Chapter 4: Structural Analysis

Topics covered typically include:

- types of structures;
- loads acting on structures;
- analysis of statically determinate structures, including diaphragm and shear wall systems;
- analysis of statically determinate trusses (similar to material covered in statics);
- internal loadings in structural members, including beam and frame shear and moment diagrams (similar to material covered in mechanics of materials);

- cable and arch structures;
- approximate analysis of statically indeterminate structures;
- deflections;
- analysis of statically indeterminate structures by force methods;
- displacement methods of analysis (slope-deflection and moment distribution); and
- influence lines (quantitative and qualitative).

Chapter 5: Reinforced Concrete Structures

Topics typically covered in an undergraduate-level reinforced concrete structural design course include:

- materials for reinforced concrete, such as concrete and reinforcing steel (Concrete as a material is discussed in more detail in Chapter 9);
- design of beams for flexure (rectangular beams, T-beams, and other configurations);
- beam design for shear (stirrups);
- development, anchorage, splicing, and other details of reinforcement;
- serviceability considerations—cracking and deflections;
- continuous beams;
- design of columns for combined axial loads and bending;
- one- and two-way slabs;
- footings; and
- punching shear.

Chapter 6: Steel Structures

Topics typically covered in an undergraduate steel structural design course include:

- steel as a structural material, including advantages and disadvantages;
- properties of structural steels, bolts, and weld metals, including ductility and fatigue performance;
- specifications, loads, and methods of design;
- analysis and design of tension members (net and gross section);
- design of axially loaded compression members;
- effective length factors for compression members;
- analysis and design of beams for flexure;
- beam shear;
- beam deflections;

- beam-columns (combined bending and axial force);
- bolted connections;
- welded connections; and
- composite beams (concrete deck slabs on steel beams).

Chapter 7: Soil Mechanics, Geotechnical Engineering, and Foundations

Topics covered in typical undergraduate geotechnical engineering and foundations courses include:

- engineering geology;
- properties and classification of soils;
- water in soils—static pressures, capillary moisture, saturated fluid flow;
- permeability and drainage;
- soil stresses;
- compressibility, consolidation, and settlement;
- shear strength;
- lateral pressures and earth-retaining structures;
- slope stability analysis;
- bearing capacity and foundations;
- improving soil conditions and properties; and
- subsurface exploration techniques.

Chapter 8: Fluid Mechanics and Hydraulics

Topics typically covered in an undergraduate fluid mechanics course include:

- fundamentals, such as density, compressibility, viscosity, surface tension, vapor pressure, specific weight, specific volume, and specific gravity;
- fluid statics: pressures and forces on submerged surfaces (generally also taught in a statics course);
- kinematics of fluid motion: velocity, acceleration, control volume, and conservation of mass (continuity equation);
- flow of an incompressible ideal fluid;
- flow of a compressible ideal fluid;
- impulse–momentum;
- fluid flow in pipes;
- fluid flow in open channels;

- fluid measurements; and
- fluid flow around immersed objects (drag and lift).

Typical topics in an undergraduate hydraulic engineering course include:

- hydrology;
- groundwater hydraulics;
- statistical analysis of hydrological data (probabilities and return periods);
- pipelines;
- open channel hydraulics;
- sediment transport;
- hydraulic machinery; and
- drainage hydraulics.

Appendix B

The ASCE Code of Ethics[1]

CIVIL ENGINEERS ARE GUIDED IN THEIR PRACTICE BY THE ASCE Code of Ethics. Other engineers have similar ethical standards, published by their respective technical societies. In addition, licensed professional engineers are subject to the codes of ethics in the jurisdictions where they are licensed and practice. These other codes, however, are often similar to the ASCE Code. The ASCE Code of Ethics is published online at https://www.asce.org/inside/codeofethics.cfm.

It is reproduced here, in full. In consists of four fundamental principles, seven fundamental canons, and a number of guidelines to practice under each fundamental canon.

---

[1] The Society's Code of Ethics was adopted on September 2, 1914, and was most recently amended on July 23, 2006. Pursuant to the Society's bylaws, it is the duty of every Society member to report promptly to the Committee on Professional Conduct any observed violation of the Code of Ethics.

Fundamental Principles[2]

Engineers uphold and advance the integrity, honor, and dignity of the engineering profession by:

1. using their knowledge and skill for the enhancement of human welfare and the environment;
2. being honest and impartial and serving with fidelity the public, their employers, and clients;
3. striving to increase the competence and prestige of the engineering profession; and
4. supporting the professional and technical societies of their disciplines.

Fundamental Canons

1. Engineers shall hold paramount the safety, health, and welfare of the public and shall strive to comply with the principles of sustainable development[3] in the performance of their professional duties.
2. Engineers shall perform services only in areas of their competence.
3. Engineers shall issue public statements only in an objective and truthful manner.
4. Engineers shall act in professional matters for each employer or client as faithful agents or trustees and shall avoid conflicts of interest.
5. Engineers shall build their professional reputation on the merit of their services and shall not compete unfairly with others.
6. Engineers shall act in such a manner as to uphold and enhance the honor, integrity, and dignity of the engineering profession and shall act with zero tolerance for bribery, fraud, and corruption.

---

[2]In April 1975, the ASCE Board of Direction adopted the fundamental principles of the Code of Ethics of Engineers as accepted by the Accreditation Board for Engineering and Technology, Inc. (ABET).

[3]In November 1996, the ASCE Board of Direction adopted the following definition of "sustainable development": "Sustainable development is the challenge of meeting human needs for natural resources, industrial products, energy, food, transportation, shelter, and effective waste management while conserving and protecting environmental quality and the natural resource base essential for future development."

7. Engineers shall continue their professional development throughout their careers and shall provide opportunities for the professional development of those engineers under their supervision.

Guidelines to Practice Under the Fundamental Canons of Ethics

CANON 1. *Engineers shall hold paramount the safety, health, and welfare of the public and shall strive to comply with the principles of sustainable development in the performance of their professional duties.*

   a. Engineers shall recognize that the lives, safety, health, and welfare of the general public are dependent upon engineering judgments, decisions, and practices incorporated into structures, machines, products, processes, and devices.
   b. Engineers shall approve or seal only those design documents, reviewed or prepared by them, which are determined to be safe for public health and welfare in conformity with accepted engineering standards.
   c. Engineers whose professional judgment is overruled under circumstances where the safety, health, and welfare of the public are endangered, or the principles of sustainable development ignored, shall inform their clients or employers of the possible consequences.
   d. Engineers who have knowledge or reason to believe that another person or firm may be in violation of any of the provisions of Canon 1 shall present such information to the proper authority in writing and shall cooperate with the proper authority in furnishing such further information or assistance as may be required.
   e. Engineers should seek opportunities to be of constructive service in civic affairs and work for the advancement of the safety, health, and well-being of their communities and the protection of the environment through the practice of sustainable development.
   f. Engineers should be committed to improving the environment by adherence to the principles of sustainable development so as to enhance the quality of life of the general public.

CANON 2. *Engineers shall perform services only in areas of their competence.*

   a. Engineers shall undertake to perform engineering assignments only when qualified by education or experience in the technical field of engineering involved.

    b. Engineers may accept an assignment requiring education or experience outside of their own fields of competence, provided their services are restricted to those phases of the project in which they are qualified. All other phases of such project shall be performed by qualified associates, consultants, or employees.
    c. Engineers shall not affix their signatures or seals to any engineering plan or document dealing with subject matter in which they lack competence by virtue of education or experience or to any such plan or document not reviewed or prepared under their supervisory control.

CANON 3. Engineers shall issue public statements only in an objective and truthful manner.

    a. Engineers should endeavor to extend the public knowledge of engineering and sustainable development and shall not participate in the dissemination of untrue, unfair, or exaggerated statements regarding engineering.
    b. Engineers shall be objective and truthful in professional reports, statements, or testimony. They shall include all relevant and pertinent information in such reports, statements, or testimony.
    c. Engineers, when serving as expert witnesses, shall express an engineering opinion only when it is founded upon adequate knowledge of the facts, upon a background of technical competence, and upon honest conviction.
    d. Engineers shall issue no statements, criticisms, or arguments on engineering matters which are inspired or paid for by interested parties, unless they indicate on whose behalf the statements are made.
    e. Engineers shall be dignified and modest in explaining their work and merit and will avoid any act tending to promote their own interests at the expense of the integrity, honor, and dignity of the profession.

CANON 4. Engineers shall act in professional matters for each employer or client as faithful agents or trustees and shall avoid conflicts of interest.

    a. Engineers shall avoid all known or potential conflicts of interest with their employers or clients and shall promptly inform their employers or clients of any business association, interests, or circumstances that could influence their judgment or the quality of their services.
    b. Engineers shall not accept compensation from more than one party for services on the same project or for services pertaining to the

same project, unless the circumstances are fully disclosed to and agreed to by all interested parties.

c. Engineers shall not solicit or accept gratuities, directly or indirectly, from contractors, their agents, or other parties dealing with their clients or employers in connection with work for which they are responsible.

d. Engineers in public service as members, advisors, or employees of a governmental body or department shall not participate in considerations or actions with respect to services solicited or provided by them or their organization in private or public engineering practice.

e. Engineers shall advise their employers or clients when, as a result of their studies, they believe a project will not be successful.

f. Engineers shall not use confidential information coming to them in the course of their assignments as a means of making personal profit if such action is adverse to the interests of their clients, employers, or the public.

g. Engineers shall not accept professional employment outside of their regular work or interest without the knowledge of their employers.

CANON 5. *Engineers shall build their professional reputation on the merit of their services and shall not compete unfairly with others.*

a. Engineers shall not give, solicit, or receive either directly or indirectly any political contribution, gratuity, or unlawful consideration in order to secure work, exclusive of securing salaried positions through employment agencies.

b. Engineers should negotiate contracts for professional services fairly and on the basis of demonstrated competence and qualifications for the type of professional service required.

c. Engineers may request, propose, or accept professional commissions on a contingent basis only under circumstances in which their professional judgments would not be compromised.

d. Engineers shall not falsify or permit misrepresentation of their academic or professional qualifications or experience.

e. Engineers shall give proper credit for engineering work to those to whom credit is due and shall recognize the proprietary interests of others. Whenever possible, they shall name the person or persons who may be responsible for designs, inventions, writings, or other accomplishments.

f. Engineers may advertise professional services in a way that does not contain misleading language or is in any other manner derogatory to the dignity of the profession. Examples of permissible advertising are as follows:

- Professional cards in recognized, dignified publications, and listings in rosters or directories published by responsible organizations, provided that the cards or listings are consistent in size and content and are in a section of the publication regularly devoted to such professional cards.
- Brochures which factually describe experience, facilities, personnel, and capacity to render service, providing they are not misleading with respect to the engineer's participation in projects described.
- Display advertising in recognized dignified business and professional publications, providing it is factual and is not misleading with respect to the engineer's extent of participation in projects described.
- A statement of the engineers' names or the name of the firm and statement of the type of service posted on projects for which they render services.
- Preparation or authorization of descriptive articles for the lay or technical press, which are factual and dignified. Such articles shall not imply anything more than direct participation in the project described.
- Permission by engineers for their names to be used in commercial advertisements, such as may be published by contractors, material suppliers, etc., only by means of a modest, dignified notation acknowledging the engineers' participation in the project described. Such permission shall not include public endorsement of proprietary products.

g. Engineers shall not maliciously or falsely, directly or indirectly, injure the professional reputation, prospects, practice, or employment of another engineer or indiscriminately criticize another's work.

h. Engineers shall not use equipment, supplies, or laboratory or office facilities of their employers to carry on outside private practice without the consent of their employers.

CANON 6. *Engineers shall act in such a manner as to uphold and enhance the honor, integrity, and dignity of the engineering profession and shall act with zero tolerance for bribery, fraud, and corruption.*

a. Engineers shall not knowingly engage in business or professional practices of a fraudulent, dishonest, or unethical nature.

b. Engineers shall be scrupulously honest in their control and spending of monies and shall promote effective use of resources through open, honest, and impartial service with fidelity to the public, employers, associates, and clients.
c. Engineers shall act with zero tolerance for bribery, fraud, and corruption in all engineering or construction activities in which they are engaged.
d. Engineers should be especially vigilant to maintain appropriate ethical behavior where payments of gratuities or bribes are institutionalized practices.
e. Engineers should strive for transparency in the procurement and execution of projects. Transparency includes disclosure of names, addresses, purposes, and fees or commissions paid for all agents facilitating projects.
f. Engineers should encourage the use of certifications specifying zero tolerance for bribery, fraud, and corruption in all contracts.

CANON 7. *Engineers shall continue their professional development throughout their careers and shall provide opportunities for the professional development of those engineers under their supervision.*

a. Engineers should keep current in their specialty fields by engaging in professional practice, participating in continuing education courses, reading in the technical literature, and attending professional meetings and seminars.
b. Engineers should encourage their engineering employees to become registered at the earliest possible date.
c. Engineers should encourage engineering employees to attend and present papers at professional and technical society meetings.
d. Engineers shall uphold the principle of mutually satisfying relationships between employers and employees with respect to terms of employment, including professional grade descriptions, salary ranges, and fringe benefits.

Appendix C

Some Cases on Video and DVD

THE HISTORY CHANNEL, AS PART OF ITS "MODERN MARVELS" series, has a series of programs entitled *Engineering Disasters*. These programs may be purchased on video or DVD from the History Channel (http://www.history.com/).

Many of the cases in these episodes are transportation and industrial cases, but structural, geotechnical, dam, and environmental problems are also addressed. Because there are typically five to seven cases discussed per 50-min episode, each case is only about 7–10 min long. As a result, they are useful for quick introductions to the cases but need to be supplemented by additional readings and discussion.

Unfortunately, there is no index of the cases featured in the various episodes. A partial list of cases is provided below, along with the chapter if they are discussed in this book. The first two videos in the series cover the most cases, but with little depth.

ENGINEERING DISASTERS
- Beauvais Cathedral (France, Middle Ages)
- Leaning Tower of Pisa (Italy, Middle Ages), discussed in more detail in *Engineering Disasters 3*
- Tacoma Narrows Bridge, 1940 (Chapter 2)
- Hyatt Regency walkway, 1981 (Chapter 2), discussed in more detail in *Engineering Disasters 11*

- Sinking of the *Titanic*
- Teton Dam, 1976 (Chapter 7)
- *Hindenburg* Zeppelin disaster
- Sioux City, Iowa, DC-10 crash
- Aloha Airlines, Honolulu, Boeing 737 fuselage fatigue failure
- TWA Flight 800 Boeing 747 crash, 1996
- Apollo 1 fire and Apollo 13 near-disaster
- Space Shuttle *Challenger* (Chapter 10)
- Hubble space telescope
- Three Mile Island nuclear power plant accident
- Chernobyl nuclear power plant disaster
- Bhopal, India, chemical poisoning, 1984, discussed in more detail in *Engineering Disasters 10*

MORE ENGINEERING DISASTERS

- Early and recent U.S. rocket failures
- The Chauchat (WWI French light machine gun)
- *USS Thresher* nuclear submarine
- Liberty ship hull failures (Chapter 9)
- *Stockholm* and *Andrea Doria* collision at sea
- Hartford Civic Center (Chapter 6)
- Kemper Arena (Chapter 4)
- St. Francis Dam
- Northeast blackout of 1965

ENGINEERING DISASTERS 3

- Leaning Tower of Pisa (Italy, Middle Ages)
- Idaho SL-1 nuclear reactor failure, 1961
- Texas Tower 4 offshore radar platform, 1961
- Soviet Soyuz 11 decompression accident, 1971
- NASA Mars Climate Orbiter and Polar Lander crashes, 1991

ENGINEERING DISASTERS 4

- MGM Grand Hotel fire, Las Vegas, 1980
- L'Ambiance Plaza collapse, 1987 (Chapter 4)
- Rocket launch failures in the USSR
- Point Pleasant Bridge collapse, 1967 (Chapter 3)
- Sinking of *Lacey v. Murrow* concrete pontoon bridge, 1990
- Design defects in Corvair and Pinto automobiles
- Comet Jet aircraft crashes (Chapter 3)

Engineering Disasters 5
- Senior Road television tower collapse, 1982
- *Exxon Valdez* oil spill, 1989
- Freeway structural collapses in earthquakes, California, 1989 and 1994
- Buffalo Creek coal waste dam failure, West Virginia, 1972
- Jefferson Island salt mine disaster, Louisiana, 1980

Engineering Disasters 6
- Harrier Jet accidents
- North Sea Piper Alpha oil platform explosion, U.K., 1988
- Willow Island cooling tower collapse, 1978 (Chapter 9)
- Communication failures in U.S. nuclear command
- Intercity express train crash, Germany, 1998

Engineering Disasters 7
- Lockheed L188 turboprop crashes
- Baldwin Hills dam failure, 1963, repeated in *Engineering Disasters 18*
- Sunshine Skyway Bridge ship collision and collapse
- Northridge earthquake—residential and reinforced and prestressed concrete failures
- Centralia, Pennsylvania, coal seam fire

Engineering Disasters 8
- Pepcon ammonium perchlorate plant explosion, Henderson, Nevada, 1988
- Mianus River Bridge collapse, 1983 (Chapter 6)
- Capsize of Ocean Ranger oil platform off Newfoundland, Canada, 1982
- Johnstown Flood, 1889 (Chapter 8)
- Learjet crash due to loss of cabin pressure, 1999 (including golfer Payne Stewart)

Engineering Disasters 9
- Grain elevator explosion, Louisiana, 1977
- Schoharie Creek Bridge, 1987 (Chapter 8)
- *Bright Field* freighter collision with riverfront mall, New Orleans, Louisiana
- Airship crashes (*Hindenburg* and R101)
- Hartford Civic Center stadium collapse, 1978 (Chapter 6)
- Rosemont Horizon Arena collapse, Illinois, 1979 (timber roof)

Engineering Disasters 10

- Hutchinson, Kansas, natural gas explosions, 2001
- *Puerto Rican* tanker explosion, 1984
- Atlantic City casino parking garage construction collapse, 2004
- Soviet White Sea Canal project, 1930s
- Bhopal, India, chemical poisoning, 1984

Engineering Disasters 11

- Liquefied natural gas tank explosion, Ohio, 1944
- Hyatt Regency walkway, 1981 (Chapter 2)
- Sinkholes in San Francisco, California, 1995, and Atlanta, Georgia, 1993
- Yangtze River flood, 1931
- Asbestos

Engineering Disasters 12

- Cryptosporidium outbreak, Wisconsin, 1993
- Texas A&M bonfire collapse, 1999
- Sherman tanks, World War II
- Denver International Airport automated baggage system, 1995

Engineering Disasters 13

- World Trade Center Building 7 fire and collapse, September 11, 2001
- Love Canal environmental disaster, 1970s
- Coconut Grove nightclub fire, 1942
- Ariane 5 rocket explosion due to software error, 1996
- Patriot Missile failure to launch due to software error, 1991
- Providence, Rhode Island, building implosion failure, 1989

Engineering Disasters 14

- El Al Boeing 747 air freighter crash, Amsterdam, Holland, 1992
- Los Angeles subway tunnel construction collapse, California, 1995
- *Sansinena* freighter explosion, Los Angeles harbor, California
- Molasses storage tank rupture, Boston, Massachusetts, 1919
- Cline Avenue bridge collapse, Indiana, 1980

Engineering Disasters 15

- Freight tunnel flood, Chicago, Illinois, 1992
- Harbour Cay Condominium, 1981 (Chapter 5)

- Delta Boeing 727 crash, Dallas–Fort Worth International Airport, Texas, 1988
- *Sultana* steamboat boiler explosion, Memphis, Tennessee, 1865
- Gotthard Road tunnel explosion and fire, Switzerland

ENGINEERING DISASTERS 16

- Guadalajara, Mexico, gasoline explosion, 1992
- MD DC-10 cargo door failures and crashes, 1970s
- Stava, Italy, mining dam failure, 1985
- London, Texas, high school natural gas explosion, 1937
- Torrey Canyon oil tanker autopilot failure, grounding, and oil spill, 1967

ENGINEERING DISASTERS 17

- Bluebird Canyon landslide, Southern California, 2005
- Tujunga Avenue sinkhole, Southern California, 2005
- Charles de Gaulle Airport Terminal 2E collapse, 2004
- Sinking of S/S *Marine Electric* modified T2 tanker bulk carrier, 1983
- Fokker F-10A crash (the crash that killed Knute Rockne), 1931
- Los Angeles Walt Disney Concert Hall glare and heat problems, 2003

ENGINEERING DISASTERS 18

- Miller Park Big Blue crane collapse, 1999
- China coal mine disasters, 2005 and earlier, and U.S. mine safety measures
- Baldwin Hills dam failure, 1963, repeated in *Engineering Disasters 7*
- X-ray shoe fitting fluoroscope machine health hazards, 1920s–1930s
- Salton Sea, California, and Aral Sea, Central Asia, environmental disasters, ongoing

ENGINEERING DISASTERS 19

- Sinking of the *SS Edmund Fitzgerald,* Lake Superior, 1975
- Boeing 737 rudder failure crashes, 1991 and 1994
- Los Angeles sodium reactor experiment nuclear accident, 1959
- Monongahela River oil tank rupture and spill, near Pittsburgh, 1988
- Galaxy 4 satellite tin whisker short circuit failure, 1998

References

Accreditation Board for Engineering and Technology, Inc. (ABET). (2007). Engineering Accreditation Commission, Criteria for Accrediting Engineering Programs, Effective for Evaluations During the 2008–2009 Accreditation Cycle, ABET, Baltimore, Md., December 4.

Allen, D. E., and Schriever, W. R. (1972). "Progressive Collapse, Abnormal Loads, and Building Codes," In *Structural Failures: Modes, Causes, Responsibilities*, ASCE, New York, pp. 21–47.

American Association of State Highway Officials. (AASHO). (1949). "Standard Specifications for Highway Bridges," AASHO, Washington, D.C.

American Concrete Institute. (ACI). (1963, 1977, 1983, 1995, 2005, 2008). *Building Code Requirements for Structural Concrete and Commentary, ACI 318-05*, American Concrete Institute, Farmington Hills, Mich.

ACI. (1989). *Avoiding Failures in Concrete Construction*, Seminar Course Manual/SCM-19 (89), Detroit.

American Institute of Steel Construction. (AISC). (1970, 1998, 2005). *Steel Construction Manual*, AISC, Chicago.

American Iron and Steel Institute. (AISI). (1996). *Cold-Formed Steel Design Manual*, AISI, Washington, D.C. with errata.

AISI. (2007). *Specification for the Design of Cold-Formed Steel Structural Members*, Washington, D.C.

American Society of Civil Engineers. (ASCE). (2004) *Civil Engineering Book of Knowledge for the 21st Century: Preparing the Civil Engineer for the Future*, ASCE, Reston, Va.

ASCE. (1968). "Causes of Silver Bridge Collapse Studied," *Civ. Eng.—ASCE,* 38(12), 87.
ASCE. (1985). "New Mianus Bridge Report Disputes Earlier Study," *Civ. Eng.—ASCE,* 55(4), 10.
ASCE. (2000). *Quality in the Constructed Project: A Guide for Owners, Designers, and Constructors, Second Edition,* ASCE Manuals and Reports on Engineering Practice No. 73, ASCE, Reston, Va.
ASCE. (2006). Code of Ethics, Web site https://www.asce.org/inside/codeofethics.cfm. Accessed September 30, 2008.
ASCE. (2007). *The New Orleans Hurricane Protection System: What Went Wrong and Why,* Report by Hurricane Katrina External Review Panel, ASCE, Reston, Va.
Arthur, H. G. (1977). "Teton Dam Failure," in *The Evaluation of Dam Safety,* Engineering Foundation Conference Proceedings, ASCE, New York, 61–68.
Bennett, J. (1973). "Metallurgical Aspects of the Failure of Point Pleasant Bridge." *J. Test. Eval.,* 1(2), 152–161.
Bignell, V., Peters, J., and Pym, C. (1977). *Catastrophic Failures,* Open University Press, Milton Keynes, New York.
Billah, Y., and Scanlan, R. (1991). "Resonance, Tacoma Narrows Bridge Failure, and Undergraduate Physics Textbooks," *American Journal of Physics,* 59(2), 118–124.
Blake, A. F. (1971). "16-Story Building Falls in Brighton," *The Boston Globe,* Boston, January 26.
Boorstin, R. O. (1987). "Bridge Collapses on the Thruway, Trapping Vehicles," *CXXXVI*(47), 101, *The New York Times,* April 6.
Bosela, P. A. (1993). "Failure of Engineered Facilities: Academia Responds to the Challenge." *J. Perform. Constr. Facil.,* 7(2), 140–144.
Bosela, P. A., and Delatte, N. J. (2006). *Forensic Engineering,* Proceedings of the Fourth Congress, October 6–9, ASCE, Reston, Va.
Bosela, P. A., Delatte, N. J., and Rens, K. L., eds. (2003). *Forensic Engineering,* Proceedings of the Third Congress, October 19–21, ASCE, Reston, Va.
Bransford, J. D., Brown, A. L., and Cocking, R. L. (1999). *How People Learn: Brain, Mind, Experience, and School,* National Academy Press, Washington, D.C.
Buchhardt, F., Magiera, G., Matthees, W., and Plank, A. (1984). "Structural Investigation of the Berlin Congress Hall Collapse," *Concrete International,* American Concrete Institute, pp. 63–68, March, reprinted in ACI (1989).
Carino, N. J., Woodward, K. A., Leyendecker, E. V., and Fattal, S. G. (1983). "A Review of the Skyline Plaza Collapse," *Concrete International,* American Concrete Institute, pp. 35–42, July, reprinted in ACI (1989).
Carper, K. L., ed. (1989). *Forensic Engineering,* Elsevier Science Publishing Co., Inc., New York. (Second edition published in 2001 by CRC Press, Boca Raton, Fla.)
Carper, K. (2001). "Why Buildings Fail," National Council of Architectural Registration Boards (www.ncarb.org), Washington, D.C.
Case Western Reserve University. (2007a). "William LeMessurier: The Fifty-Nine-Story Crisis: A Lesson in Professional Behavior," Online Ethics Center for Engi-

neering and Science, http://www.onlineethics.org/cms/8888.aspx. Accessed September 30, 2008.

Case Western Reserve University. (2007b). "Roger Boisjoly—The *Challenger* Disaster," Online Ethics Center for Engineering and Science, http://www.onlineethics.org/cms/7123.aspx. Accessed September 30, 2008.

Center for Science, Mathematics, and Engineering Education, National Research Council. (1996). *From Analysis to Action,* National Academy Press, Washington, D.C.

Charleston [West Virginia] *Daily Mail.* (1999). "State Disasters Come in All Sizes," February 12, http://www.dailymail.com/static/specialsections/lookingback/lb0212.htm (accessed May 16, 2007).

Cleveland Plain Dealer. (2007). "Epoxy Called Factor in Tunnel Collapse," July 11, p. A9, from the Associated Press.

Commission. (2007). Commission of Inquiry into the Collapse of a Portion of the de la Concorde Overpass: October 3, 2006–October 15, 2007. *Report,* Government of Quebec, Canada. Available online at http://www.cevc.gouv.qc.ca/UserFiles/File/Rapport/report_eng.pdf. Accessed September 30, 2008.

Committee on Science and Technology. (1986). *Investigation of the* Challenger *Accident, October 29, 1986,* House of Representatives. U.S. Government Printing Office, Washington, D.C. Available online at http://history.nasa.gov/rogersrep/51lcover.htm and http://www.gpoaccess.gov/challenger/64_420.pdf. Accessed September 30, 2008.

Committee on Undergraduate Science Education. (1999). *Transforming Undergraduate Education in Science, Mathematics, Engineering, and Technology,* Center for Science, Mathematics, and Engineering Education, National Research Council, Washington, D.C.

Culver, C. G. (2002). "Discussion of 'Another Look at the L'Ambiance Plaza Collapse,'" *J. Perf. of Constr. Fac., 16*(1), 42.

Cuoco, D. A., Peraza, D. B., Scarangello, T. Z. (1992). "Investigation of L'Ambiance Plaza Building Collapse," *J. Perf. of Constr. Fac., 6*(4), 211–231.

Delatte, N. J. (1997). "Integrating Failure Case Studies and Engineering Ethics in Fundamental Engineering Mechanics Courses." *J. Prof. Issues Eng. Educ. Pract., 123*(3), 111–116.

Delatte, N. J. (2000). "Using Failure Case Studies in Civil Engineering Education," in *Forensic Engineering,* K. L. Rens, O. Rendon-Herrero, and P. A. Bosela, eds., ASCE, Reston, Va., 430–440.

Delatte, N. J. (2002). "Agricultural Product Loads and Warehouse Failures," *J. Perf. of Constr. Fac., 16*(3), 116–123.

Delatte, N. J. (2003). "Using Failure Case Studies in Civil Engineering Courses," *Proceedings of the 2003 American Society for Engineering Education Annual Conference & Exposition,* Nashville, Tennessee, 22–25 June 2003.

Delatte, N. J. (2005). "'Failure of Cold-Formed Steel Beams during Concrete Placement,' technical note." *J. Perf. of Constr. Fac., 19*(2), 178–181.

Delatte, N. J. (2006). "Learning from Failures," *Civil Engineering Practice, 21*(2), 21–38.

Delatte, N. (2008). "Using Failure Case Studies to Address Civil Engineering Program and BOK Criteria," In *Proceedings of the 2008 American Society for Engineering Education Annual Conference and Exposition*, Pittsburgh, Penn., June 22–25. ASEE, Washington, D.C.

Delatte, N. J., and Rens, K. L. (2002). "Forensics and Case Studies in Civil Engineering Education: State of the Art," *J. Perf. of Constr. Fac.*, 16(3), 98–109.

Delatte, N., Sutton, R., Beasley, W., and Bagaka's, J. (2007). "Assessing the Impact of Case Studies on the Civil Engineering and Engineering Mechanics Curriculum," *Proceedings of the 2007 American Society for Engineering Education Annual Conference and Exposition*, Honolulu, Hawaii, June 24–27.

Delatte, N., Bosela, P., Sutton, R., Beasley, W., and Bagaka's, J. (2008) "Assessing the Impact of Case Studies on the Civil Engineering and Engineering Mechanics Curriculum: Phase II," *Proceedings of the 2008 American Society for Engineering Education Annual Conference and Exposition*, Pittsburgh, June 22–25.

Dietz, D., and Preston, D. (2007) "Records Find Insurers Undercut Policy Promises," *Plain Dealer*, Cleveland, Ohio, from Bloomberg News, August 3.

Engineering News Record. (ENR). (1941). "Why the Tacoma Narrows Bridge Failed." *ENR*, May, 8.

ENR. (1967). "Collapse May Never Be Solved." *ENR*, December, 21, 69–71.

ENR. (1968). "Systems Built Apartments Collapse." *ENR*, May 23, 1968, p. 23.

ENR. (1970). "Britain Tightens Building Standards, Moves to Stem 'Progressive Collapse.'" *ENR*, April 16, 1970, p. 12.

ENR. (1971). "Cause of Fatal Collapse Unknown." *ENR*, February 4, 1971, p. 13.

ENR. (1978a). "Design Flaws Collapsed Steel Space Frame Roof." *ENR*, April, 6.

ENR. (1978b). "Collapsed Space Truss Roof Had a Combination of Flaws." *ENR*, June, 22.

ENR. (1978c). "Collapsed Roof Design Defended." *ENR*, June, 29.

ENR. (1979a). "New Theory on Why Hartford Roof Fell." *ENR*, June, 14.

ENR. (1979b). "Hartford Collapse Blamed on Weld." *ENR*, June, 24.

Fattal, S. G., and Lew, H. S. (1980). *Analysis of Construction Conditions Affecting the Structural Response of the Cooling Tower at Willow Island, West Virginia*, National Bureau of Standards, Washington, D.C., July.

Federal Emergency Management Agency. (FEMA). (2002). *World Trade Center Building Performance Study: Data Collection, Preliminary Observations, and Recommendations*, FEMA 403, Federal Emergency Management Agency, Washington D.C. Available online at http://www.fema.gov/rebuild/mat/wtc study.shtm. Accessed September 30, 2008.

Feld, J. (1964). *Lessons from Failures of Concrete Structures*, American Concrete Institute (ACI) Monograph No. 1, Detroit, Michigan, reprinted in ACI (1989).

Feld, J. (1978). *Failure Lessons in Concrete Construction*, A Collection of Articles from *Concrete Construction* magazine. Updated February 1978, Addison, Illinois, reprinted in ACI (1989).

Feld, J., and Carper, K. (1997). *Construction Failure*, 2nd ed., John Wiley & Sons, New York.

Fitzgerald, M. (2007). "Roth Testifies before House Subcommittees, Conveys ASCE's Support for Levee and Dam Safety Programs," *ASCE News, 32*(6), 8–9.

Frank, W. S. (1988). "The Cause of the Johnstown Flood: A New Look at the Historic Johnstown Flood of 1889," *Civ. Engrg.*, May, 63–66. Also available at http://smoter.com/flooddam/johnstow.htm. Accessed September 30, 2008.

Freiman, F. L., and Schlager, N. (1995a). *Failed Technology: True Stories of Technological Disasters, Vol. 1,* UXL, Gale Research, Detroit.

Freiman, F. L., and Schlager, N. (1995b). *Failed Technology: True Stories of Technological Disasters, Vol. 2,* UXL, Gale Research, Detroit.

Fuller, R. (1975). "Industrialized Concrete Construction for HUD." *Industrialization in Concrete Building Construction.* American Concrete Institute, Detroit, Mich.

Gardner, N. J., Huh, J., and Chung, L. (2002). "Lessons from the Sampoong Department Store Collapse," *Cement Concr. Compos., 24,* 523–529.

Genevois, R., and Ghirotti, M. (2005). "The 1963 Vaiont Landslide," *Giornale di Geologia Applicata I* (Journal of Applied Geology I), 41–52.

Ghosh, S. K., Fanella, D. A., and Rabbat, B. G. (1995). *Notes on ACI 318-95 Building Code Requirements for Structural Concrete with Design Applications,* Portland Cement Association, Skokie, Ill.

Gillum, J. D. (2000). "The Engineer of Record and Design Responsibility." *J. Perf. of Constr. Fac., 14*(2), 67–70.

Granger, R. O., Peirce, J. W., Protze, H. G., Tobin, J. J., and Lally, F. J. (1971). *The Building Collapse at 2000 Commonwealth Avenue, Boston, Massachusetts, on January 25, 1971,* Report of the Mayor's Investigating Commission, The City of Boston.

Greene, B. H., and Christ, C. A. (1998). "Mistakes of Man: The Austin Dam Disaster of 1911," *Pennsylvania Geology, 29*(2/3), 7–14, online at http://www.dcnr.state.pa.us/topogeo/pub/pageolmag/pageolonline.aspx. Accessed September 30, 2008.

Griffiths, H., Pugsley, A. G., and Saunders, O. (1968). *Report of the Inquiry into the Collapse of Flats at Ronan Point, Canning Town,* Her Majesty's Stationery Office, London.

Hanna, A. (1999). *Concrete Formwork Systems,* Marcel Dekker, Inc., New York.

Heger, F. J. (1991). "Public-Safety Issues in Collapse of L'Ambiance Plaza," *J. Perf. of Constr. Fac., 5*(2) 92–112.

Hendron, A. J., and Patton, F. D. (1985). *"The Vaiont Slide: A Geotechnical Analysis Based on New Geological Observations of the Failure Surface, Vol. I, Main Text,* Technical report GL-85-5, Department of the Army, U.S. Army Corps of Engineers, U.S. Army Engineer Waterways Experiment Station, Vicksburg, Mississippi, June.

Hinman, E. E., and Hammond, D. J. (1997). *Lessons from the Oklahoma City Bombing,* Defensive Design Techniques, ASCE Press, New York.

Holgate, H., Derry, John G. G., and Galbraith, J. (1908). "Royal Commission Quebec Bridge Inquiry Report," Sessional Paper No. 154, S. E. Dawson, printer to the King, Ottawa.

Holt, R., and Hartmann, J. (2008). *Adequacy of the U10 & L11 Gusset Plate Designs for the Minnesota Bridge No. 9340 (I-35W over the Mississippi River): Interim Report*, Federal Highway Administration, Turner-Fairbank Highway Research Center Report, January 11, 2008; also available at http://www.ntsb.gov/dockets/Highway/HWY07MH024/default.htm. Accessed September 30, 2008.

Houser, M. (2007). "Convention Center Joint Did Not Conform to Steel Industry Guidelines," *Pittsburgh Tribune-Review*, February 24.

Houser, M., and Ritchie, J. (2007). "Convention Center Collapse Blamed on Bolt Connection," *Pittsburgh Tribune-Review*, February 22.

Houston, B. (1991). "Structural Engineer Gives Advice on Handling Fallout of Disaster," *Hartford Courant*, October 28.

Huber, F. (1991). "Update: Bridge Scour." *Civ. Engrg.—ASCE*, 61(9), 62–63.

Hurd, M. K. (1995). *Formwork for Concrete*, American Concrete Institute Committee 347, Formwork for Concrete, Special Publication Number 4, 6th ed., American Concrete Institute, Farmington Hills, Mich.

Independent Panel. (1976). Independent Panel to Review Cause of Teton Dam Failure. "Report to the U.S. Department of the Interior and State of Idaho on Failure of Teton Dam," Idaho Falls, Idaho, December.

Interagency Performance Evaluation Task Force. (IPET). (2007). "Performance Evaluation of the New Orleans and Southeast Louisiana Hurricane Protection System," Final Report of the Interagency Performance Evaluation Task Force, Vol. I—Executive Summary and Overview (Interim Final), U.S. Army Corps of Engineers, March 26; also available at https://ipet.wes.army.mil/. Accessed September 30, 2008.

Johnstown Flood National Memorial. (2008). Available at http://www.nps.gov/archive/jofl/home.htm). Accessed September 30, 2008.

Kaminetzky, D. (1991). *Design and Construction Failures: Lessons from Forensic Investigations*, McGraw-Hill, New York.

King, S., and Delatte, N. J. (2004). "Collapse of 2000 Commonwealth Avenue: Punching Shear Case Study." *J. Perf. of Constr. Fac.*, 18(1), 54–61.

Korman, R. (1987). "Flawed Connection Detail Triggered Fatal L'Ambiance Plaza Collapse," *ENR*, October, 29.

Korman, R. (1988). "Mediated Settlement Seeks to Close the Book on L'Ambiance Plaza," *ENR*, November, 24.

Korman, R. (1995). "Critics Grade Citicorp Confession," *ENR*, 234(21), 10.

Kremer, E. (2002). "(Re)Examining the Citicorp Case: Ethical Paragon or Chimera," *Cross Currents*, http://www.crosscurrents.org/kremer2002.htm, Association for Religious and Intellectual Life, Fall 2002, Vol. 52, No. 3. Accessed September 30, 2008.

Lee, G. C., Ketter, R. L., and Hsu, T. L. (1981). *The design of single story rigid frames*, Metal Building Manufacturer's Association, Cleveland. Ohio

Leonards, G. A. (1982). "Investigation of Failures," *J. Geotech. Engrg. Div.*, 108(2), 185–246.

Lev Zetlin Associates. (LZA). (1978). "Report of the Engineering Investigation Concerning the Causes of the Collapse of the Hartford Coliseum Space Truss Roof on January 18, 1978," submitted to the city of Hartford, Connecticut, June 12.

Levy, M., and Salvadori, M. (1992). *Why Buildings Fall Down: How Structures Fail*, W.W. Norton, New York.

Lew, H. S. (1979). *Investigation of Construction Failure of Reinforced Concrete Cooling Tower at Willow Island, West Virginia*, NBSSIR 78-158, National Bureau of Standards, Washington, D.C.

Lew, H. S. (1980). "West Virginia Cooling Tower Collapse Caused by Premature Form Removal," *Civ. Engrg.—ASCE*, 50(2), 62–67.

Lew, H. S., Carino, N. J., Fattal, S. G., and Batts, M. E. (1981). "Investigation of Construction Failure of Harbour Cay Condominium in Cocoa Beach, Florida," Publication No. NBSIR 81-2374, National Bureau of Standards, Washington D.C.

Lew, H. S., Carino, N. J., and Fattal, S. G. (1982). "Cause of the Condominium Collapse in Cocoa Beach, Florida." *Concrete International*, August.

Leyendecker, E. V., and Fattal, S. G. (1977). "Investigation of the Skyline Plaza Collapse in Fairfax County, Virginia," Building Science Series No. 94, National Bureau of Standards, Washington, D.C.

Lichtenstein, A. G. (1993). "The Silver Bridge Collapse Recounted," *J. Perf. of Constr. Fac.*, 7(4), 249–261.

Litle, William A. (1972). "Boston Collapse," In *Structural Failures: Modes, Causes, Responsibilities*, ASCE, New York, p. 99.

Luth, G. P. (2000). "Chronology and Context of the Hyatt Regency Collapse," *J. Perf. of Constr. Fac.*, 14(2), 51–61.

Macauley, D. (2000). *Building Big*, Houghton Mifflin Company, New York.

Martin, R., and Delatte, N. J. (2000). "Another Look at the L'Ambiance Plaza Collapse," *J. Perf. of Constr. Fac.*, 14(4), 160–165.

Martin, R., and Delatte, N. J. (2001). "Another Look at Hartford Civic Center Coliseum Collapse," *J. Perf. of Constr. Fac.*, 15(1), 31–36.

McCullough, D. (1968). *The Johnstown Flood*, Simon & Schuster, New York.

McGrath, P., and Foote, D. (1981). "What Happened at the Hyatt?" *Newsweek*, p. 26, August 3.

McGuire, W. (1992). "Comments on L'Ambiance Plaza Lifting Collar/Shearheads," *J. Perf. of Constr. Fac.*, 6(2), 78–85.

McKaig, T. (1962). *Building Failures: Case Studies in Construction and Design*, McGraw-Hill, New York.

Mencken, H. L. (1949) "The Divine Afflatus," *A Mencken Chrestomathy*, chapter 25, p. 443.

Metal Building Manufacturer's Association. (MBMA). (1996). *Low Rise Building Systems Manual*, Metal Building Manufacturer's Association, Cleveland, Ohio.

Middleton, W. D. (2001). *Bridge at Quebec*, Indiana University Press, Bloomington, Ind.

Ministère de l'Agriculture. (1960). *Final Report of the Investigating Committee of the Malpasset Dam*, translated from the French and published by the U.S. Department of the Interior, National Science Foundation, and Israel Program for Scientific Translations, published in two volumes in 1963.

Mlakar, P. E., Dusenberry, D. O., Harris, J. R., Haynes, G., Phan, L. T., and Sozen, M. A. (2003). *The Pentagon Building Performance Report*, ASCE Press, Reston, Va.

Modjeski, R., Borden, H. P., and Monsarrat, C. N. (1919). *The Quebec Bridge over the St. Lawrence River*, Report of the Government Board of Engineers, Canada Department of Railways and Canals.

Moncarz, P. D., Hooley, R., Osteraas, J. D., and Lahnert, B. J. (1992). "Analysis of stability of L'Ambiance Plaza lift-slab towers." *J. Perf. Constr. Fac., ASCE,* 6(4), 232–245.

Moncarz, P. D., and Taylor, R. K. (2000). "Engineering Process Failure—Hyatt Walkway Collapse," *J. Perf. of Constr. Fac.,* 14(2), 46–50.

Morganstern, J. (1997). "The Fifty-Nine Story Crisis," *J. Profl. Issues in Engrg. Educ. and Pract.,* January, 23–29. The article was previously published in *The New Yorker* magazine on May 29, 1995.

Morley, J. (1996). "'Acts of God': The Symbolic and Technical Significance of Foundation Failures." *J. Perf. of Constr. Fac.,* 10(1), 23–31.

Morrison, A. (1980). "Willow Island Aftermath: The Limits of OSHA," *Civ. Engrg.—ASCE,* 50(3), 68–73.

National Bureau of Standards (NBS). (1982). *Investigation of the Kansas City Hyatt Regency Walkways Collapse*, NBSIR 82-2465, NBS, National Institute of Standards and Technology, U.S. Department of Commerce, Washington, D.C.

National Transportation Safety Board. (NTSB). (1970). "Collapse of U.S. 35 Highway Bridge, Point Pleasant, West Virginia, December 15, 1967," Highway Accident Report, NTSB, Washington, D.C.

National Transportation Safety Board. (NTSB). (1983). "Collapse of Suspended Span of Route 95 Highway Bridge over the Mianus River, Greenwich, Connecticut, June 28, 1983," Highway Accident Report, NTSB, Washington, D.C.

National Transportation Safety Board. (NTSB). (1988). "Collapse of New York Thruway (I-90) Bridge over the Schoharie Creek, near Amsterdam, New York, April 5, 1987," Highway Accident Report NTSB/HAR-88/02, NTSB, Washington, D.C.

National Transportation Safety Board. (NTSB). (2007). "Ceiling Collapse in the Interstate 90 Connector Tunnel, Boston, Massachusetts, July 10, 2006," Highway Accident Report NTSB/HAR-07/02, NTSB, Washington, D.C. Available at http://www.ntsb.gov/publictn/2007/HAR0702.pdf. Accessed September 30, 2008.

Nawy, E. G. (2006). *Prestressed Concrete: A Fundamental Approach*, Prentice-Hall, Upper Saddle River, N.J.

Neville, A. M. (1995). *Properties of Concrete*, 4th ed., Prentice-Hall, Upper Saddle River, N.J.

Newman, K. (1986). "Common Quality in Concrete Construction," *Concrete International*, pp. 37–48, March, reprinted in ACI (1989).

Nicastro, D. H. (1996). "Annotated Bibliography of Forensic Engineering." *J. Perform. Constr. Facil.*, 10(1), 2–4.

Nishanian, J., and Wiles, E. (1970). *Reassembly of Point Pleasant Bridge Documentation of Structural Damage and Identification of Laboratory Specimens*, Federal Highway Administration, Washington, D.C.

Nordlund, R. L., and Deere, D. U. (1970). "Collapse of Fargo Grain Elevator," *J. Soil Mech. and Found. Div.*, 96(2), 585–607.

O'Connell, H. M., Dexter, R. J., and Bergson, P. M. (2001). *Fatigue Evaluation of the Deck Truss of Bridge 9340*, Final Report 2001-10, Minnesota Department of Transportation, March.

Osteraas, J. D. (2006). "Murrah Building Bombing Revisited: A Qualitative Assessment of Blast Damage and Collapse Patterns," *J. Perf. of Constr. Fac.*, 20(4), 330–335.

Palmer, R., and Turkiyyah, G. (1999). "CAESAR: An Expert System for Evaluation of Scour and Stream Stability." Report 426, National Cooperative Highway Research Program (NCHRP), Washington D.C.

Pearson, C., and Delatte, N. (2003). "Lessons from the Progressive Collapse of the Ronan Point Apartment Tower," In *Forensic Engineering*, Proceedings of the Third Congress, edited by Paul A. Bosela, Norbert J. Delatte, and Kevin L. Rens, ASCE, Reston, Va., pp. 190–200.

Pearson, C., and Delatte, N. (2005). "Ronan Point Apartment Tower Collapse and Its Effect on Building Codes," *J. Perf. of Constr. Fac.*, 19(2), 172–177.

Peck, R. B., and Bryant, F. G. (1953). "The Bearing-Capacity Failure of the Transcona Elevator," *Geotechnique*, 3, 201–208.

Peterson, I. (1978a). "51 Killed in Collapse of Scaffold at Power Plant in West Virginia," *New York Times*, April 28, 1.

Peterson, I. (1978b). "Onlookers Relive Horror at Plunge of 51 on Scaffold", *New York Times*, April 29, 1.

Petroski, H. (1985). *To Engineer Is Human*, St. Martins Press, New York.

Petroski, H. (1991). "Still Twisting." *Am. Sci.*, Sept/Oct.

Petroski, H. (1994). *Design Paradigms: Case Histories of Error and Judgment in Engineering*, Cambridge University Press, New York.

Petroski, H. (1995). *Engineers of Dreams: Great Bridge Builders and the Spanning of America*, Vintage Books, New York.

Petroski, H. (1996). *Invention by Design*, Harvard University Press, Cambridge, Mass.

Petroski, H. (1997) *When Engineering Fails*, video. Films for the Humanities and Sciences, Princeton, N.J.

Pfatteicher, S. K. A. (2000). "'The Hyatt Horror:' Failure and Responsibility in American Engineering." *J. Perf. of Constr. Fac.*, 14(2), 62–66.

Phelps, E. H. (1969). "A Review of the Stress Corrosion Behavior of Steels with High Yield Strength." In *Proceedings of Conference. Fundamental Aspects of Stress Corrosion Cracking,* National Association of Corrosion Engineers, Houston, Texas, pp. 398–410.

Popescu, A., and Popescu, R. (2003). "Building Research Skills: Course-Integrated Training Methods." *J. Prof. Issues Engrg. Educ. Pract.,* 129(1), 40–43.

Poston, R. W., Feldmann, G. C., and Suarez, M. G. (1991). "Evaluation of L'Ambiance Plaza Post-Tensioned Floor Slabs," *J. Perf. of Constr. Fac.,* 5(2), 75–91.

Rendon-Herrero, O. (1993a). "Including Failure Case Studies in Civil Engineering Courses." *J. Perform. Constr. Facil.,* 7(3), 181–185.

Rendon-Herrero, O. (1993b). "Too Many Failures: What Can Education Do?" *J. Perform. Constr. Facil.,* 7(2), 133–139.

Rendon-Herrero, O. (1994). "Discussion of 'Investigation of L'Ambiance Plaza Building Collapse,'" *J. Perf. of Constr. Fac.,* 8(2), 162–164.

Rens, K., ed. (1997). *Forensic Engineering,* ASCE, New York.

Rens, K. L., Rendon-Herrero, O., Bosela, P. A., eds. (2000a). *Forensic Engineering,* Proceedings of the Second Conference, ASCE, Reston, Va.

Rens, K. L., Rendon-Herrero, O., and Clark, M. J. (2000b). "Failure of Constructed Facilities in the Civil Engineering Curricula." *J. Perform. Constr. Facil.,* 4(1), 27–37.

Ritchie, J., and Houser, M. (2007). "Investigators Descend on Convention Center Collapse Site," *Pittsburgh Tribune-Review,* February 7.

Roddis, W. M. K. (1993). "Structural Failures and Engineering Ethics," *J. Struct. Engrg.,* 119(5), 1539–1555.

Ross, S. (1984). *Construction Disasters: Design Failures, Causes, and Prevention,* McGraw-Hill, New York.

Rubin, R. A., and Banick, L. A. (1987). "The Hyatt Regency Decision: One View," *J. Perf. of Constr. Fac.,* 1(3), 161–167.

Russell, H. G., and Rowe, T. J. (1985). "Collapse of Ramp C," *Concrete International,* American Concrete Institute, pp. 32–37, December, reprinted in ACI (1989).

Scheffey, C. (1971). "Pt. Pleasant Bridge Collapse: Conclusions of the Federal Study," *Civ. Eng.,* 41(7), 41–45.

Schlager, N. (1994). *When Technology Fails: Significant Technological Disasters, Accidents, and Failures of the Twentieth Century,* Gale Research, Detroit.

Scott, R. (2001). *In the Wake of Tacoma: Suspension Bridges and the Quest for Aerodynamic Stability,* ASCE Press, Reston, Va.

Semenza, E., and Ghirotti, M. (2000). "History of the 1963 Vaiont Slide: The Importance of Geological Factors," *Bulletin of Engineering Geology and the Environment,* 59(2), 87–97.

Shepherd, R., and Frost, J. D. (1995). *Failures in Civil Engineering: Structural, Foundation, and Geoenvironmental Case Studies,* ASCE, New York.

Shermer, C. (1968). "Eye-Bar Bridges and the Silver Bridge Disaster," *Engineers Joint Council,* IX(1), 20–31.

Solava, S., and Delatte, N. (2003). "Lessons from the Failure of the Teton Dam," In *Forensic Engineering*, Proceedings of the Third Congress, edited by Paul A. Bosela, Norbert J. Delatte, and Kevin L. Rens, ASCE, Reston, Va., pp. 178–189.

Standard Building Code. (SBC). (1997). *Southern Building Code Congress (SBC)*, Birmingham, Alabama.

Space Shuttle *Challenger* Disaster. (2007). Texas A&M University, http://ethics.tamu.edu/ethics/shuttle/shuttle1.htm. Accessed September 30, 2008.

Steinman, D. B. (1924). "Design of Florianopolis Suspension Bridge." *ENR*, November, 13, 780–782.

Tarkov, J. A. (1986). "A Disaster in the Making," *Am. Herit. Invent. Technol.*, Spring, 10–17.

Tassava, C. J. (2003). "Weak Seams: Controversy over Welding Theory and Practice in American Shipyards, 1938–1946." *Hist. Technol.*, 19(2), 87–108.

Thornton, C. H., Tomasetti, R. L., and Joseph, L. M. (1988). "Lessons From Schoharie Creek," *Civ. Eng.—ASCE*, 58(5), 46–49.

Thornton-Tomasetti, P.C. (1987). "Overview Report Investigation of the New York State Thruway Schoharie Creek Bridge Collapse," Prepared for New York State Disaster Preparedness Commission, December.

U.S. Army Corps of Engineers. (USACE). (1968). *After-Action Report, Silver Bridge Collapse, Point Pleasant West Virginia, 15 December 1967*, Department of the Army, Huntington, W.V.

U.S. Bureau of Reclamation. (USBR). (2001). "The Failure of Teton Dam," http://www.pn.usbr.gov/news/01new/dcoped.html. Accessed May 6, 2003.

Valiani, A., Caleffi, V., and Zanni, A. (2002). "Case Study: Malpasset Dam-Break Simulation Using a Two-Dimensional Finite Volume Method," *J. Hyd. Engrg.*, 128(5), 460–472.

Varney, R. (1971). "Vibration Studies Relating to the Failure of the Point Pleasant Bridge," *Public Roads*, 36(8), 161–166.

Velivasakis, E. E. (1997). "The Willow Island Cooling Tower Scaffold Collapse: America's Worst Construction Accident," In *Forensic Engineering*, edited by Kevin L. Rens, ASCE, Reston, Va., pp. 94–105.

Wearne, P. (2000). *Collapse: When Buildings Fall Down*, TV Books, L.L.C. (www.tvbooks.com), New York. (This book is a companion to The Learning Channel's television series "Collapse.")

Willcut, H. M., Mayfield, W. D., and Valco, T. D. (undated). "Cottonseed Storage," Cotton Incorporated, 4505 Creedmore Road, Raleigh, N.C.

Wiss, Janney, Elstner Associates, Inc. (WJE). (2008). David L. Lawrence Convention Center: Investigation of the 5 February 2007 Collapse, Pittsburgh, PA, Final Report, prepared for the Sports and Exhibition Authority of Pittsburgh and Allegheny County, 4 February. Report online at http://www.pgh-sea.com/images/DLLCCCollapseFinalReportFeb%2008.pdf. Accessed September 30, 2008.

Wiss, Janney, Elstner Associates, Inc., and Mueser Rutledge Consulting Engineers. (WJE and MRCE). (1987). "Collapse of Thruway Bridge at Schoharie Creek," Final Report, Prepared for New York State Thruway Authority, November.

Yu, W.-W. (1991). *Cold-Formed Steel Design*, 2nd ed., Wiley-Interscience, New York.

Zallen, R. M., and Peraza, D. B. (2004). *Engineering Considerations for Lift-Slab Construction*, ASCE, Reston, Va.

Index

2000 Commonwealth Avenue, 133–144
  causes of failure, review of, 140–141
  collapse, 134–136
  Commission investigation, the, 136–139
  design and construction, 133–134
  design and detailing concerns, 141–142
  procedural concerns, 142–143
  punching shear, mechanism, the, 139–140

Agricultural product warehouse failure, 43, 90–97, 219, 256
  design and construction, 91–95
  failure hypotheses, 96
  inspection, 90–91
Air Force warehouse shear failures, 130–133
  lessons learned, 132–133
  Robins Air Force Base, 131–132
  Wilkins Air Force depot, 130–131
Aircraft impacts, 38–42

Austin concrete dam failure, 44–48, 256, 299
Autoroute 19 de la Concorde Overpass, 170–171

Bombing of the Oklahoma City Murrah Federal Building, 155–162
  attack and recovery, 157–161
  Murrah Federal Building, The, 156–157
  technical lessons learned, 161–162

Citicorp Tower, 126, 219, 333–345
  alternative point of view, an, 340–343
  alternatives, the, 338
  intervening two decades, the, 343–345
  plan, the, 338–339
  publicity, 339
  repair, executing the, 339–340
  repercussions, 340

404 BEYOND FAILURE

Cleveland lift-slab parking garage, 121–124
Cold-formed steel beam construction failure, 188–195
  collapse, 189–190
  description of the structure, 189
  investigation, 190–193
  reconstruction, 193–194
Comet jet aircraft crashes, 81–86
  aircraft designs, changes to, 85–86
  crashes the, 83–84
  investigations, 84–85

Dee Bridge, 215–217

Engineering design, 1–3

Fluid mechanics and hydraulics, 257–299

Harbour Cay Condominium, 149–155
  design versus construction error, 153
  lessons learned, 154–155
  NBS findings, summary of, 152–153
  professional aspects, 153–154
  structure and construction, description of, 152
Hartford Civic Center Stadium collapse, 174–184
  collapse, 177
  design and construction, 174–177
  educational aspects, 183
  ethical aspects, 182–183
  failure, causes of, 177–181
  professional and procedural aspects, 181–182
  technical aspects, 181
Hyatt Regency walkway, 8–25, 86–87, 218
  collapse, 9–10
  design and construction, 8–9
  educational aspect, 23
  essential reading, 25
  ethical concerns, 19–21
  events leading up to the collapse, 15–18
  failure, causes of, 11–14
  legal repercussions, 18
  lessons learned, 24
  lessons not learned, 24–25
  procedural concerns, 19
  technical concerns, 19
  the human factor, 21–23

Johnstown flood, 255–267

Kemper Arena, 124–126
  collapse, 124–125
  educational aspects, 126
  investigation results, 125–126
  structure and roof, the, 124

L'Ambiance Plaza collapse, 48, 87, 107–121, 167, 218
  collapse, 109
  design and construction, 107–108
  ethical aspects, 119
  failure, causes of, 109–118
  human factor, the, 119–120
  legal repercussions, 118
  professional and procedural aspects, 119
  technical aspects, 118

Malpasset Dam, 267–277
  design and construction, 269
  failure of the dam, 269–272
  investigations and repercussions, 272–274
  lessons learned, 274–276
Management, ethics, and professional issues, 333–360

Mechanics of materials, 51–87
Mianus River Bridge collapse, 42–43, 184–188
  investigation, 185–186
  lessons learned, 187–188
  standards and procedures, changes to, 187
Minneapolis I-35 W Bridge collapse, 211–215

New Orleans Hurricane Katrina levee failures, 256, 287–299

Pentagon attack, the, 162–167
  attack and building performance, 163
  building design and construction, 162–163
  investigation and analysis, 163–165
  technical lessons learned, 166
Pittsburgh Convention Center expansion joint failure, 87, 206–211
  collapse, 207
  David L. Lawrence Convention Center, The, 206–207
  expansion joint detailing and forces, 209–211
  investigations, 207–209
Point Pleasant Bridge, 70–82
  bridge inspection, changes to, 80
  collapse, 74, 218
  design and construction, 71–73
  educational aspects, 80–81
  fracture, corrosion, and fatigue, evidence of, 77–79
  investigation, 74
  standards, 79–80
Professional engineer license, misuse of the, 357–359
Property loss investigations, 359–360

Quebec Bridge, 43–44, 51–70, 217–218
  aftermath, 68
  capacity of compression members, 66–67
  collapse, 59
  conception, design, and construction, 52–55
  ethical aspects, 67
  events leading up to the collapse, 55–59
  failure, causes of, 61–64
  procedural and professional aspects, 64–65
  Royal Commission report, 59–61
  second bridge, the, 68

Reinforced concrete structures, 129–171
Ronan Point, 97–106, 167
  collapse, 100–101
  design and construction, 98–100
  ethical aspects, 105
  failure, causes of, 101–102
  professional and procedural aspects, 103–105
  technical aspects, 102–103

Sampoong Superstore, 169–170, 352–357
  collapse, 353–354
  design and construction, 353
  investigations and conclusions, 354–357
Schoharie Creek Bridge, 167–168, 256, 277–287
Skyline Plaza in Bailey's Crossroads, 144–149
  estimated concrete strength, 147
  framework, 145–147
  legal repercussions, 148
  lessons learned, 149

Skyline Plaza in Bailey's Crossroads—*continued*
  NBS findings, summary of, 147–148
  structure and construction, description of, 145
Soil mechanics, geotechnical engineering, and foundations, 221–256
Space shuttle *Challenger*, 345–352
  investigations, 351
  leaks in the primary seal, 348–349
  meeting before the launch, the, 350
  shuttle explosion, the, 350–351
  space shuttle design and operation, 346–348
  whistle blowing, 351
Statics and dynamics, 7–50
Steel structures, 173–219
Structural analysis, 89–127

Tacoma Narrows Bridge, 26–38
  causes of failure, 30–32
  collapse, the 27–30
  earlier problems with suspension bridges, 32–33
  educational aspects, 36–37
  ethical concerns, 34
  evolution of bridge aerodynamics, 34–36
  later problems with suspension bridges, 33–34
  technical concerns, 32

Teton Dam, 49–50, 223–224
  educational aspects, 233
  failure, the, 224–226
  investigating panel, 226–229
  panel investigation and results, 229–232
  professional and procedural aspects, 232–233
  technical aspects, 232
The World Trade Center attack, 195–206
  design and construction, 195–201
  attack, the, 201–204
  Federal Emergency Management Agency study findings, overall, 204–205
  recommendations to improve robustness and safety, 205–206
Transcona and Fargo grain elevators, The, 249–255
  Transcona grain elevator, the, 249–251
  Fargo grain elevator, 251–255
  lessons learned, 255

Vaiont dam reservoir slope stability failure, 48, 234–248, 299

Willow Island Cooling Tower collapse, 167

About the Author

NORBERT J. DELATTE JR., PH.D., P.E., is a professor in the Department of Civil and Environmental Engineering at Cleveland State University. He is a past chair of the executive committee of ASCE's Technical Council on Forensic Engineering, as well as a past chair of the council's education committee. Previously, he served on the faculty of the University of Alabama at Birmingham and of the U.S. Military Academy, West Point, New York. Before entering academia, he was an officer in the U.S. Army Corps of Engineers, serving on active duty in South Korea and the Middle East as well as a number of stateside postings. He received his bachelor's degree from the Citadel, his master's from the Massachusetts Institute of Technology, and his doctorate from the University of Texas at Austin. He is a licensed professional engineer in Alabama, Ohio, and Virginia.

VERMONT STATE COLLEGES
0 0003 0827398 6

Hartness Library
Vermont Technical College
One Main St.
Randolph Center, VT 05061